Springer Series in Electrophysics
Volume 16

Edited by Theodor Tamir

Springer Series in Electrophysics

Editors: Günter Ecker   Walter Engl   Leopold B. Felsen

David Schieber

Electromagnetic Induction Phenomena

With 88 Figures

Springer-Verlag Berlin Heidelberg New York
London Paris Tokyo

Professor Dr. David Schieber

Technion, Israel Institute of Technology, Department of Electrical Engineering
Technion City, Haifa 32000, Israel

Guest Editor:

Professor Theodor Tamir, Ph. D.

981 East Lawn Drive, Teaneck, NJ 07666, USA

Series Editors:

Professor Dr. Günter Ecker

Ruhr-Universität Bochum, Theoretische Physik, Lehrstuhl I,
Universitätsstrasse 150, D-4630 Bochum-Querenburg, Fed. Rep. of Germany

Professor Dr. Walter Engl

Institut für Theoretische Elektrotechnik, Rhein.-Westf. Technische Hochschule,
Templergraben 55, D-5100 Aachen, Fed. Rep. of Germany

Professor Leopold B. Felsen, Ph.D.

Polytechnic Institute of New York, 333 Jay Street, Brooklyn, NY 11201, USA

ISBN-13:978-3-642-71017-9   e-ISBN-13:978-3-642-71015-5
DOI: 10.1007/978-3-642-71015-5

Library of Congress Cataloging-in-Publication Data. Schieber, David, 1938- Electromagnetic induction
phenomena. (Springer series in electrophysics ; v. 16) Bibliography: p. 1. Induction, Electromagnetic. I. Title.
II. Series. QC638.S35 1986   537   86-3813

Typsetting: K+V Fotosatz GmbH, 6124 Beerfelden

2153/3150-543210

Foreword

From an engineering perspective, Electrodynamics is the province of two cultures. The most easily identified of the two is primarily concerned with phenomena in which the propagation of electromagnetic waves is crucial. Included are the designers of microwave circuits, of antennae and of many-wavelength communication channels. The interests of the second group focus on dynamical processes associated with the evolution of field sources, whether these be electrons and holes migrating in a semiconductor, or currents diffusing in a moving metal.

Because the second culture is primarily concerned with the interaction between electromagnetic fields and media, where the latter are often responsible for the dominant dynamical processes, it addresses applications that are more widely ranging. A few from a very long list would include electrostatic printing, rotating machines, power transmission apparatus, the electromagnetics of biological systems and physical electronics. Whether by nature or by design, the phenomena of interest are generally electroquasistatic or magnetoquasistatic in this second branch of electrodynamics. It is tempting to say that the two branches of electrodynamics can be distinguished by the frequency range, but electron-beam and microwave-magnetic devices, with their respective plasma oscillations and spin waves, are examples where the frequencies can be in the GHz range while the fundamental interactions are quasistatic. By design, so also are those that determine the frequency response of a transistor.

Because of the diversity of applications for the second branch, advances that are fundamental to the field of electrodynamics tend to be poorly communicated to others working within that branch. Communication between the two cultures is even more difficult to secure. This book, which will receive the interest of both cultures, will serve to encourage others to take the author's perspective.

James R. Melcher
Stratton Professor of Electrical Engineering and Physics,
Department of Electrical Engineering and Computer Science,
Massachusetts Institute of Technology

Preface

This book treats an area in applied electrodynamics with which I was strongly involved during the decade 1971–1981; whereas the writing of this volume was contemplated for several years, the final decision to implement that idea was taken some months after the passing away of Professor Franz Ollendorff in December 1981.

Professor Ollendorff was my highly revered preceptor, who initiated me into the more advanced aspects of applied electrodynamics. For sixteen years I had the unique privilege of studying and working with him, every single discussion being an unforgettable experience of unsurpassed scientific beauty. With his all-embracing intellect and razor-sharp critique – invariably presented in scholarly terms – he would often point out a weak spot in the analysis proper or in the phrasing of a problem; time and again these comments taught me something new and I was always awed by his keen insight and intuition. In this context, I would like to quote Professor Chen-To Tai's remark originally referring to Herman Minkowski: "As a youngster, I was puzzled by a Chinese inscription on a tombstone saying 'Bequeath Fragrance Hundred Generations'. I certainly appreciate the significance of this inscription now".

As an undergraduate, Franz Ollendorff had studied under M. Planck and M. von Laue; he used to recall that the dominant approach was then still largely that of H. Hertz and H. A. Lorentz, and he liked to recount von Laue's use, on the blackboard, of a moving and deforming loop to illustrate Faraday's law as extended to moving media. However, the loop approach was later replaced by the Einstein-Minkowski approach, which eliminated the artificial borderline (in existence even today for some) between "transformer emf" and "motion-induced emf", thus avoiding pointless questions with regard to unipolar induction.

Although Professor Ollendorff had always maintained a lively interest – and had published quite a lot – in the relevant relativistic electrodynamics (he often used to stress Einstein's saying that Maxwell's equations ... "can be grasped ... only by way of the special theory of relativity"), he was able to devote sufficient attention and time to this topic only quite late in his scientific career. His contribution was reflected *inter alia* in a comprehensive relativistic theory of unipolar induction machines, and later in a series of papers on tubular motors which appeared in the Archiv für Elektrotechnik in the seventies.

The representation of practical problems in terms of abstract models was one of Professor Ollendorff's strongest points — an art that he practiced for about forty years at the Technion. Some of it is reflected in these pages; however, let me immediately add that, in addition to the "secret" of how to solve such problems, he also taught me that the path followed on the way to the final results leads to an inner freedom which is not affected by external conditions.

Following his example, the present text resorts to some engineering aspects of restricted relativity when dealing with the analysis of moving media: not only are Maxwell's equations thus quite naturally extended, but any tacit physical premises otherwise involved are dispensed with by taking this secure path; furthermore the analytical treatment leading to relatively simple expressions is then also greatly facilitated. In other sections, dealing with stationary media, an attempt is similarly made to provide as far as possible closed-form solutions, so as to obtain additional insight into certain problems of applied electromagnetism.

It goes without saying that all errors and mistakes are obviously mine — not counting the oversights which can hardly be avoided in such a text as this.

I use this opportunity to express my appreciation to:

- My alma mater the Technion and our Electrical Engineering Department, for providing technical assistance;
- Ing. Eliezer Goldberg, resident Scientific Editor at the Technion, who unsparingly contributed his expertise and advice;
- Mr. Armand Dankner, for his devoted help with the graphics;
- Mrs. Ann Gerzon, who typed the successive versions of the manuscript with exemplary skill and patience.

A special debt of gratitude is due to the Editor of this volume, Professor Theodor Tamir of the Polytechnic Institute of New York, without whose impetus, friendship and encouragement the book would have never been completed.

Since I view this volume as a modest homage to the memory of Professor Ollendorff, a great scholar and teacher, let me conclude with the traditional Aramaic prayer line

Yith'gaddal ve'yith'kaddash Sh'mei Rabba.

Haifa, Mt. Carmel   *D. Schieber*
January 1986

Contents

1. Introduction

Electrodynamics is an integral component of electrical engineering. Although the area of electrical engineering has by now become an interdisciplinary field and its definition is quite diffuse, electrodynamics remains one of its mainstays. Its applications in satellite communications, microwave heating, tomography, biotechnology, geophysical remote-sensing, electromechanical energy-conversion and power systems, small-scale electromagnetic motors, modern optics, vehicular technology and elsewhere, render it indispensable to the student, and it is for this reason that electromagnetism is a basic component of the relevant core curriculum.

Obviously, a course in electromagnetism must today be sufficiently broad to encompass as many as possible of the potential directions that the aspiring electrical engineer may choose to take; it must therefore be taught in depth, without losing sight of the surrounding reality. Traditional modes of presentation whereby the student is treated to a wealth of impractical problems in electrostatics and magnetostatics are both obsolete and self-defeating because they may detract from his motivation and leave him totally unprepared for the more advanced topics of waves, distributed systems, microwaves, antennae, optics, etc. In the author's experience, an approach based upon the axiomatic introduction of Maxwell's equations and subsequent branching off to the relevant subsystems (i.e., electro-quasistatics, magneto-quasistatics, stationary current fields and finally dynamics) both broadens and deepens the student's interest in electromagnetism. As just noted, this approach must be supplemented by illustrative examples drawn from the realities of the profession, many of which are necessarily at graduate level.

The present volume purposely deals with a few such examples, most of them encountered — albeit in much less abstract terms — by the author in the course of his career. The relevant situations obviously had to be modelled so as to permit analytic treatment.

Before proceeding with the examples proper, a few brief general remarks on applied electrodynamics are in order.

1.1 Electrodynamics of Moving Media

The vast literature on phenomenological electrodynamics of moving media falls into three broad categories:

a) Minkowski's electrodynamics, based upon Einstein's restricted theory of relativity;
b) Covariant description in arbitrary frames of reference, based upon Einstein's general theory of relativity;
c) Observation of moving electrified matter from the inertial laboratory.

1.1.1 Minkowski Electrodynamics [1]

Within the framework of Minkowski's electrodynamics, Maxwell's equations relate the electric field E, the magnetic field H, the inductions D and B and the free electric charge and current densities ρ and j at a certain instant of time t by means of

$$\nabla \times H = j + \frac{\partial D}{\partial t} \; , \tag{1.1.1}$$

$$\nabla \times E = -\frac{\partial B}{\partial t} \; , \tag{1.1.2}$$

$$\nabla \cdot D = \rho \; , \tag{1.1.3}$$

$$\nabla \cdot B = 0 \; . \tag{1.1.4}$$

These equations, valid for a point at rest with respect to the inertial frame of reference x, y, z, t of the laboratory, may be used for the determination of E and H once ρ and j are known, provided constitutive laws relating D and B to E and H, respectively, are available.

For a perfectly nonconducting, linear medium — again at rest — with relative scalar permittivity ε_r (ε_0 in vacuum) and relative scalar permeability μ_r (μ_0 in vacuum), the constitutive equations are postulated to read

$$D = \varepsilon_r \varepsilon_0 E; \quad B = \mu_r \mu_0 H \; . \tag{1.1.5}$$

Assuming now that the medium moves with respect to the laboratory frame at a speed v along the x-axis, we mentally attach to it a primed frame of reference x', y', z', t', in turn taken to be linked to the unprimed one by means of the Lorentz transformation [2]

$$x' = \frac{x - \dfrac{v}{c} ct}{\sqrt{1 - \beta^2}} \; ; \quad y' = y; \quad z' = z; \quad ct' = \frac{ct - \dfrac{v}{c} x}{\sqrt{1 - \beta^2}} \; ; \quad \beta^2 = \frac{v^2}{c^2} \; . \tag{1.1.6}$$

---

[1] Rationalized MKSA units (SI) are used here throughout.
[2] In order to obtain even approximate, low-speed transformation formulae, it is simplest to resort to exact relativistic formulae holding at all speeds.

Here c stands for the *in-vacuo* phase velocity of a monochromatic plane light wave.

On application of the Lorentz transformation, we find [1.1] that the primed field and induction components (subscripts x, y, z) are related to the unprimed ones through

$$E'_x = E_x \; ; \qquad B'_x = B_x \; ;$$

$$E'_y = \frac{E_y - vB_z}{\sqrt{1 - \beta^2}} \; ; \quad B'_y = \frac{B_y + \dfrac{v}{c^2}E_z}{\sqrt{1 - \beta^2}} \; ; \qquad (1.1.7)$$

$$E'_z = \frac{E_z + vB_y}{\sqrt{1 - \beta^2}} \; ; \quad B'_z = \frac{B_z - \dfrac{v}{c^2}E_y}{\sqrt{1 - \beta^2}} \; ,$$

along with

$$H'_x = H_x \; ; \qquad D'_x = D_x \; ;$$

$$H'_y = \frac{H_y + vD_z}{\sqrt{1 - \beta^2}} \; ; \quad D'_y = \frac{D_y - \dfrac{v}{c^2}H_z}{\sqrt{1 - \beta^2}} \; ; \qquad (1.1.8)$$

$$H'_z = \frac{H_z - vD_y}{\sqrt{1 - \beta^2}} \; ; \quad D'_z = \frac{D_z + \dfrac{v}{c^2}H_y}{\sqrt{1 - \beta^2}} \; .$$

Replacing v by $(-v)$, the inverse equations result, namely

$$E_x = E'_x \; ; \qquad B_x = B'_x \; ;$$

$$E_y = \frac{E'_y + vB'_z}{\sqrt{1 - \beta^2}} \; ; \quad B_y = \frac{B'_y - \dfrac{v}{c^2}E'_z}{\sqrt{1 - \beta^2}} \; ; \qquad (1.1.9)$$

$$E_z = \frac{E'_z - vB'_y}{\sqrt{1 - \beta^2}} \; ; \quad B_z = \frac{B'_z + \dfrac{v}{c^2}E'_y}{\sqrt{1 - \beta^2}} \; ,$$

and

$$H_x = H'_x \; ; \qquad\qquad D_x = D'_x \; ;$$

$$H_y = \frac{H'_y - v D'_z}{\sqrt{1 - \beta^2}} \; ; \quad D_y = \frac{D'_y + \dfrac{v}{c^2} H'_z}{\sqrt{1 - \beta^2}} \; ; \qquad\qquad (1.1.10)$$

$$H_z = \frac{H'_z + v D'_y}{\sqrt{1 - \beta^2}} \; ; \quad D_z = \frac{D'_z - \dfrac{v}{c^2} H'_y}{\sqrt{1 - \beta^2}} \; .$$

Substituting (1.1.9, 10) in (1.1.5), we obtain the "shorthand" expressions, in which v stands for the generalized, three-dimensional velocity:

$$D' - \frac{1}{c^2} v \times H' = \varepsilon_r \varepsilon_0 (E' - v \times B') \; , \qquad\qquad (1.1.11)$$

$$B' + \frac{1}{c^2} v \times E' = \mu_r \mu_0 (H' + v \times D') \; . \qquad\qquad (1.1.12)$$

Expressing now the inductions as dependent on field intensities only, we find

$$D' = \frac{1}{1 - \varepsilon_r \mu_r \beta^2} \left\{ \varepsilon_r \varepsilon_0 \left[E'(1 - \beta^2) + \frac{v(v \cdot E')}{c^2}(1 - \varepsilon_r \mu_r) \right] \right.$$

$$\left. + \frac{v \times H'}{c^2}(1 - \varepsilon_r \mu_r) \right\} \; , \qquad\qquad (1.1.13)$$

$$B' = \frac{1}{1 - \varepsilon_r \mu_r \beta^2} \left\{ \mu_r \mu_0 \left[H'(1 - \beta^2) + \frac{v(v \cdot H')}{c^2}(1 - \varepsilon_r \mu_r) \right] \right.$$

$$\left. - \frac{v \times E'}{c^2}(1 - \varepsilon_r \mu_r) \right\} \; , \qquad\qquad (1.1.14)$$

and reverting once more to the unprimed frame of reference, the new constitutive equations result, i.e.,

$$D = \frac{1}{1 - \varepsilon_r \mu_r \beta^2} \left\{ \varepsilon_r \varepsilon_0 \left[E(1 - \beta^2) + \frac{v(v \cdot E)}{c^2}(1 - \varepsilon_r \mu_r) \right] \right.$$

$$\left. - \frac{v \times H}{c^2}(1 - \varepsilon_r \mu_r) \right\} \; , \qquad\qquad (1.1.15)$$

$$B = \frac{1}{1 - \varepsilon_r \mu_r \beta^2} \left\{ \mu_r \mu_0 \left[H(1 - \beta^2) + \frac{v(v \cdot H)}{c^2}(1 - \varepsilon_r \mu_r) \right] \right.$$

$$\left. + \frac{v \times E}{c^2}(1 - \varepsilon_r \mu_r) \right\} . \tag{1.1.16}$$

Let us emphasize that (1.1.15, 16) represent the relations that an observer at rest, with respect to the unprimed inertial frame of reference, determines with regard to a medium receding at the instantaneous velocity v.

1.1.2 Covariant Description in Arbitrary Frames of Reference

This approach is based upon the formalism of general relativity [1.2–4]. From the engineering point of view, it is deficient in that we do not yet have a theory encompassing both the elastic and electromagnetic aspects of the constitutive laws in highly accelerated frames of reference; in fact, we do not yet even have enough experimental data as a substitute for such a theory; hence, its application is at present only rarely warranted in applied electromagnetism.

1.1.3 Observation of Moving Electrified Matter

We mention here two different approaches:

The Minkowski approach as given by Lorentz and Pauli [Ref. 1.1, pp. 285 to 287]. With

$$E^* \equiv E + v \times B , \tag{1.1.17}$$

$$H^* \equiv H - v \times D , \tag{1.1.18}$$

the field equations at a point moving with an arbitrary velocity v with respect to the inertial laboratory now read

$$\nabla \times H^* = j + \frac{\partial D}{\partial t} - \nabla \times (v \times D) , \tag{1.1.19}$$

$$\nabla \times E^* = -\frac{\partial B}{\partial t} + \nabla \times (v \times B) , \tag{1.1.20}$$

$$\nabla \cdot D = \rho , \tag{1.1.21}$$

$$\nabla \cdot B = 0 . \tag{1.1.22}$$

Although (1.1.19, 20) are largely formal, they possess a distinct practical advantage when the "observation" approach is extended to an integral formulation of Maxwell's laws [1.5].

The Chu approach. About twenty-five years ago, *Chu* and co-workers [1.6] presented two formulations of macroscopic electrodynamics, based upon a description of magnetic dipoles by magnetic charges and by current loops, respectively. One of the assumptions underlying these formulations is that moving and deforming material bodies contribute to the electromagnetic fields as sources; they use only two macroscopic fields – different from those of Minkowski – and obey free space-like equations while dispensing with constitutive equations.

Introducing the volume magnetization M and volume polarization P, the E, H ("non-Amperian") Chu equations read

$$\nabla \times H - \frac{\partial}{\partial t}\, \varepsilon_0 E = j + \frac{\partial P}{\partial t} + \nabla \times (P \times v) \ , \tag{1.1.23}$$

$$\nabla \times E + \frac{\partial}{\partial t}\, \mu_0 H = - \frac{\partial}{\partial t}\, (\mu_0 M) - \nabla \times (\mu_0 M \times v) \ , \tag{1.1.24}$$

$$\nabla \cdot \varepsilon_0 E = \rho - \nabla \cdot P \ , \tag{1.1.25}$$

$$\nabla \cdot \mu_0 H = - \nabla \cdot \mu_0 M \ . \tag{1.1.26}$$

The approach was criticized inter alia on force considerations (see, e.g., [1.7]), but subsequent work [1.8–10] proved these criticisms unjustified and later analysis [1.11] provided some additional insight into it.

In the present volume, we have chosen to adhere to the Einstein-Minkowski electrodynamics, its simplicity [1.12] strongly arguing in its favour, in particular where the material constants are given in regions of space or where the excitation sources are embedded in vacuo.

[Let us parenthetically remark that relativity is, in fact, making ever deeper inroads into modern electrical engineering: with the advent of charge accelerators (for cancer therapy, etc.), electron microscopes, microwave and high-voltage television tubes – analysis by means of restricted relativity is indispensable.]

1.2 Some Remarks on Electrodynamics of Moving Media and Restricted Relativity

The electrodynamics of moving media becomes readily accessible when treated in conjunction with restricted relativity: in fact, Einstein's first paper on relativity was called "On the Electrodynamics of Moving Bodies" and its starting point was the phenomenon of unipolar induction. Even those approaches that are based upon viewing of electrified bodies by a single observer gain in scope and insight by this means [Ref. 1.6, pp. 453–504].

Without restricted relativity, the explanation, e.g., of motion-induced emf is artificial, requiring auxiliary assumptions, or at best – under some of the formerly mentioned approaches – rather laborious [Ref. 1.6, pp. 394–407]. However, once it is used, the true mathematical structure of induction emerges as "in a mountain landscape when the fog lifts" [Ref. 1.1, p. 212]. It is quite strange that some are still reluctant to take advantage of restricted relativity, objecting that it is too complicated or that laboratory speeds are too low to warrant its use. These objections are, however, irrelevant. As regards the first objection, the comprehension of restricted relativity requires at most one year of college physics and mathematics, plus an open mind. Concerning the second objection – it is downright wrong: in electrodynamics, even at very low speeds, restricted relativity has a distinct advantage over Galileo's relativity; not only does it leave Maxwell's equations invariant, but it also yields a deeper insight into physical effects that are otherwise quite difficult to visualize.

Thus, turning to the formerly mentioned phenomenon of induction, let us consider two frames of reference interconnected as per (1.1.6); in the primed frame of reference, existence of the three components V_x, V_y, V_z of the magnetic vector potential [3] V is postulated along with that of the electric scalar potential ϕ. As usual,

$$E = -\frac{\partial V}{\partial t} - \nabla \phi , \tag{1.2.1}$$

$$B = \nabla \times V . \tag{1.2.2}$$

Imposing upon V and ϕ the condition

$$\nabla \cdot V + \frac{1}{c^2} \frac{\partial \phi}{\partial t} = 0 , \tag{1.2.3}$$

we find (see also Sect. 5.1, this section, below, and [Ref. 1.1, p. 213]) that V_x, V_y, V_z and ϕ/c constitute a four-vector; hence (1.1.6) also prescribes the transformation law for the primed potentials, i.e.,

$$V_x' = \frac{V_x - \frac{v}{c}\frac{\phi}{c}}{\sqrt{1-\beta^2}} \ ; \quad V_y' = V_y; \quad V_z' = V_z; \quad \frac{\phi'}{c} = \frac{\frac{\phi}{c} - \frac{v}{c}V_x}{\sqrt{1-\beta^2}} \ . \tag{1.2.4}$$

---

[3] Our deviation from the standard notation practice in using V for the magnetic vector potential instead of the regular A is not entirely arbitrary: in analogy to the linear momentum of a particle of mass m moving at a velocity v, which is given by the product mv, the additional electromagnetic momentum of a charge q in a region pervaded by a magnetic vector potential, is given by the product qV. On the strength of this formal analogy, we resort to V, reserving A for the surface current density.

At speeds $v \ll c$ we may approximate

$$x' = x - vt; \quad y' = y; \quad z' = z; \quad t' = t , \tag{1.2.5}$$

along with

$$V_x' = V_x; \quad V_y' = V_y; \quad V_z' = V_z . \tag{1.2.6}$$

However, there still remains

$$\phi' = \phi - v V_x , \tag{1.2.7}$$

illustrating how, thanks to restricted relativity, the interplay between the velocity and the magnetic field results directly.

When dealing with electromagnetic induction, the problem is very often of a magneto-quasistatic nature; how can quasistatics, in which the finite speed of light plays no specific role, be reconciled with restricted relativity?

Before considering this, let us review our treatment of quasistatics.

1.3 Some Remarks on Quasistatics

A direct derivation of quasistatic electrodynamics is not often found in literature, and even the whole subsets of relevant equations are quite seldom quoted (see, however, [1.13] and [Ref. 1.8, pp. 45 – 47]).

The formal approach is somewhat handicapped by the need to incorporate considerations not explicitly stated by differential laws. For example (Fig. 1.1) two perfectly conducting plates (dimensions b, l, and spacing h) excited, e.g., by a low-frequency sinusoidal power supply, represent a magneto-quasistatic approximation so long as they are short-circuited at one end, but when they are open-circuited, an electro-quasistatic situation results; further, a quite different situation is to be expected if the open- or short-circuit is replaced by a finite conductivity [1.14].

The usual definitions of quasistatics are thus somewhat loose, often necessitating intuition and experience, and may give rise to certain questions of principle; thus, one approach to quasistatics is through the *in-vacuo* wave length λ of the considered field which oscillates at frequency f. If the characteristic dimension l of the system under investigation is much smaller than the said wavelength, the exchange mechanism between electric and magnetic energy may be disregarded and it is a case of quasistatics; the relevant inequality reads

$$\frac{c}{f} \gg l . \tag{1.3.1}$$

A somewhat different interpretation is called for if this expression is rewritten in the form

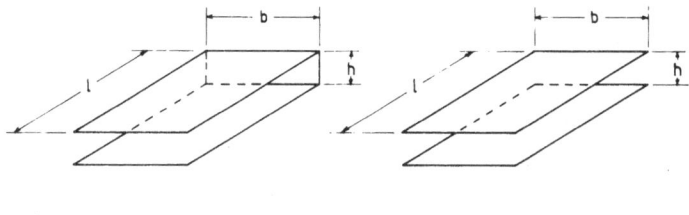

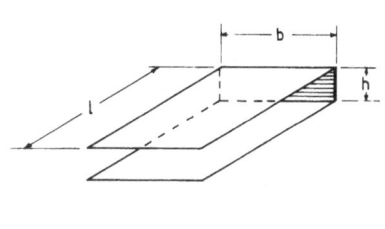

Fig. 1.1a – c. Highly conducting plates
(a) Short-circuit; (b) Open-circuit; (c) Resistive
termination

$$\frac{l}{c} \ll T \ , \tag{1.3.2}$$

where the period T is defined as usual by

$$T = \frac{1}{f} \ . \tag{1.3.3}$$

Quasistatics may now be said to ensue if the time required for a plane light
wave to propagate along the system is much shorter than the characteristic
period T. Still, both points of view are often somewhat hazy in practice. More
specific approaches are then resorted to, and quasistatics is applied if certain
processes are slow compared, e.g., to the magnetic diffusion time τ_d, or to the
electric charge relaxation time τ_r. For a system of electric conductivity σ,
permeability μ_0, permittivity ε_0 and characteristic length l, the relevant rela-
tionships read

$$\tau_d = \mu_0 \sigma l^2 \quad \text{and} \tag{1.3.4}$$

$$\tau_r = \frac{\varepsilon_0}{\sigma} \ , \tag{1.3.5}$$

the time needed for propagation of a plane light wave along l being

$$\sqrt{\tau_d \tau_r} = \frac{l}{c} \equiv \tau_{em} \ . \tag{1.3.6}$$

Interpretation of quasistatics in terms of a time-rate parameter expansion
is now possible [Ref. 1.6, pp. 213 – 240]; however, in lossless systems there are

no characteristic times distinguishing magneto- and electro-quasistatics, and insight and intuition are again called for.

Considerations of this kind open up a still different path for the formal derivation and interpretation of quasistatics — via series expansion of the fields in terms of the angular supply frequency ω. The usefulness — and definition — of quasistatics now lies in the fact that if ω is "low enough", higher-order terms may be safely neglected [Ref. 1.6, pp. 329 – 331]. As a rule of thumb, "low enough" means much less than the lowest angular frequency of resonance ω_0 of the system considered [1.15].

At this juncture, however, a new difficulty arises: how can quasistatics be reconciled with transient phenomena whose high-frequency Fourier spectrum is unbounded? It should be borne in mind that, on the one hand, Kirchhoff's laws, in fact, follow from Maxwell's equations as quasistatic approximations and, on the other, that transients are considered as a matter of routine in circuit theory[4]. How then can *low-frequency* quasistatics be reconciled with transient phenomena comprising a *high-frequency* spectrum?

We thus have to deal with *two* questions, namely that of the end of Sect. 1.2, and the present one. Before trying to find the answers, let us quote the quasistatic equations. Based upon the *Minkowski* formulation for bodies at rest — see (1.1.1 – 4) — the relevant approximations read:

Electro-Quasistatics

$$\nabla \times H = j + \frac{\partial D}{\partial t} \;, \tag{1.3.7}$$

$$\nabla \times E = 0 \;, \tag{1.3.8}$$

$$\nabla \cdot D = \rho \;, \tag{1.3.9}$$

$$\nabla \cdot B = 0 \;. \tag{1.3.10}$$

While electric charge conservation, i.e.,

$$\nabla \cdot j + \frac{\partial \rho}{\partial t} = 0 \tag{1.3.11}$$

follows from (1.3.7), Eq. (1.3.10), which signifies absence of free magnetic charges, is of no immediate use here.

Magneto-Quasistatics

$$\nabla \times H = j \;, \tag{1.3.12}$$

$$\nabla \times E = -\frac{\partial B}{\partial t} \;, \tag{1.3.13}$$

---

[4] Omission of the capacitance when dealing with inductive components, or of the inductance when dealing with capacitive components, reduces in effect the number of degrees of freedom of the investigated system [1.16].

$$\nabla \cdot D = 0 \ , \tag{1.3.14}$$

$$\nabla \cdot B = 0 \ . \tag{1.3.15}$$

In the present case, (1.3.14) expressing absence of free electric charges is of little practical value: charge conservation is now symbolized by closed current paths, i.e.,

$$\nabla \cdot j = 0 \ , \tag{1.3.16}$$

in turn implied by (1.3.12).

A different approach to quasistatics is provided by the *Chu* equations (bodies at rest):

Electro-Quasistatics

$$\nabla \times H - \frac{\partial}{\partial t}(\varepsilon_0 E) = j + \frac{\partial P}{\partial t} \ , \tag{1.3.17}$$

$$\nabla \times E = 0 \ , \tag{1.3.18}$$

$$\nabla \cdot \varepsilon_0 E = \rho - \nabla \cdot P \ , \tag{1.3.19}$$

$$\nabla \cdot \mu_0 H = - \nabla \cdot \mu_0 M \ . \tag{1.3.20}$$

Magneto-Quasistatics

$$\nabla \times H = j \tag{1.3.21}$$

$$\nabla \times E + \frac{\partial}{\partial t}(\mu_0 H) = - \frac{\partial}{\partial t}(\mu_0 M) \ , \tag{1.3.22}$$

$$\nabla \cdot \varepsilon_0 E = - \nabla \cdot P \ , \tag{1.3.23}$$

$$\nabla \cdot \mu_0 H = - \nabla \cdot \mu_0 M \ . \tag{1.3.24}$$

We have purposely chosen to quote these approximations within the framework of two different formulations in order to underline different points of view.

Now, although (1.3.10 and 14) are redundant within the context of the relevant simplified Minkowskian sets of equations, they are not void of content; in fact, it is the physical character of the inductions D and B which provides an answer to the formerly mentioned two questions, as well as to a formal approach to quasistatics. To this end we now consider the so-called "Cerenkov electrodynamics".

1.4 "Cerenkov Electrodynamics"

The concept of "Cerenkov electrodynamics" was introduced for heuristic reasons in order to reconcile − from the point of view of the theory of knowledge − the theory of electro- and magneto-quasistatics on the one hand, with relativity and high-frequency phenomena on the other hand; it is still open to research and even to discussion, and in the present text we use it only as a background in our attempt to remove formal inconsistencies. The approach proper is outlined in two papers by *Ollendorff* [1.17]; a few notions will be reiterated here.

For this purpose we consider the electromagnetic field in a non-conducting, charge-free homogeneous and isotropic material of relative permittivity ε_r and relative permeability μ_r; once more, the material moves at constant speed v along the $+x$-axis of a right-handed Cartesian frame of reference x, y, z in which time t is measured by a suitably distributed array of stationary clocks, and again a frame of reference x', y', z', t' is assumed to be "rigidly" attached to the moving material.

Plane homogeneous electromagnetic waves propagate within the mentioned medium − of infinite extent − at phase velocity $\bar{c}$, which is n times smaller than the free space velocity c:

$$\bar{c} = \frac{c}{n} = \frac{c}{\sqrt{\varepsilon_r \mu_r}} \ . \tag{1.4.1}$$

Instead of resorting now to the transformation (1.1.6), a so-called "Cerenkov transformation" is postulated: it results from the Lorentz transformation by replacing in the former c by $\bar{c}$, i.e.,

$$x' = \frac{x - \dfrac{v}{\bar{c}} \bar{c} t}{\sqrt{1 - \gamma^2}} ; \quad y' = y; \quad z' = z; \quad \bar{c} t' = \frac{\bar{c} t - \dfrac{v}{\bar{c}}}{\sqrt{1 - \gamma^2}} ; \quad \gamma^2 = \frac{v^2}{\bar{c}^2} \ . \tag{1.4.2}$$

Imposing upon the magnetic vector potential V and upon the electric scalar potential ϕ the condition

$$\nabla \cdot V + \frac{n^2}{c^2} \frac{\partial \phi}{\partial t} = 0 \ , \tag{1.4.3}$$

as opposed to (1.2.3), we find that each of them obeys at charge- and current-free points in space-time, the three-dimensional wave equation

$$\nabla^2 V = \frac{n^2}{c^2} \frac{\partial^2 V}{\partial t^2}; \quad \nabla^2 \phi = \frac{n^2}{c^2} \frac{\partial^2 \phi}{\partial t^2} \ . \tag{1.4.4}$$

Thus, introducing the "world coordinates"

$$x_1 \equiv x; \quad x_2 \equiv y; \quad x_3 \equiv z; \quad x_4 = i\bar{c}t; \quad i = \sqrt{-1} \ , \tag{1.4.5}$$

we are able to rewrite (1.4.3) in the form

$$\frac{\partial V_x}{\partial x_1} + \frac{\partial V_y}{\partial x_2} + \frac{\partial V_z}{\partial x_3} + \frac{\partial}{\partial x_4}\left(i\,\frac{n\phi}{c}\right) = 0 \ . \tag{1.4.6}$$

The components of a four-dimensional "Cerenkov vector" now obey, per definition, the transformation (1.4.2); further on the basis of the relation

$$\frac{\partial}{\partial x_j'} = \sum_{k=1}^{4} \frac{\partial x_k}{\partial x_j'}\,\frac{\partial}{\partial x_k}; \quad j = 1,2,3,4 \ , \tag{1.4.7}$$

the operators $\partial/\partial x_j$ formally constitute the components

$$\Box_1 = \frac{\partial}{\partial x_1} = \frac{\partial}{\partial x}; \quad \Box_2 = \frac{\partial}{\partial x_2} = \frac{\partial}{\partial y} \ ; \tag{1.4.8}$$

$$\Box_3 = \frac{\partial}{\partial x_3} = \frac{\partial}{\partial z}; \quad \Box_4 = \frac{\partial}{\partial x_4} = \frac{1}{i}\frac{n}{c}\frac{\partial}{\partial t} \ , \tag{1.4.9}$$

of a symbolic four-vector $\vec{\Box}$. Hence (1.4.6) implies existence of a four-vector Ω whose components now read

$$\Omega_1 = V_x; \quad \Omega_2 = V_y; \quad \Omega_3 = V_z; \quad \Omega_4 = i\,\frac{n\phi}{c} \ . \tag{1.4.10}$$

As the differential operator

$$\Box^2 = (\vec{\Box})^2 = \frac{\partial^2}{\partial x_1^2} + \frac{\partial^2}{\partial x_2^2} + \frac{\partial^2}{\partial x_3^2} + \frac{\partial^2}{\partial x_4^2} \ , \tag{1.4.11}$$

is invariant under a Cerenkov transformation — so are the wave equations (1.4.4), a fact which is of basic importance for the approach considered here. The primed counterparts of Ω are now directly obtained from (1.4.2), i.e.,

$$V_x' = \frac{V_x - (vn^2/c^2)\phi}{\sqrt{1 - n^2v^2/c^2}}; \quad V_y' = V_y; \quad V_z' = V_z;$$

$$n\frac{\phi'}{c} = \frac{n\dfrac{\phi}{c} - n\dfrac{v}{c}V_x}{\sqrt{1 - n^2v^2/c^2}} \ . \tag{1.4.12}$$

When Cerenkov invariance is imposed upon (1.2.1,3), the so-called "Cerenkov electrodynamics" results. We shall not further elaborate on this topic (see again [1.17]); let us only point out that, within this framework, the phenomenological numbers ε_r, μ_r — pertaining to the investigated medium — may be set arbitrarily. In particular, fictitious media with $n = 0$ may now be defined without violating the laws of electrodynamics[5].

---

[5] The same formal approach is sometimes adopted in the Minkowski electrodynamics, too, e.g., in a plasma where the frequency of excitation equals the so-called plasma frequency; the dielectric constant formally vanishes in this case. However, the underlying "philosophy" is totally different from that pertaining to "Cerenkov electrodynamics".

In contrast to Minkowski electrodynamics, the following transformations between the primed and unprimed field components now result:

$$E'_x = E_x; \qquad\qquad B'_z = B_z ,$$

$$E'_y = \frac{E_y - vB_z}{\sqrt{1 - n^2 v^2/c^2}}; \qquad B'_y = \frac{B_y + v(n^2/c^2)E_z}{\sqrt{1 - n^2 v^2/c^2}} , \qquad (1.4.13)$$

$$E'_z = \frac{E_z + vB_y}{\sqrt{1 - n^2 v^2/c^2}}; \qquad B'_z = \frac{B_z - v(n^2/c^2)E_y}{\sqrt{1 - n^2 v^2/c^2}} ,$$

along with

$$H'_x = H_x; \qquad\qquad D'_x = D_x ,$$

$$H'_y = \frac{H_y + vD_z}{\sqrt{1 - n^2 v^2/c^2}}; \qquad D'_y = \frac{D_y - v(n^2/c^2)H_z}{\sqrt{1 - n^2 v^2/c^2}} , \qquad (1.4.14)$$

$$H'_z = \frac{H_z - vD_y}{\sqrt{1 - n^2 v^2/c^2}}; \qquad D'_z = \frac{D_z + v(n^2/c^2)H_y}{\sqrt{1 - n^2 v^2/c^2}} .$$

Inversely, we directly obtain

$$E_x = E'_x; \qquad\qquad B_z = B'_z ,$$

$$E_y = \frac{E'_y + vB'_z}{\sqrt{1 - n^2 v^2/c^2}}; \qquad B_y = \frac{B'_y - v(n^2/c^2)E'_z}{\sqrt{1 - n^2 v^2/c^2}} , \qquad (1.4.15)$$

$$E_z = \frac{E'_z - vB'_y}{\sqrt{1 - n^2 v^2/c^2}}; \qquad B_z = \frac{B'_z + v(n^2/c^2)E'_y}{\sqrt{1 - n^2 v^2/c^2}} ,$$

and

$$H_x = H'_x; \qquad\qquad D_x = D'_x ,$$

$$H_y = \frac{H'_y - vD'_z}{\sqrt{1 - n^2 v^2/c^2}}; \qquad D_y = \frac{D'_y + v(n^2/c^2)H'_z}{\sqrt{1 - n^2 v^2/c^2}} , \qquad (1.4.16)$$

$$H_z = \frac{H'_z + vD'_y}{\sqrt{1 - n^2 v^2/c^2}}; \qquad D_z = \frac{D'_z - v(n^2/c^2)H'_y}{\sqrt{1 - n^2 v^2/c^2}} .$$

Resorting now to (1.1.5) we find, as opposed to (1.1.11 and 12), that

$$D' - \frac{n^2}{c^2} v \times H' = \varepsilon_r \varepsilon_0 (E' - v \times B') \ , \tag{1.4.17}$$

$$B' + \frac{n^2}{c^2} v \times E' = \mu_r \mu_0 (H' + v \times D') \ , \tag{1.4.18}$$

which are identically satisfied by

$$D' = \varepsilon_r \varepsilon_0 E' \ , \tag{1.4.19}$$

$$B' = \mu_r \mu_0 H' \ , \tag{1.4.20}$$

compare: (1.1.13, 14).
Reverting to the laboratory, (1.4.17 and 18) yield

$$D + \frac{n^2}{c^2} v \times H = \varepsilon_r \varepsilon_0 (E + v \times B) \ , \tag{1.4.21}$$

$$B - \frac{n^2}{c^2} v \times E = \mu_r \mu_0 (H - v \times D) \ , \tag{1.4.22}$$

i.e.

$$D = \varepsilon_r \varepsilon_0 E \ , \tag{1.4.23}$$

$$B = \mu_r \mu_0 H \ . \tag{1.4.24}$$

Hence, the *constitutive* equations are *invariant* under a "Cerenkov transformation".

The fact that (1.4.23, 24) − instead of (1.1.15, 16) − hold even in the presence of motion, is crucial for the following:

Proceeding from the definition of a fictitious medium with

$$n = 0 \ , \tag{1.4.25}$$

we obtain at first the Galilean transformation

$$x' = x - vt; \quad y' = y; \quad z' = z; \quad t' = t \ . \tag{1.4.26}$$

Further, (1.4.25) is obviously satisfied if we define a fictitious "anti-electric" medium [1.18]

$$\varepsilon_r = 0; \quad \mu_r \neq 0 \ ; \tag{1.4.27}$$

hence, see (1.4.19, 23),

$$D = 0; \quad D' = 0 \ , \tag{1.4.28}$$

on the one hand, whereas on the other − the pre-Maxwell equations

$$\nabla \times H = j \ ,$$

(1.4.29)

$$\nabla \times E = -\frac{\partial B}{\partial t} \ ,$$

(1.4.30)

ensue, immediately yielding the magneto-quasistatic approximation.

Finally, field and induction components transform according to

$$H' = H; \quad B' = B \ ,$$

(1.4.31)

$$E'_x = E_x \ ,$$

(1.4.32)

$$E'_y = E_y - v B_z \ ,$$

(1.4.33)

$$E'_z = E_z + v B_y \ .$$

(1.4.34)

Obviously, (1.4.25) is also satisfied by a fictitious "anti-magnetic" medium [1.19], in which

$$\varepsilon_r \neq 0; \quad \mu_r = 0 \ ,$$

(1.4.35)

so that

$$B = 0; \quad B' = 0 \ .$$

(1.4.36)

For this case the Maxwell-Minkowski expressions (1.1.1, 2) simplify to

$$\nabla \times H = \frac{\partial D}{\partial t} + j \ ,$$

(1.4.37)

$$\nabla \times E = 0 \ ,$$

(1.4.38)

yielding thus the electro-quasistatic set; field and induction components now transform according to the prescription

$$E' = E; \quad D' = D \ ,$$

(1.4.39)

$$H'_x = H_x \ ,$$

(1.4.40)

$$H'_y = H_y + v D_z \ ,$$

(1.4.41)

$$H'_z = H_z - v D_y \ .$$

(1.4.42)

Let us point out that the initial assumption of a charge-free material is — according to (1.4.28) — a requirement for the existence of magneto-quasistatic fields; on the other hand, existence of free charge densities and their attendant convection currents is obviously compatible with (1.4.37, 38).

It is now clear that the quasistatic approximations are formally obtainable by judicious choice of an anti-electric or anti-magnetic medium in which the field resides; moreover, while quasistatic Minkowskian electrodynamics, see

(1.1.7,8), implies existence of inductions in a primed frame of reference even when totally non-existent in the unprimed one, this is remedied by the outlined Ollendorff-Cerenkov approach in which, see (1.4.28,36), vanishing of inductions in one frame of reference assures their non-existence in any other as well.

Finally let us point out that derivation, e.g., of E from a potential function ϕ now leads to a *time-dependent* Poisson equation

$$\nabla^2 \phi(x, y, z, t) = -\frac{\rho(x, y, z, t)}{\varepsilon_r \varepsilon_0} , \qquad (1.4.43)$$

which, however, nowhere implies "slowness".

Hence the two posed questions have found their answers.

Let us once more emphasize that the Ollendorff-Cerenkov approach[6] does not pretend to develop new theories of relativity [1.20]; nor is it involved with electron theory [1.21]: it is purely an artificial means which introduces fictitious media of either negligible specific magnetic energy density w_m with respect to the relevant electric counterpart w_e, or vice versa.

A simple criterion for quasistatics therefore ensues:

with

$$w_m = \tfrac{1}{2}\mu_r \mu_0 (H)^2; \qquad w_e = \tfrac{1}{2}\varepsilon_r \varepsilon_0 (E)^2 , \qquad (1.4.44)$$

magneto-quasistatics is obtained if

$$w_m \gg w_e , \qquad (1.4.45)$$

and electro-quasistatics − if

$$w_m \ll w_e . \qquad (1.4.46)$$

Furthermore, as regards consistency, the relation $\phi' = \phi - v V_x$, e.g., is still obtained from relativity with a finite value of c, even though a quasistatic problem may be considered.

Let us once more stress that the expressions

$$D' - \frac{n^2}{c^2} v \times H' = \varepsilon_r \varepsilon_0 (E' - v \times B') \qquad (1.4.47)$$

$$B' + \frac{n^2}{c^2} v \times E' = \mu_r \mu_0 (H' + v \times D') , \qquad (1.4.48)$$

and their implications, namely

$$D' = \varepsilon_r \varepsilon_0 E' , \qquad (1.4.49)$$

$$B' = \mu_r \mu_0 H' , \qquad (1.4.50)$$

---

[6] which still ascribes, *in vacuo*, a finite value of $c = 1/\sqrt{\mu_0 \varepsilon_0}$.

are basic in avoiding the inconsistencies inherent in the application of restricted relativity to quasistatics: in contrast to (1.4.49, 50), Minkowski's electrodynamics yields (1.1.13, 14), so that the simple omission of the electric induction (magneto-quasistatics), or of the magnetic induction (electro-quasistatics) in one frame of reference does not eliminate it in the other – admittedly a source of inconsistency [Ref. 1.8, p. 44].

1.5 Further Remarks on Quasistatics

We supplement our review of quasistatics by a short digression to stationary charge flow, i.e., stationary current. For such fields, Maxwell's laws reduce to

$$\nabla \times H = j \ , \tag{1.5.1}$$

$$\nabla \times E = 0 \ , \tag{1.5.2}$$

being often supplemented by Ohm's law. Obviously, Ollendorff-Cerenkov electrodynamics with $n = \sqrt{\mu_r \varepsilon_r}$ formally incorporates this case as well, with $\mu_r = \varepsilon_r = 0$.

Stationary fields arise in practical cases in which the wave-propagational properties of the fields are irrelevant, yet quasistatics is adequate. Such a situation may be visualized by considering, e.g., Fig. 1.1 c, where the conducting plates terminate in a resistive sheet. With dc excitation, a stationary current arises in the configuration; E and H fields coexist. However, when the overall load resistance R of the resistive sheet is low, the E-field is "weak" and the configuration models an inductor, whereas when it is high the H-field is relatively unimportant and a model for a capacitor ensues. Where is the boundary between these two cases?

As already hinted, when the specific magnetic energy density w_m greatly outweighs the electric specific energy density w_e, it is the *inductor* case, whereas when $w_m \ll w_e$, it is the *capacitor* case; finally, when

$$w_m = w_e \ , \tag{1.5.3}$$

an intermediate case ensues (we are still dealing with a dc situation).

Thus, in Fig. 1.1 c, approximate expressions for w_m and w_e read

$$w_m \simeq \frac{1}{2}\mu_0 \left(\frac{I_0}{b}\right)^2 \ , \tag{1.5.4}$$

$$w_e \simeq \frac{1}{2}\varepsilon_0 \left(\frac{I_0 R}{h}\right)^2 \ , \tag{1.5.5}$$

so that (1.5.3), in fact, implies

$$\sqrt{\frac{\mu_0}{\varepsilon_0}} \simeq \frac{b}{h} R \ . \tag{1.5.6}$$

Or, with b and h of the same order −

$$\sqrt{\frac{\mu_0}{\varepsilon_0}} \simeq R \ . \tag{1.5.7}$$

Conversely, it is clear that the condition for magneto-quasistatics may now be rewritten in the form

$$\frac{1}{2} \mu_0 \left(\frac{I_0}{b}\right)^2 \gg \frac{1}{2} \varepsilon_0 \left(\frac{I_0 R}{h}\right)^2 \ , \tag{1.5.8}$$

i.e.,

$$\sqrt{\frac{\mu_0}{\varepsilon_0}} \gg \frac{b}{h} R \ , \quad \text{or} \tag{1.5.9}$$

$$\sqrt{\frac{\mu_0}{\varepsilon_0}} \gg R \ , \tag{1.5.10}$$

and that for electro-quasistatics in the form

$$\frac{1}{2} \mu_0 \left(\frac{I_0}{b}\right)^2 \ll \frac{1}{2} \varepsilon_0 \left(\frac{I_0 R}{h}\right)^2 \ , \tag{1.5.11}$$

leading towards

$$\sqrt{\frac{\mu_0}{\varepsilon_0}} \ll \frac{b}{h} R \ , \tag{1.5.12}$$

i.e.

$$\sqrt{\frac{\mu_0}{\varepsilon_0}} \ll R \ . \tag{1.5.13}$$

Under a different point of view − which we very briefly mention − we can try to determine up to what (angular) frequency the system of Fig. 1.1c (analyzed for dc) still represents a "resistor", i.e., the power dissipated at the load greatly exceeds the magnetic and/or electric energy stored in the volume (blh) per cycle. Thus the first case prevails as long as

$$\frac{1}{2} \mu_0 \left(\frac{I_0}{b}\right)^2 (blh) \omega \ll I_0^2 R \ , \tag{1.5.14}$$

i.e., dispensing with the factor 1/2 which is redundant in our far-reaching approximations, as long as

$$\omega \ll \frac{R}{\mu_0 \dfrac{hl}{b}} \; . \tag{1.5.15}$$

For the second case we have

$$\frac{1}{2} \varepsilon_0 \left(\frac{I_0 R}{h} \right)^2 (blh)\, \omega \ll I_0^2 R \; , \tag{1.5.16}$$

or

$$\omega \ll \frac{1}{\left(\varepsilon_0 \dfrac{bl}{h} \right) R} \; . \tag{1.5.17}$$

Obviously, $\mu_0 hl/b$ and $\varepsilon_0 bl/h$ represent, respectively, the relevant static plate inductance and capacitance. We shall not, however, pursue this topic further.

1.6 Summary of Approaches

Following the above remarks, the approach underlying the analysis in the present volume is readily summarized:

- wherever possible, quasistatics is resorted to on the basis of dimensionless criteria validated à posteriori by analytical results, or experimentally;
- solutions proceed by investigating electric, magnetic or stationary subsystems, or alternatively primary ("excitation") and secondary ("reaction") fields;
- electrodynamics of moving media is used with relativity in the background: a minimum of assumptions is thus required and possible epistemological inconsistencies are smoothed out by the Ollendorff-Cerenkov approach, likewise in the background.

1.7 Outline of Contents

Chapter 2 deals with some idealized engineering aspects of electromagnetism in the presence of moving armatures, e.g., unipolar induction, braking phenomena and linear machines. While R&D or design engineers nowadays have direct and routine access to relevant computer packages, a basic grasp of the underlying electromagnetic principles is still very often a must for them: with the advent of modern specialized machines used in the computer industry (spooling), in cryogenics, in missiles, in computed tomography, in linear traction and elsewhere, any approach exclusively based on experimentation in conjunction with man-machine interaction is quite limited in scope and must be augmented by real insight derived from Maxwell's equations. It is only this latter insight that directly contributes to better understanding among engineers and applied physicists engaged in design and production of special purpose magneto-electric converters.

Other sections dealing with electromagnetic induction, transient phenomena and screening, are again highly idealized versions of engineering problems. Thus, in Chap. 3 some configurations of induction phenomena for bodies at rest are analyzed; eddy currents, reaction fields, proximity effects and a variant of spiral coils are considered through straightforward application of Maxwell's equations. Calculation of eddy currents in engineering applications often calls for more advanced approaches and the attendant boundary-value problems may be quite complicated. Some practical tools are accordingly outlined in this chapter, and wherever possible, closed-form solutions are sought, so as to permit direct visualization of the rôle of all the parameters in the final result. While closed-form treatment admittedly entails some extra analytical work (which it is fashionable to shirk today, on the pretext of availability of the computer), the fact that the system parameters appear in a single expression, which may even prove in the end to be a well-known function, is a distinct advantage.

In Chap. 4 some transient phenomena are investigated. A model of a grounding system is presented and field build-up under an electrical surge is analyzed. Another section of the chapter deals with field penetration into saturable cores; experiment shows that build-up of saturation in power equipment is relatively slow and the calculated results substantiate this empirical finding. Further, transients in a highly idealized version of a super-conducting component are calculated. With the expected advent of superconductors in power cables, in signal transmission lines as well as in the construction of electrical generators, this topic will acquire practical interest. For the moment, however, only some very preliminary results pertaining to switching phenomena in superconducting material are put forward in the present text.

The chapter concludes with electromechanical transients in liquid metal; MHD generation and pumping in general, and linear pumps in particular are of interest in present-day technology and this fact motivated the analysis of this section.

The final (fifth) chapter is concerned with some dynamic phenomena; the first two sections deal with radiation pressure as well as with interaction between a traveling wave and a moving dielectric slab. Although engineering applications (e.g., laser jet engines) are still in the future, micromanipulation of particles, optical levitation and force spectroscopy, among others, are already practised.

The section on a moving dielectric slab not only models a dielectric motor, but also provides insight into some wave-kinematic aspects of rapidly expanding gas columns probed by bouncing-off of electromagnetic waves.

Finally, two topics dealing with relatively low-frequency screening are taken up; the relevant problems are presented in highly modelled versions, and analyses are performed with the aid of the Hertz and Fitzgerald (Ref. 1.18, p. 200 and Ref. 1.22) electromagnetic superpotentials.

2. Principles of Magneto-Electric Interactions

This chapter treats some aspects of electrodynamics of moving media. The problem of induction is tackled in two examples through application of the Fitzgerald superpotential; whereas the first example is primarily of academic interest, the second has practical value because it outlines the applications of the superpotential to electromagnetic coils, in general, and to linear induction devices, in particular. Subsequent sections deal with unipolar induction, a phenomenon of interest in itself (e.g., for high-current machines) as well as because it opened up the path for restricted relativity in Einstein's first paper on the topic. In the early days of applied electromagnetics this topic puzzled also Faraday, who − obviously unaware of relativity − was not sure "where ... the current (is) generated: in the wire ... or in the magnet ...". The final two sections are concerned with aspects of electromagnetic braking and emphasize some points of view inherent in linear braking devices.

2.1 Faraday's Law in the Presence of Moving Media

It is common knowledge that the practical application of Faraday's law to moving media is attended by difficulties even today. The student, the engineer, even the scientist is not always certain how to apply the so-called "flux-rule" − and/or whether the "induced electromagnetic force" is to be measured, in the frame of reference of the laboratory or in that attached to the moving body: it is a fact that, although more than eighty years have passed since Einstein published his epoch-making paper "On the Electrodynamics of Moving Bodies" (with Faraday's law particularly in mind), its principles have apparently been less well assimilated than Maxwell's approach to bodies at rest.

2.1.1 Overview and Basic Remarks

The difficulties just outlined are greatly enhanced by the familiar attempts to extend Maxwell's equations to moving bodies. For example, many well-known textbooks present the pre-relativity formula:

$$\nabla \times E' = -\frac{\partial B}{\partial t} + \nabla \times (v \times B) \ ,$$

where E' denotes the electric field measured in the reference frame of the medium in motion, v is the velocity of the medium with respect to the laboratory, B is the magnetic induction as recorded in the laboratory and t is the time parameter, also as measured in the laboratory (the basic meaning assigned to time in different frames of reference thus being completely lost).

The main drawbacks of this approach are as follows:

- Fields are measured in two distinct frames of reference, which entails a considerable amount of complication in interpretation and results (see also Sect. 2.1.5).
- The equation is valid, in principle, only for velocities $v^2 \ll c^2$, c denoting the phase velocity of a plane wave of electromagnetic radiation in vacuum.

The situation is somewhat improved if a new field E^* is introduced (Sect. 1.1.3 and [2.1]), again such that

$$\nabla \times E^* = -\frac{\partial B}{\partial t} + \nabla \times (v \times B) \ ,$$

but now all the field components are "measured" in the laboratory. In this case, the equation is not at variance with relativity, but its significance is largely formal and the casual reader may be expected to have trouble grasping the real difference between E' and E^*, especially because the right-hand sides of both equations are identical.

It should be noted that, whereas the above equations are often derived through application of the Helmholtz-Lorentz derivative [2.1, 2]

$$\dot{B} = \frac{\partial B}{\partial t} + v(\nabla \cdot B) - \nabla \times (v \times B) \ ,$$

(with $\nabla \cdot B = 0$, which is the case if B represents the magnetic induction), sometimes the "substantial" derivative

$$\frac{DB}{Dt} = \frac{\partial B}{\partial t} + (v \cdot \nabla)B \ ,$$

is mentioned under other approaches to the induction law [2.3]. This duality in tackling the same phenomenon is evidently also a source of puzzlement.

We shall not attempt, within the scope of this section, to cover the whole complex of these problems; still, comments by famous physicists on Faraday's law in the present context are highly instructive. Thus, for instance, *Landau* and *Lifshitz* [2.3], when advocating the use of coordinates *fixed* to the induced body, indicate that equivalence between the moving body and the body at rest is complete, irrespective of Galileo's or Einstein's principle of relativity; this is due to the fact "that the electromagnetic induction is independent of the cause of the change in the magnetic flux". *Feynman* et al. [2.4] considered the "transformer" term $\partial B / \partial t$ vs. the "motion-induced" term $\nabla \times (v \times B)$ and stressed the fact that "we know of no other place in physics where such a simple and accurate general principle requires for its real understanding an analysis in terms of *two different phenomena*" (i.e., that the flux rule is linked to both). *Sommerfeld* [Ref. 2.1, p. 363], despite the fact that electromagnetic induction has found its widest practical application in rotating machinery, is careful to emphasize that the Maxwell-Minkowski equations are not "directly applicable to problems involving rotation", although he expressly points out on two different occasions [Ref. 2.1, p. 277 and 280] that the velocity may be almost arbitrary.

In view of the above, it seems that it is much easier to replace – at least in principle – the various applied analyses by an approach based from the outset on a relativistic foundation. Such an approach is adequate for both uniform and "slowly" accelerated motion [2.5]; in the latter, however, a test for "slow" acceleration should be applied and the formalism of general relativity [2.6] resorted to if required.

For simplicity, we confine ourselves in this section to non-accelerated, translatory motion of a solid body, our aim being illustration of an analysis of electromagnetic induction in terms of special relativity.

2.1.2 Fitzgerald's Superpotential and Its Use in Different Frames of Reference

Our operative problem reads:

How does one, *in principle*, tackle the induction phenomena of Maxwell's equations within the framework of restricted relativity?

In an attempt to answer this, we introduce an inertial frame of reference characterized by the right-handed Cartesian system of coordinates x, y, z in which the running time t is continuously recorded by suitably distributed stationary clocks. For matter-free space (vacuum permeability $\mu_0 = 4\pi \times 10^{-7}$ henry/m, vacuum permittivity $\varepsilon_0 \simeq 8.85 \times 10^{-12}$ farad/m), Maxwell's equations relating the electric field E to its magnetic counter-part H read:

$$\nabla \times H = \frac{\partial}{\partial t} \varepsilon_0 E \ , \tag{2.1.1}$$

$$\nabla \times E = -\frac{\partial}{\partial t} \mu_0 H \ . \tag{2.1.2}$$

In perfectly empty space, these relations are supplemented by

$$\nabla \cdot \mu_0 H = 0 \ , \tag{2.1.3}$$

$$\nabla \cdot \varepsilon_0 E = 0 \ . \tag{2.1.4}$$

By virtue of (2.1.4), we may introduce the electric vector potential C such that

$$\varepsilon_0 E = \nabla \times C \ . \tag{2.1.5}$$

Hence, reverting to (2.1.1), the magnetic field is readily obtained in terms of C and of an additional magnetic scalar potential χ:

$$H = \frac{\partial C}{\partial t} - \nabla \chi \ , \tag{2.1.6}$$

and recalling the induction law (2.1.2), we have

$$\nabla \times (\nabla \times C) = -\mu_0 \varepsilon_0 \frac{\partial}{\partial t} \left(\frac{\partial C}{\partial t} - \nabla \chi \right) \ . \tag{2.1.7}$$

Expanding the curl curl operator in a Cartesian system of coordinates and introducing the notation $c = (\mu_0 \varepsilon_0)^{-1/2}$ for the phase velocity of a plane light wave in vacuum, mentioned earlier, we obtain

$$\nabla (\nabla \cdot C) - \nabla^2 C = -\frac{1}{c^2} \left(\frac{\partial^2 C}{\partial t^2} - \nabla \frac{\partial \chi}{\partial t} \right) \ . \tag{2.1.8}$$

In view of (2.1.5), only the curl of C is specified; we are thus at liberty to choose

$$\nabla \cdot C = \frac{1}{c^2} \frac{\partial \chi}{\partial t} \ , \tag{2.1.9}$$

whereby (2.1.8) reduces to

$$\nabla^2 C = \frac{1}{c^2} \frac{\partial^2 C}{\partial t^2} \ . \tag{2.1.10}$$

Let us add, for the sake of completeness, that a similar equation is obtained for χ, namely

$$\nabla^2 \chi = \frac{1}{c^2} \frac{\partial^2 \chi}{\partial t^2} \ . \tag{2.1.11}$$

With the Fitzgerald vector superpotential F introduced through the defining relations:

$$C = \frac{1}{c^2} \frac{\partial F}{\partial t} \ , \tag{2.1.12}$$

$$\chi = \nabla \cdot F \ , \tag{2.1.13}$$

Eq. (2.1.9) becomes an identity and the differential equation for F follows directly either from (2.1.10,12), or from (2.1.11,13):

$$\nabla^2 F = \frac{1}{c^2} \frac{\partial^2 F}{\partial t^2} \ . \tag{2.1.14}$$

Recourse to four-dimensional electrodynamics, together with our exclusive use here of Cartesian coordinates (which obviates the need for distinction between covariant and contravariant vector and tensor components), will considerably simplify the subsequent stages of our analysis. Accordingly, instead of the three-dimensional vector F with components F_x, F_y, F_z, we introduce[1] a matrix representation of the tensor $[F]$, specified by the array

$$[F] = \begin{vmatrix} 0 & 0 & 0 & \dfrac{i}{c}F_x \\[2mm] 0 & 0 & 0 & \dfrac{i}{c}F_y \\[2mm] 0 & 0 & 0 & \dfrac{i}{c}F_z \\[2mm] -\dfrac{i}{c}F_x & -\dfrac{i}{c}F_y & -\dfrac{i}{c}F_z & 0 \end{vmatrix} ; \quad i = \sqrt{-1} . \qquad (2.1.15)$$

The reasoning behind this step is as follows: if (2.1.9) is to hold (by virtue of its form alone) in every inertial frame of reference, then the three components C_x, C_y, C_z of C on the one hand[2], and χ on the other, must form a four-vector Ω [Ref. 2.5, p. 145; p. 535], i.e., in our notation

$$\Omega = \left(C_x, C_y, C_z, -\frac{i}{c}\chi \right) , \quad \text{with} \qquad (2.1.16)$$

$$\frac{\partial C_x}{\partial x} + \frac{\partial C_y}{\partial y} + \frac{\partial C_z}{\partial z} + \frac{1}{ic\,\partial t}\left(-\frac{i}{c}\chi \right) = 0 . \qquad (2.1.17)$$

The four-vector is directly obtainable from (2.1.15) through application of the tensor divergence:

$$\text{Div}\,[F] \equiv \frac{\partial F_{jk}}{\partial x_k} ; \quad j,k = 1,2,3,4 , \qquad (2.1.18)$$

where the notation refers to Einstein's summation rule, and moreover

$$x_1 = x; \quad x_2 = y; \quad x_3 = z; \quad x_4 = ict . \qquad (2.1.19)$$

Hence, it is readily seen that

$$\Omega = \text{Div}\,[F] . \qquad (2.1.20)$$

---

[1] In analogy with Sommerfeld's representation of the Hertzian tensor, see [Ref. 2.1, p. 221].
[2] We do not differentiate, within the present context, between three-dimensional vectors and pseudo-vectors.

We thus have, for the first row of the matrix:

$$\frac{\partial F_{jk}}{\partial x_k} = \frac{\partial F_{1k}}{\partial x_k} = \frac{\partial F_{11}}{\partial x} + \frac{\partial F_{12}}{\partial y} + \frac{\partial F_{13}}{\partial z} + \frac{\partial F_{14}}{\partial \mathrm{i}ct}$$

$$= \frac{1}{\mathrm{i}c} \frac{\partial}{\partial t} \left(\frac{\mathrm{i}}{c} F_x \right) = \frac{1}{c^2} \frac{\partial F_x}{\partial t} \ . \tag{2.1.21}$$

Similar relations hold for the second and third rows, while for the fourth:

$$\frac{\partial F_{4k}}{\partial x_k} = \frac{\partial F_{41}}{\partial x} + \frac{\partial F_{42}}{\partial y} + \frac{\partial F_{43}}{\partial z} + \frac{\partial F_{44}}{\partial(\mathrm{i}ct)}$$

$$= \frac{\partial}{\partial x} \left(-\frac{\mathrm{i}}{c} F_x \right) + \frac{\partial}{\partial y} \left(-\frac{\mathrm{i}}{c} F_y \right) + \frac{\partial}{\partial z} \left(-\frac{\mathrm{i}}{c} F_z \right)$$

$$= -\frac{\mathrm{i}}{c} \nabla \cdot F \ . \tag{2.1.22}$$

The possibility of setting up a matrix as per (2.1.15) which, in turn, implies (2.1.20), depends upon the physical character of the four relevant components – here C_x, C_y, C_z, χ – which must obviously be known a priori.

We next supplement the inertial frame of reference x, y, z, t, by a primed frame x', y', z', t' attached to a medium moving with respect to the laboratory with constant speed v along the x-axis. The two frames are interrelated through the continuous and homogeneous Lorentz transformation:

$$x' = \frac{x - (v/c)ct}{\sqrt{1 - \beta^2}} ; \quad y' = y; \quad z' = z;$$

$$ct' = \frac{ct - (v/c)x}{\sqrt{1 - \beta^2}} ; \quad \beta^2 = \frac{v^2}{c^2} \ . \tag{2.1.23}$$

Or, using a matrix $[\alpha_{jk}]$ defined through:

$$[\alpha_{jk}] = \begin{vmatrix} \dfrac{1}{\sqrt{1 - \beta^2}} & 0 & 0 & \dfrac{\mathrm{i}\beta}{\sqrt{1 - \beta^2}} \\ 0 & 1 & 0 & 0 \\ 0 & 0 & 1 & 0 \\ \dfrac{-\mathrm{i}\beta}{\sqrt{1 - \beta^2}} & 0 & 0 & \dfrac{1}{\sqrt{1 - \beta^2}} \end{vmatrix} , \tag{2.1.24}$$

and resorting once more to Einstein's summation rule, we may write for short

$$x'_j \doteq \alpha_{jk} x_k; \quad j, k = 1, 2, 3, 4 \ . \tag{2.1.25}$$

Eq. (2.1.25) represents the transformation law for a four-vector, while that for a tensor of second rank T_{mn} into a primed tensor T'_{jk} reads

$$T'_{jk} = \alpha_{jm} \alpha_{kn} T_{mn} \ .\tag{2.1.26}$$

Further, we introduce a variant of (2.1.15), namely a square array comprising *only* the component F_z:

$$[F] = \begin{vmatrix} 0 & 0 & 0 & 0 \\ 0 & 0 & 0 & 0 \\ 0 & 0 & 0 & \dfrac{i}{c} F_z \\ 0 & 0 & -\dfrac{i}{c} F_z & 0 \end{vmatrix} \ .$$

Thus, in the primed frame of reference we find for the components F'_{jk} of $[F']$, the primed counterpart of $[F]$, the expressions:

$$[F'] = \begin{vmatrix} 0 & 0 & \dfrac{(v/c^2)F_z}{\sqrt{1-\beta^2}} & 0 \\ 0 & 0 & 0 & 0 \\ \dfrac{-(v/c^2)F_z}{\sqrt{1-\beta^2}} & 0 & 0 & \dfrac{i}{c}\dfrac{F_z}{\sqrt{1-\beta^2}} \\ 0 & 0 & -\dfrac{i}{c}\dfrac{F_z}{\sqrt{1-\beta^2}} & 0 \end{vmatrix} \ .\tag{2.1.27}$$

The primed four-vector $\boldsymbol{\Omega}'$, comprising the three components C'_x, C'_y, C'_z of the electric vector potential C' and also the primed magnetic scalar potential χ', is accordingly given by the new equation:

$$\boldsymbol{\Omega}' = \mathrm{Div}\,'[F'] \ ,\tag{2.1.28}$$

i.e.,

$$C'_x = \frac{\partial}{\partial z'}\left(\frac{(v/c^2)F_z}{\sqrt{1-\beta^2}}\right) = \frac{(v/c^2)}{\sqrt{1-\beta^2}}\frac{\partial F_z}{\partial z'} \ ,\tag{2.1.29}$$

$$C'_y = 0 \ ,\tag{2.1.30}$$

$$C_z' = \frac{\partial}{\partial x'}\left(\frac{-(v/c^2)F_z}{\sqrt{1-\beta^2}}\right) + \frac{\partial}{\partial ict'}\left(\frac{i}{c}\frac{F_z}{\sqrt{1-\beta^2}}\right) ,$$

$$= \frac{1}{\sqrt{1-\beta^2}}\left(-\frac{v}{c^2}\frac{\partial F_z}{\partial x'} + \frac{1}{c^2}\frac{\partial F_z}{\partial t'}\right) , \quad \text{and} \tag{2.1.31}$$

$$-\frac{i}{c}\chi' = \frac{\partial}{\partial z'}\left(-\frac{i}{c}\frac{F_z}{\sqrt{1-\beta^2}}\right) = -\frac{i}{c}\frac{1}{\sqrt{1-\beta^2}}\frac{\partial F_z}{\partial z'} . \tag{2.1.32}$$

The primed potential functions have thus directly been derived in terms of F_z. Reverting now to (2.1.5), we explicitly have:

$$\varepsilon_0 E_x = \frac{\partial C_z}{\partial y} - \frac{\partial C_y}{\partial z} , \tag{2.1.33}$$

$$\varepsilon_0 E_y = \frac{\partial C_x}{\partial z} - \frac{\partial C_z}{\partial x} , \tag{2.1.34}$$

$$\varepsilon_0 E_z = \frac{\partial C_y}{\partial x} - \frac{\partial C_x}{\partial y} . \tag{2.1.35}$$

By virtue of the principle of relativity, we similarly find for the primed field E' (components E_x', E_y', E_z') the expressions:

$$\varepsilon_0 E_x' = \frac{\partial C_z'}{\partial y'} - \frac{\partial C_y'}{\partial z'} , \tag{2.1.36}$$

$$\varepsilon_0 E_y' = \frac{\partial C_x'}{\partial z'} - \frac{\partial C_z'}{\partial x'} , \tag{2.1.37}$$

$$\varepsilon_0 E_z' = \frac{\partial C_y'}{\partial x'} - \frac{\partial C_x'}{\partial y'} . \tag{2.1.38}$$

An identical procedure, based on (2.1.6) and its primed counterpart $H' = (\partial C'/\partial t') - \nabla'\chi'$, yields the magnetic field in the moving frame of reference.

The above is the framework around which a sound approach to the electrodynamics of bodies in motion is to be constructed; to show that there is more to it than barren formality, we apply it in a practical example, namely, a "sheet" gliding past an exciting electromagnet.

2.1.3 Statement of Practical Problem and Basic Assumptions

An iron-core electromagnet (Fig. 2.1) is excited by a low frequency (say 50 or 60 Hz) power source; the magnetic poles (only one of which is shown in the diagram) are assumed − for simplicity − to be made of fictitious material ex-

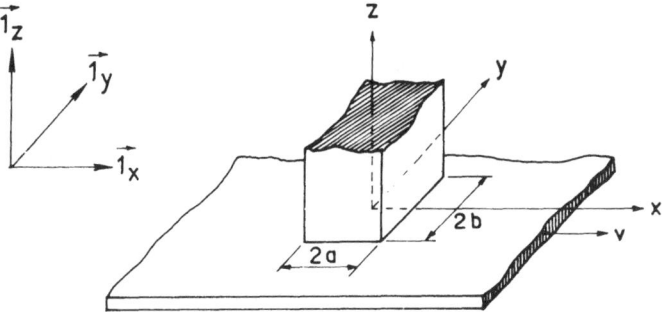

Fig. 2.1. Geometry of sheet gliding past electromagnet

hibiting an extremely high relative magnetic permeability μ_r (in the limit, $\mu_r \to \infty$) and, by contrast, a dielectric constant and relative electric conductivity (the latter, e.g., with respect to copper) both tending to zero.

For clarity we now assume that the imposed magnetic field comprises a *uniform* axial component H_z only, wholly localized to the region between the pole pieces; by contrast, the magnetic field exterior to the poles is assumed to be *vanishingly small*. This situation, in spite of its patently fictional character, is fairly closely approximated in practice by well-designed pole pieces with a *very narrow* air gap: in keeping with this model we mentally separate the interior region from the exterior one by a fictitious "magnetic wall", impenetrable in the direction perpendicular to the z-axis, and along which we assume a discontinuity in H_z. Apart from the very narrow region along the pole perimeter, which we exclude from our analysis (see below and also Sect. 2.3.3), the magnetic field is therefore, under this approximation, curl-free. These simplifying assumptions greatly reduce the computational labor without, however, impairing our insight into the main problem of motion-induced emf.

The excitation coils of the electromagnet (not shown) are taken to be quite remote from the air gap; it is further assumed that the magnetic circuit closes upon itself at a similarly remote location.

In the very narrow air gap, which interrupts also the "magnetic wall" whose width is much smaller than each horizontal pole dimension $2a$, $2b$, a nonconducting, nonmagnetic and uncharged sheet is gliding with a regulated constant velocity v along the x-axis of a right-hand Cartesian system of coordinates x, y, z (unit vectors $1_x, 1_y, 1_z$).

Finally, in order to obviate any additional complications due to immaterial "end-effects" entailed by finite dimensions, we also assume that the sheet is "practically infinite", again compared to the linear dimensions of the pole cross section.

What is the electromagnetic field intensity recorded on the sheet?

2.1.4 Solution

We introduce the vector F having a z-component F_z only

$$F = F_z 1_z \ . \tag{2.1.39}$$

For strictly monochromatic excitation – obviously, also an idealization – with angular supply frequency ω and complex amplitude $\bar{F}_z$, we have

$$F_z = \bar{F}_z e^{-i\omega t} , \tag{2.1.40}$$

where the linear operator Re (real part only!) has been suppressed. Real infiniteness is obviously ruled out; in concrete terms, say that the width (along the y-axis) is about ten times the dimension $2b$, while the "infiniteness" along the x-axis is simulated through the mental image of a conveyor belt propelled on two pulleys fixed at $x = \pm l$ with, say, $l \simeq 20 \times (2a)$.

Resorting to (2.1.14), we find that $\bar{F}_z$ satisfies the Helmholtz (partial) differential equation

$$\nabla^2 \bar{F}_z + \frac{\omega^2}{c^2} \bar{F}_z = 0 . \tag{2.1.41}$$

However, on account of its rather specialized geometry, our fictitious magnetic field is vortex-free, i.e.,

$$\nabla \times H = 0 . \tag{2.1.42}$$

Accordingly, (2.1.41) reduces to the much simpler Laplace equation

$$\nabla^2 \bar{F}_z = 0 . \tag{2.1.43}$$

A) Imposed Magnetic Field – Laboratory Frame of Reference

In order to be able to express the magnetic field through simple analytic functions, we somewhat relax at this stage the stringent spatial assumptions which wholly confine it to the very narrow gap, assuming it to be in fact confined in the $z = 0$ *plane* only to the rectangular region $|x| < a; |y| < b$.

Any resulting finite vortex densities of H will be relegated (see below) to a two-dimensional, *excluded* singularity, of no further interest. Denoting now by $B^{(e)}$ the imposed (= exciting, superscript e) magnetic induction, we can write in the $z = 0$ plane that $B^{(e)} = 0 \mathbf{1}_x + 0 \mathbf{1}_y + B^{(e)} \mathbf{1}_z$; introducing the complex amplitude $\bar{B}^{(e)}$, we have

$$B^{(e)} = \begin{cases} \bar{B}^{(e)} e^{-i\omega t}; & |x| < a; \quad |y| < b \\ 0 ; & |x| > a; \quad |y| > b \end{cases}; \quad z = 0 . \tag{2.1.44}$$

With the variables of integration m, n, ξ and η, the induction component (2.1.44) may compactly be expressed through the Fourier integral

$$B^{(e)} = \bar{B}^{(e)} e^{-i\omega t} \frac{1}{2\pi} \int_{-\infty}^{\infty} e^{imx} dm \int_{-a}^{a} e^{-im\xi} d\xi \frac{1}{2\pi} \int_{-\infty}^{\infty} e^{iny} dn \int_{-b}^{b} e^{-in\eta} d\eta$$

$$= \bar{B}^{(e)} e^{-i\omega t} \frac{4ab}{(2\pi)^2} \int_{-\infty}^{\infty} \int_{-\infty}^{\infty} \frac{\sin ma}{ma} e^{imx} \frac{\sin nb}{nb} e^{iny} dm \, dn . \tag{2.1.45}$$

As remarked, for the quasistatic approach, the wave equation describing the behavior of the Fitzgerald superpotential reduces to the Laplace equation. Although in our case this equation cannot be solved by separation of variables so as to satisfy the boundary conditions on the pole pieces, such a solution may nevertheless be postulated if the poles are reduced to "vortex threads", namely:

$$\lim_{\substack{|B^{(e)}| \to \infty \\ a \to 0 \\ b \to 0}} B^{(e)}(4ab) = \bar{\phi}^{(e)} , \tag{2.1.46}$$

$\bar{\phi}^{(e)}$ being finite and standing for the excitation flux. The fiction of the vortex thread implies, in mathematical terms, that the spatial distribution of $B^{(e)}$ in the $z = 0$ plane has been replaced by a two-dimensional δ-function: thus, for $a \to 0$, $b \to 0$, we have

$$B^{(e)} = \bar{\phi}^{(e)} e^{-i\omega t} \frac{1}{(2\pi)^2} \int_{-\infty}^{\infty} \int_{-\infty}^{\infty} e^{imx} e^{iny} dm \, dn ; \quad z = 0 . \tag{2.1.47}$$

Now, the expression

$$\lim_{\substack{m = -\alpha \\ \alpha \to \infty}} \int_{}^{\alpha} e^{ixm} dm = 2\pi\delta(x) \tag{2.1.48}$$

represents the Dirac delta-function, so that the "vortex thread" effectively defines, in turn, the distribution

$$B^{(e)} = \bar{\phi}^{(e)} e^{-i\omega t} \delta(x)\delta(y); \quad z = 0 . \tag{2.1.49}$$

Reverting now to (2.1.43), we are able to write down the appropriate solution for the primary Fitzgerald component $F^{(e)} \equiv F_z$ as per (2.1.40). Thus, introducing a still unknown spectral density $\bar{f}(m, n)$, we have

$$F^{(e)} = e^{-i\omega t} \int_{-\infty}^{\infty} \int_{-\infty}^{\infty} \bar{f}(m, n) e^{i(mx + ny)} \cosh(\sqrt{m^2 + n^2} z) \, dm \, dn . \tag{2.1.50}$$

How do we determine $\bar{f}(m, n)$? Eqs. (2.1.6, 12 – 14 and 39), in general, yield the components H_x, H_y, H_z of the magnetic field H,

$$H_x = -\frac{\partial^2 F_z}{\partial x \, \partial z} , \tag{2.1.51}$$

$$H_y = -\frac{\partial^2 F_z}{\partial y \, \partial z} , \tag{2.1.52}$$

$$H_z = \frac{\partial^2 F_z}{\partial x^2} + \frac{\partial^2 F_z}{\partial y^2} . \tag{2.1.53}$$

Hence, in the $z = 0$ plane, we find $-$ in our case $-$ the magnetic induction component $B^{(e)} = \mu_0 H_z^{(e)}$ via (2.1.50), i.e.,

$$B^{(e)} = -\mu_0 e^{-i\omega t} \int_{-\infty}^{\infty} \int_{-\infty}^{\infty} \bar{f}(m,n) e^{i(mx+ny)} (m^2+n^2) \, dm \, dn; \quad z = 0 \ . \quad (2.1.54)$$

From (2.1.45, 54), we immediately obtain

$$\bar{B}^{(e)} \frac{4ab}{(2\pi)^2} \frac{\sin ma}{ma} \frac{\sin nb}{nb} = -\mu_0 (m^2+n^2) \bar{f}(m,n) \qquad (2.1.55)$$

where reference should be made to (2.1.46). Hence

$$F^{(e)} = -\frac{1}{\mu_0} \frac{\bar{\phi}^{(e)}}{(2\pi)^2} e^{-i\omega t} \int_{-\infty}^{\infty} \int_{-\infty}^{\infty} \frac{\sin ma}{ma} \frac{\sin nb}{nb} e^{i(mx+ny)}$$

$$\cdot \frac{\cosh \sqrt{m^2+n^2} z}{m^2+n^2} \, dm \, dn \ , \qquad (2.1.56)$$

or, finally, as $a \to 0$ and $b \to 0$:

$$F^{(e)} = -\frac{1}{\mu_0} \frac{\bar{\phi}^{(e)}}{(2\pi)^2} e^{-i\omega t} \int_{-\infty}^{\infty} \int_{-\infty}^{\infty} e^{i(mx+ny)} \frac{\cosh \sqrt{m^2+n^2} z}{m^2+n^2} \, dm \, dn \ . \qquad (2.1.57)$$

The pole region has thus been replaced by a "vortex filament" of magnetic excitation and any finite-field vortex densities are now wholly confined to the very narrow air gap localized at $a \to 0$, $b \to 0$.

It goes without saying that the relevance of the proposed approximation to the considered problem improves with increasing distance from the region of singularity.

B) Electromagnetic Field $-$ Moving Frame of Reference

We now turn our attention to the primed frame of reference, in which the gliding sheet is at rest. In view of (2.1.36 $-$ 38), and (2.1.29 $-$ 31), there results, in general,

$$\varepsilon_0 E_x' = \frac{\partial}{\partial y'} \left[\frac{1}{\sqrt{1-\beta^2}} \left(-\frac{v}{c^2} \frac{\partial F_z}{\partial x'} + \frac{1}{c^2} \frac{\partial F_z}{\partial t'} \right) \right] , \qquad (2.1.58)$$

$$\varepsilon_0 E_y' = \frac{\partial}{\partial z'} \left[\frac{1}{\sqrt{1-\beta^2}} \left(\frac{v}{c^2} \frac{\partial F_z}{\partial z'} \right) \right]$$

$$- \frac{\partial}{\partial x'} \left[\frac{1}{\sqrt{1-\beta^2}} \left(-\frac{v}{c^2} \frac{\partial F_z}{\partial x'} + \frac{1}{c^2} \frac{\partial F_z}{\partial t'} \right) \right] , \qquad (2.1.59)$$

$$\varepsilon_0 E_z' = -\frac{\partial}{\partial y'}\left[\frac{1}{\sqrt{1-\beta^2}}\left(\frac{v}{c^2}\frac{\partial F_z}{\partial z'}\right)\right].\tag{2.1.60}$$

We once more resort to the transformation (2.1.23), finding

$$\frac{1}{\sqrt{1-\beta^2}}\left(-\frac{v}{c^2}\frac{\partial}{\partial x'}+\frac{1}{c^2}\frac{\partial}{\partial t'}\right)=\frac{1}{c^2}\frac{\partial}{\partial t},\tag{2.1.61}$$

along with

$$\frac{\partial}{\partial x'}=\frac{1}{\sqrt{1-\beta^2}}\left(\frac{\partial}{\partial x}+\frac{v}{c^2}\frac{\partial}{\partial t}\right); \quad \frac{\partial}{\partial y'}=\frac{\partial}{\partial y}; \quad \frac{\partial}{\partial z'}=\frac{\partial}{\partial z}.\tag{2.1.62}$$

Hence, the primed electric field components are expressed as follows:

$$\varepsilon_0 E_x' = \frac{\partial}{\partial y}\left(\frac{1}{c^2}\frac{\partial F_z}{\partial t}\right),\tag{2.1.63}$$

$$\varepsilon_0 E_y' = \frac{1}{\sqrt{1-\beta^2}}\frac{v}{c^2}\frac{\partial^2 F_z}{\partial z^2}$$
$$-\frac{1}{\sqrt{1-\beta^2}}\left(\frac{\partial}{\partial x}+\frac{v}{c^2}\frac{\partial}{\partial t}\right)\left(\frac{1}{c^2}\frac{\partial F_z}{\partial t}\right),\tag{2.1.64}$$

$$\varepsilon_0 E_z' = -\frac{1}{\sqrt{1-\beta^2}}\frac{v}{c^2}\frac{\partial^2 F_z}{\partial y\,\partial z}.\tag{2.1.65}$$

Although (2.1.63 – 65) may perhaps be found *at this stage* aesthetically "wanting" – they have primed quantities on the left and unprimed ones on the right – there nevertheless is a distinct practical advantage to this mode of presentation, because it brings out with clarity the link between the field components and F_z.

It should be borne in mind that, while the above results were obtained via the special theory of relativity (whereby the field intensities underwent transformations based on physical reality), the first equation in Sect. 2.1.1 is – on the other hand – due merely to *kinematic* considerations. The fundamental difference between the approaches – albeit primed *and* unprimed coordinates figure in both – is now quite obvious.

Regrouping the terms in (2.1.64), we arrive, see Sect. 2.1.6A(a), at

$$\varepsilon_0 E_y' = -\frac{1}{\sqrt{1-\beta^2}}\frac{1}{c^2}\frac{\partial^2 F_z}{\partial x\,\partial t}-\frac{v}{c^2}\frac{1}{\sqrt{1-\beta^2}}\left(\frac{\partial^2 F_z}{\partial x^2}+\frac{\partial^2 F_z}{\partial y^2}\right),\tag{2.1.66}$$

so that finally the electric field components read:

$$E_x' = \mu_0 \frac{\partial^2 F_z}{\partial y\, \partial t} \ ,$$
(2.1.67)

$$E_y' = -\frac{1}{\sqrt{1-\beta^2}}\, \mu_0 \frac{\partial^2 F_z}{\partial x\, \partial t} - \frac{1}{\sqrt{1-\beta^2}}\, \mu_0 v \left(\frac{\partial^2 F_z}{\partial x^2} + \frac{\partial^2 F_z}{\partial y^2} \right) \ ,$$
(2.1.68)

$$E_z' = -\frac{1}{\sqrt{1-\beta^2}}\, \mu_0 v \frac{\partial^2 F_z}{\partial y\, \partial z} \ .$$
(2.1.69)

Resorting to (2.1.57) for $F_z = F^{(e)}$, we obtain *in the z = 0 plane*

$$E_x' = -\frac{\bar{\phi}^{(e)}}{(2\pi)^2}\, \omega e^{-i\omega t} \int_{-\infty}^{\infty} \int_{-\infty}^{\infty} e^{i(mx+ny)} \frac{n}{m^2+n^2}\, dm\, dn \ ,$$
(2.1.70)

$$E_y' = \frac{1}{\sqrt{1-\beta^2}}\, \frac{\bar{\phi}^{(e)}}{(2\pi)^2}\, \omega e^{-i\omega t} \int_{-\infty}^{\infty} \int_{-\infty}^{\infty} e^{i(mx+ny)} \frac{m}{m^2+n^2}\, dm\, dn$$

$$-\frac{1}{\sqrt{1-\beta^2}}\, \frac{\bar{\phi}^{(e)}}{(2\pi)^2}\, v e^{-i\omega t} \int_{-\infty}^{\infty} \int_{-\infty}^{\infty} e^{i(mx+ny)}\, dm\, dn \ ,$$
(2.1.71)

whereas E_z' vanishes.

Turning our attention to the magnetic field, we have from (2.1.6), as applied in the *primed* frame of reference:

$$H_x' = \frac{\partial C_x'}{\partial t'} - \frac{\partial \chi'}{\partial x'} \ ,$$
(2.1.72)

$$H_y' = \frac{\partial C_y'}{\partial t'} - \frac{\partial \chi'}{\partial y'} \ ,$$
(2.1.73)

$$H_z' = \frac{\partial C_z'}{\partial t'} - \frac{\partial \chi'}{\partial z'} \ ,$$
(2.1.74)

i.e., with (2.1.29 – 32):

$$H_x' = \frac{\partial}{\partial t'} \left(\frac{v}{c^2}\, \frac{1}{\sqrt{1-\beta^2}}\, \frac{\partial F_z}{\partial z'} \right) - \frac{\partial}{\partial x'} \left(\frac{1}{\sqrt{1-\beta^2}}\, \frac{\partial F_z}{\partial z'} \right) \ ,$$
(2.1.75)

$$H_y' = \qquad\qquad\qquad\qquad -\frac{\partial}{\partial y'} \left(\frac{1}{\sqrt{1-\beta^2}}\, \frac{\partial F_z}{\partial z'} \right) \ ,$$
(2.1.76)

$$H_z' = \frac{\partial}{\partial t'} \left[\frac{1}{\sqrt{1-\beta^2}} \left(-\frac{v}{c^2} \frac{\partial F_z}{\partial x'} + \frac{1}{c^2} \frac{\partial F_z}{\partial t'} \right) \right]$$

$$- \frac{\partial}{\partial z'} \left(\frac{1}{\sqrt{1-\beta^2}} \frac{\partial F_z}{\partial z'} \right) . \tag{2.1.77}$$

Now, as $\partial/\partial z = \partial/\partial z'$, we find for $F_z = F^{(e)}$ in the $z = 0$ plane that $H_x' = 0$; $H_y' = 0$.

Further, see also Sect. 2.1.6A(b), we find

$$H_z' = \frac{1}{\sqrt{1-\beta^2}} \frac{v}{c^2} \frac{\partial^2 F_z}{\partial x \partial t} + \frac{1}{\sqrt{1-\beta^2}} \left(\frac{\partial^2 F_z}{\partial x^2} + \frac{\partial^2 F_z}{\partial y^2} \right) , \tag{2.1.78}$$

or i.e., see (2.1.57),

$$H_z' = -\frac{1}{\sqrt{1-\beta^2}} \frac{v}{c^2 \mu_0} \frac{\bar{\phi}^{(e)}}{(2\pi)^2} \omega e^{-i\omega t} \int_{-\infty}^{\infty} \int_{-\infty}^{\infty} e^{i(mx+ny)} \frac{m}{m^2+n^2} dm\, dn$$

$$+ \frac{1}{\sqrt{1-\beta^2}} \frac{1}{\mu_0} \frac{\bar{\phi}^{(e)}}{(2\pi)^2} e^{-i\omega t} \int_{-\infty}^{\infty} \int_{-\infty}^{\infty} e^{i(mx+ny)} dm\, dn; \quad z = 0 . \tag{2.1.79}$$

After integration (Sect. 2.1.6B), in which the notation $r = \sqrt{x^2+y^2}$; $\alpha = \tan^{-1}(y/x)$ has been used, the field components may conveniently be listed as follows:

$$E_x' = -\frac{i\omega \bar{\phi}^{(e)}}{2\pi r} \sin \alpha e^{-i\omega t} , \tag{2.1.80}$$

$$E_y' = \frac{1}{\sqrt{1-\beta^2}} \frac{i\omega \bar{\phi}^{(e)}}{2\pi r} \cos \alpha e^{-i\omega t}$$

$$- \frac{1}{\sqrt{1-\beta^2}} v \bar{\phi}^{(e)} e^{-i\omega t} \delta(x) \delta(y) , \tag{2.1.81}$$

$$E_z' = 0 , \quad \text{and} \tag{2.1.82}$$

$$H_x' = 0 , \tag{2.1.83}$$

$$H_y' = 0 , \tag{2.1.84}$$

$$H_z' = -\frac{1}{\sqrt{1-\beta^2}} \frac{i\omega \bar{\phi}^{(e)}}{2\pi r} \frac{v}{\mu_0 c^2} \cos \alpha e^{-i\omega t}$$

$$+ \frac{1}{\sqrt{1-\beta^2}} \frac{\bar{\phi}^{(e)}}{\mu_0} e^{-i\omega t} \delta(x) \delta(y) . \tag{2.1.85}$$

It should be noted once again that these field components are recorded in the *primed* frame of reference, even if expressed in terms of the *unprimed* coordinates, which may themselves be stated in terms of the primed ones. On the other hand, when the fields in the unprimed frame of reference are sought, a *physical* rather than kinematic transformation is called for, using (Sect. 1.1.1 and [2.7])

$$E_x = E_x' , \tag{2.1.86}$$

$$E_y = \frac{E_y' + v\mu_0 H_z'}{\sqrt{1-\beta^2}} , \tag{2.1.87}$$

$$E_z = \frac{E_z' - v\mu_0 H_y'}{\sqrt{1-\beta^2}} , \tag{2.1.88}$$

$$\mu_0 H_x = \mu_0 H_x' \tag{2.1.89}$$

$$\mu_0 H_y = \frac{\mu_0 H_y' - \dfrac{v}{c^2} E_z'}{\sqrt{1-\beta^2}} , \tag{2.1.90}$$

$$\mu_0 H_z = \frac{\mu_0 H_z' + \dfrac{v}{c^2} E_y'}{\sqrt{1-\beta^2}} , \tag{2.1.91}$$

Obviously, as $F_z = F^{(e)}$ is known, the unprimed fields are obtainable directly in the laboratory frame of reference. On conceptual grounds, however, the importance of the "relativistic" factors is here emphasized even for very low velocities: thus, for instance, see (2.1.81, 85, 87),

$$E_y = \frac{1}{\sqrt{1-\beta^2}} \left[\frac{1}{\sqrt{1-\beta^2}} \frac{i\omega\bar{\phi}^{(e)}}{2\pi r} \cos\alpha e^{-i\omega t} - \frac{1}{\sqrt{1-\beta^2}} v\bar{\phi}^{(e)} e^{-i\omega t} \delta(x)\delta(y) \right.$$

$$\left. - \frac{v\mu_0}{\sqrt{1-\beta^2}} \frac{i\omega\bar{\phi}^{(e)}}{2\pi r} \frac{v}{\mu_0 c^2} \cos\alpha e^{-i\omega t} + \frac{v\mu_0}{\sqrt{1-\beta^2}} \frac{\bar{\phi}^{(e)}}{\mu_0} e^{-i\omega t} \delta(x)\delta(y) \right]$$

$$= \frac{i\omega\bar{\phi}^{(e)}}{2\pi r} \cos\alpha e^{-i\omega t} , \tag{2.1.92}$$

and the interplay of the various terms is quite clear. For the sake of completeness, we list below the unprimed field components in the sheet plane:

$$E_x = -\frac{i\omega \bar{\phi}^{(e)}}{2\pi r} \sin \alpha e^{-i\omega t} , \tag{2.1.93}$$

$$E_y = \frac{i\omega \bar{\phi}^{(e)}}{2\pi r} \cos \alpha e^{-i\omega t} , \tag{2.1.94}$$

$$E_z = 0 , \tag{2.1.95}$$

along with

$$H_x = 0 , \tag{2.1.96}$$

$$H_y = 0 , \tag{2.1.97}$$

$$H_z = \frac{\bar{\phi}^{(e)}}{\mu_0} e^{-i\omega t} \delta(x) \delta(y) . \tag{2.1.98}$$

Evidently, for low velocities ($\beta^2 \ll 1$) and *not* directly beneath the vortex thread (i.e., $x \neq 0$, $y \neq 0$), the primed and unprimed electromagnetic field components are indistinguishable in the realm of magneto-quasistatics – a fact which explains why the "flux rule" yields correct results; strictly speaking, however, only the Maxwell-Lorentz-Einstein electrodynamics provide the correct conceptual basis for motion-induced electromagnetism without additional assumptions not implied directly by Maxwell's equations for bodies at rest [3].

2.1.5 Discussion

The approach outlined above, based upon restricted relativity, permits simple insight into the so-called "creation" of motion-induced fields; the rather lengthy solution has been put forward, *on grounds of principle* only, as in practice the introduced refinements are not really warranted by numerical results differing significantly from those obtained through application of the "flux rule". Whenever doubt arises, relativity provides an unambiguous answer.

The results quoted above cover also the phenomenon of unipolar induction which, while dating back to the days of Faraday, has been directly associated with relativity only much later [2.9 – 11]. The topic will, however, be tackled in more detail below (Sect. 2.3).

At this stage we are in a position to understand better the E'/E^* duality in the introduction: whereas E' formally yields the approximate field ($\beta^2 \ll 1$) in

---

[3] At this point, the following quotation [2.8] may once more be mentioned: "The special theory of relativity owes its origin to Maxwell's equations of the electromagnetic field. *Inversely, the latter can be grasped formally in satisfactory fashion only by way of the special theory of relativity*". (Our italics).

the moving frame of reference, E^* is an artificial concept which is useful in relating Faraday's law – by means of the Helmholtz derivative – to moving media as *viewed from the laboratory*, namely,

$$\oint_{(l)} E^* \cdot dl = -\frac{d}{dt} \int_{(a)} B \cdot da , \tag{2.1.99}$$

integration being here performed over a closed contour of area a (vector differential da) and length l (vector differential dl).

Turning now to the question of measurement, we may envisage two modes:

a) *"Slip rings"*, i.e. a pair of contacts fixed in the laboratory, across which the sheet is gliding; for activation of the voltmeter, assumed to be of high sensitivity, we ascribe to the sheet an extremely small but finite conductivity σ_s (e.g., using once more the conductivity of copper σ_c as reference: $\sigma_s \ll \sigma_c$ by many order of magnitude). In these circumstances, the resulting current – if any – would not disturb the imposed primary magnetic field, while providing for measurement.

As the slip rings are firmly fixed in the laboratory, they enable the field to be recorded at a point $x = $ const, $y = $ const. We then have, for example, for the y-component:

$$E_y' \simeq \frac{i\omega \bar{\phi}^{(e)}}{2\pi} \frac{\cos\alpha}{r} = \frac{i\omega \bar{\phi}^{(e)}}{2\pi} \frac{x}{x^2+y^2} e^{-i\omega t} , \tag{2.1.100}$$

i.e., the amplitude, say $\bar{\phi}^{(e)} = \phi^{(e)}e^{i0}$, is time-independent.

b) *Expanding leads,* i.e., a pair of elastic leads firmly attached to the gliding sheet, so that in this case $x' = $ const, $y' = $ const. Resorting, for simplicity, to the Galileo transformation we have $x' = x - vt$; $y' = y$; $z' = z'$; $t' = t$, so that now

$$E_y' \simeq \frac{i\omega \bar{\phi}^{(e)}}{2\pi} \frac{x}{x^2+y^2} e^{-i\omega t} = \frac{i\omega \bar{\phi}^{(e)}}{2\pi} \frac{x'+vt'}{(x'+vt')^2+y'^2} e^{-i\omega t'} . \tag{2.1.101}$$

Hence, the field amplitude is time-dependent.

Of the two modes, only the first is encountered in practice (the slip rings attached to the wound rotor of an induction motor are a common example), and the second method was mentioned only in order to underline the meaning of E_y'. Obviously, in principle neither of them can be regarded as satisfactory: in (a) we do not yet know exactly how to allow for physical effects across the sliding boundaries, whereas in (b) only inadequate information is available on the behaviour of expanding conductors.

2.1.6 Appendix

A) Field Expressions

a) For better clarity we derive (2.1.68) step by step. Keeping in mind that in principle F_z obeys the partial differential equation

$$\frac{\partial^2 F_z}{\partial x^2} + \frac{\partial^2 F_z}{\partial y^2} + \frac{\partial^2 F_z}{\partial z^2} = \frac{1}{c^2} \frac{\partial^2 F_z}{\partial t^2} , \tag{2.1.102}$$

even though we resorted – in view of the rather specialized field – to the approximation

$$\nabla \times H = 0 , \tag{2.1.103}$$

we find, see (2.1.64), that

$$\varepsilon_0 E_y' = \frac{v}{c^2} \frac{1}{\sqrt{1-\beta^2}} \frac{\partial^2 F_z}{\partial z^2} - \frac{1}{\sqrt{1-\beta^2}} \left(\frac{\partial}{\partial x} + \frac{v}{c^2} \frac{\partial}{\partial t} \right) \frac{1}{c^2} \frac{\partial F_z}{\partial t}$$

$$= \frac{v}{c^2} \frac{1}{\sqrt{1-\beta^2}} \frac{\partial^2 F_z}{\partial z^2} - \frac{1}{c^2} \frac{1}{\sqrt{1-\beta^2}} \left(\frac{\partial^2 F_z}{\partial x \partial t} + \frac{v}{c^2} \frac{\partial^2 F_z}{\partial t^2} \right)$$

$$= \frac{v}{c^2} \frac{1}{\sqrt{1-\beta^2}} \frac{\partial^2 F_z}{\partial z^2} - \frac{1}{c^2} \frac{1}{\sqrt{1-\beta^2}}$$

$$\times \left[\frac{\partial^2 F_z}{\partial x \partial t} + v \left(\frac{\partial^2 F_z}{\partial x^2} + \frac{\partial^2 F_z}{\partial y^2} + \frac{\partial^2 F_z}{\partial z^2} \right) \right] , \tag{2.1.104}$$

i.e.,

$$\varepsilon_0 E_y' = -\frac{1}{\sqrt{1-\beta^2}} \frac{1}{c^2} \frac{\partial^2 F_z}{\partial x \partial t}$$

$$-\frac{1}{\sqrt{1-\beta^2}} \frac{1}{c^2} v \left(\frac{\partial^2 F_z}{\partial x^2} + \frac{\partial^2 F_z}{\partial y^2} \right) . \tag{2.1.105}$$

Now, as $\mu_0 \varepsilon_0 = 1/c^2$, we have

$$E_y' = -\frac{1}{\sqrt{1-\beta^2}} \mu_0 \frac{\partial^2 F_z}{\partial x \partial t} - \frac{1}{\sqrt{1-\beta^2}} \mu_0 v \left(\frac{\partial^2 F_z}{\partial x^2} + \frac{\partial^2 F_z}{\partial y^2} \right) . \tag{2.1.106}$$

b) Rewriting the expression (2.1.77) for H_z' by means of (2.1.61,62), one obtains

$$H_z' = \frac{\partial}{\partial t'} \left(\frac{1}{c^2} \frac{\partial F_z}{\partial t} \right) - \frac{1}{\sqrt{1-\beta^2}} \frac{\partial^2 F_z}{\partial z^2} . \tag{2.1.107}$$

Or, applying (2.1.61) in the reverse, i.e. with

$$\frac{1}{c^2} \frac{\partial}{\partial t'} = \frac{1}{\sqrt{1-\beta^2}} \left(+ \frac{v}{c^2} \frac{\partial}{\partial x} + \frac{1}{c^2} \frac{\partial}{\partial t} \right) , \tag{2.1.108}$$

we find

$$H'_z = \frac{1}{\sqrt{1-\beta^2}} \left(\frac{v}{c^2} \frac{\partial}{\partial x} + \frac{1}{c^2} \frac{\partial}{\partial t} \right) \frac{\partial F_z}{\partial t} - \frac{1}{\sqrt{1-\beta^2}} \frac{\partial^2 F_z}{\partial z^2} . \qquad (2.1.109)$$

Resorting once more to (2.1.102), we finally have

$$H'_z = \frac{1}{\sqrt{1-\beta^2}} \frac{v}{c^2} \frac{\partial^2 F_z}{\partial x \partial t} + \frac{1}{\sqrt{1-\beta^2}} \left(\frac{\partial^2 F_z}{\partial x^2} + \frac{\partial^2 F_z}{\partial y^2} \right) . \qquad (2.1.110)$$

B) Performance of Integrations

We evaluate the following integrals, see (2.1.70, 71, 79),

a) $\displaystyle \Sigma_1 = \int\limits_{-\infty}^{\infty} \int\limits_{-\infty}^{\infty} e^{i(mx+ny)} \frac{n}{m^2+n^2} \, dm \, dn$,   and $\qquad (2.1.111)$

b) $\displaystyle \Sigma_2 = \int\limits_{-\infty}^{\infty} \int\limits_{-\infty}^{\infty} e^{i(mx+ny)} \frac{m}{m^2+n^2} \, dm \, dn$. $\qquad (2.1.112)$

Substituting

$$x = r\cos\alpha, \quad m = \lambda\cos\gamma , \qquad (2.1.113)$$

$$y = r\sin\alpha, \quad n = \lambda\sin\gamma , \qquad (2.1.114)$$

we have:

$$mx + ny = \lambda r \cos\gamma\cos\alpha + \lambda r \sin\gamma\sin\alpha = \lambda r \cos(\gamma-\alpha) , \qquad (2.1.115)$$

and moreover

$$\frac{n}{m^2+n^2} = \frac{\lambda\sin\gamma}{\lambda^2} = \frac{\sin\gamma}{\lambda} ; \qquad \frac{m}{m^2+n^2} = \frac{\lambda\cos\gamma}{\lambda^2} = \frac{\cos\gamma}{\lambda} ;$$

$$dm \, dn = \lambda \, d\lambda \, d\gamma . \qquad (2.1.116)$$

Hence:

a) $\displaystyle \Sigma_1 = \int\limits_{\lambda=0}^{\infty} \int\limits_{\gamma=0}^{2\pi} e^{i\lambda r\cos(\gamma-\alpha)} \sin\gamma \, d\lambda \, d\gamma$. $\qquad (2.1.117)$

Or, with

$$\gamma - \alpha \equiv \gamma' \qquad \lambda r \equiv u , \qquad (2.1.118)$$

$$\Sigma_1 = \frac{1}{r} \int_{u=0}^{\infty} \int_{\gamma'=0}^{2\pi} e^{iu\cos\gamma'} \sin(\gamma'+\alpha)\, du\, d\gamma'$$

$$= \frac{1}{r} \left(\sin\alpha \int_0^{\infty} \int_0^{2\pi} e^{iu\cos\gamma'} \cos\gamma'\, du\, d\gamma' \right.$$

$$\left. + \cos\alpha \int_0^{\infty} \int_0^{2\pi} e^{iu\cos\gamma'} \sin\gamma'\, du\, d\gamma' \right) . \tag{2.1.119}$$

Now the Bessel function of the first kind, order n and argument z, may be expressed [Ref. 2.12, p. 149] in terms of the integral representation:

$$J_n(z) = \frac{i^{-n}}{2\pi} \int_0^{2\pi} e^{iz\cos\phi} e^{in\phi}\, d\phi . \tag{2.1.120}$$

Hence, in our case, we find

$$\int_0^{2\pi} e^{iu\cos\gamma'} \cos\gamma'\, d\gamma' = 2\pi i J_1(u) \quad \text{and} \tag{2.1.121}$$

$$\int_0^{2\pi} e^{iu\cos\gamma'} \sin\gamma'\, d\gamma' = 0 , \tag{2.1.122}$$

so that

$$\Sigma_1 = \frac{\sin\alpha}{r} i2\pi \int_0^{\infty} J_1(u)\, du = \frac{\sin\alpha}{r} i2\pi , \tag{2.1.123}$$

i.e., finally, see (2.1.70),

$$E_x' = -\frac{\bar{\phi}^{(e)}}{(2\pi)^2} \omega e^{-i\omega t} i \frac{\sin\alpha}{r} 2\pi$$

$$= -\frac{i\omega}{2\pi r} \bar{\phi}^{(e)} e^{-i\omega t} \sin\alpha . \tag{2.1.124}$$

b) $$\Sigma_2 = \int_{-\infty}^{\infty} \int_{-\infty}^{\infty} e^{i(mx+ny)} \frac{m}{m^2+n^2}\, dm\, dn . \tag{2.1.125}$$

Proceeding as in (a), we find

$$\Sigma_2 = \int_{\lambda=0}^{\infty} \int_{\gamma=0}^{2\pi} e^{i\lambda r\cos(\gamma-\alpha)} \cos\gamma\, d\lambda\, d\gamma , \tag{2.1.126}$$

and again with $\gamma - \alpha = \gamma'$; $\lambda r = u$:

$$\Sigma_2 = \frac{1}{r} \int\limits_{u=0}^{\infty} \int\limits_{\gamma'=0}^{2\pi} e^{iu\cos\gamma'} \cos(\gamma'+\alpha)\,du\,d\gamma'$$

$$= \frac{1}{r}\left(\cos\alpha \int\limits_{0}^{\infty} \int\limits_{0}^{2\pi} e^{iu\cos\gamma'}\cos\gamma'\,du\,d\gamma' \right.$$

$$\left. - \sin\alpha \int\limits_{0}^{\infty}\int\limits_{0}^{2\pi} e^{iu\cos\gamma'}\sin\gamma'\,du\,d\gamma'\right) . \tag{2.1.127}$$

With the results from (a), this reduces to

$$\Sigma_2 = \frac{1}{r}\cos\alpha \int\limits_{0}^{\infty} i2\pi J_1(u)\,du = \frac{\cos\alpha}{r} i2\pi . \tag{2.1.128}$$

Or, finally, see (2.1.71, 79), and also the defining expression (2.1.48),

$$E_y' = \frac{1}{\sqrt{1-\beta^2}}\frac{i\omega}{2\pi r}\bar{\phi}^{(e)}e^{-i\omega t}\cos\alpha - \frac{1}{\sqrt{1-\beta^2}}v\bar{\phi}^{(e)}e^{-i\omega t}\delta(x)\delta(y) , \tag{2.1.129}$$

and

$$H_z' = -\frac{1}{\sqrt{1-\beta^2}}\frac{i\omega}{2\pi r}\bar{\phi}^{(e)}e^{-i\omega t}\frac{v}{\mu_0 c^2}\cos\alpha$$

$$+ \frac{1}{\sqrt{1-\beta^2}}\frac{\bar{\phi}^{(e)}}{\mu_0}e^{-i\omega t}\delta(x)\delta(y) . \tag{2.1.130}$$

2.2 Linear Induction Devices

In the preceding section we discussed the induction law and the Fitzgerald superpotential. While not yielding new concrete results, it was pointed out that relativity does permit a completely new insight into the nature of the electromagnetic field in the presence of a moving part of an electrotechnical device. The resulting fundamental picture enables the engineer to feel confidence in that approach; however, once the underlying principles and procedures have been mastered, the wider frame of analysis may no longer be fully needed in practice. Accordingly, the present section deals with such a practical problem via a simpler approach, with relativistic principles hovering in the background.

The specific practical device under investigation is the linear motor, which has been known in one form or another for quite some time, e.g., its use in launching aircraft from carrier decks was investigated and even applied in practice toward the end of World War II and subsequently. However, it did not attract too much interest until *Laithwaite* drew attention to its various versions in an extensive series of papers and books (for example [2.13 – 15]). Investigation of magneto-hydrodynamic generation and liquid metal pumps imparted additional impetus to the topic. During the seventies, much of the interest was motivated by the search for pollution-free and more efficient urban and intercity transportation, and it seemed that the linear motor might be suitable for traction of the high-speed rolling stock having no wheels; however, development was hampered mainly by lack of adequate funds. In the meantime, research on linear induction devices yielded a high output of relevant literature which included applications, on the one hand, and theoretical treatment of principles and performance characteristics, on the other.

To the best of the author's knowledge, no work up-to-date resorted to the Fitzgerald superpotential in analysing the linear motor. This "oversight" is remedied in the present treatment, which makes use of the latter and of the fact that it is a "disguised six-vector", with its z-component F_z invariant for $\beta^2 \ll 1$, see (2.1.27).

2.2.1 Definition of Problem

A schematic presentation of a double-sided short-stator linear induction motor is shown in Fig. 2.2; now while the distribution of Ampere-turns across the stator may actually be prescribed – or constrained – with relative ease, this is not the case with the magnetic flux density. However, for simplicity, we *assume* along the stator the existence of a specialized winding which yields a sinusoidal induction in the air gap; in other words, the principle of the linear induction motor is here formulated in *analogy* with its conventional counterpart, with the rotating squirrel cage replaced by a gliding continuous sheet of conducting material.

In order to proceed with the mathematical analysis, we consider the following: a very thin, homogeneous and isotropic sheet of nonmagnetic material of thickness Δ and electrical conductivity σ is moving rectilinearly with regulated constant velocity v in the air gap of width $2b$ of an electromagnet of rectangular cross section, having the dimensions $2h$, $2l$ (Fig. 2.2). The sheet is assumed to be very long ("infinite") in the direction of motion, but of finite width $2a$ in the perpendicular direction. The electrical properties of the sheet are specified by the limiting process

$$\lim_{\substack{\sigma \to \infty \\ \Delta \to 0}} (\sigma \Delta) = \text{finite} .$$

The stator iron is replaced by fictitious *slotless* material of infinite relative permeability ($\mu_r \to \infty$) and vanishingly small electrical conductivity. Its winding (of vanishingly small electrical resistance) is connected to a balanced

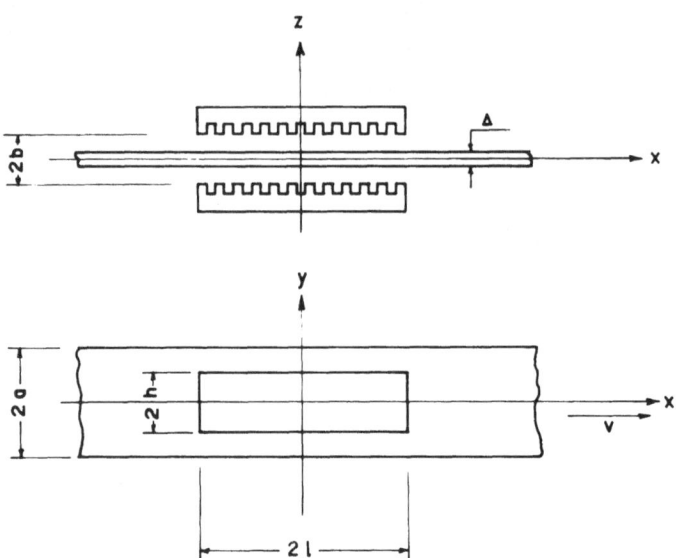

Fig. 2.2. Schematic representation of linear motor

multiphase ac voltage and excites in the region $|x| < l$, $|y| < h$ of the $z = 0$ plane a (quasi-)homogeneous magnetic flux density directed mainly along the z-axis and comprising vanishingly small horizontal components. The angular frequency ω of this field is sufficiently low for the time rate of the electric induction to be disregarded.

What is the electromagnetic field of the structure?

Once more [see also remark preceding (2.1.46)] the given shape of the exciting electromagnet (= stator) rules out the exact solution of the Laplace equation by means of known (tabulated) functions; accordingly, we replace the original ferromagnetic material of finite horizontal dimensions by a fictitious stator of infinite horizontal extension, retaining nevertheless the requirement that the *primary field remain confined* to the original region $|x| < l$, $|y| < h$. Although the influence of the secondary field is thus exaggerated, this mathematical device will enable us to express the resulting electromagnetic fields in a closed analytic form, while retaining the salient conditions of the problem, i.e., a rectangularly confined exciting field and a moving strip of finite width.

2.2.2 General Approach to Solution

For the system under consideration as above, we once more introduce two distinct frames of reference:

- Right-handed Cartesian coordinates x, y, z with intrinsic time t; these coordinates are attached to the stator and their origin is located in the sheet plane $z = 0$, at the center of the normal projection of the region to which the field is confined.

- Right-handed Cartesian coordinates x', y', z' with intrinsic time t', attached to the moving sheet which is now linked, however, to the unprimed coordinates through the *Galilean transformation* $x' = x - vt$; $y' = y$; $z' = z$; $t' = t$.

As hinted already, once the relativistic point of view has been outlined – entailing here specifically the invariance of F_z – we may resort, via Cerenkov electrodynamics, to Galilean relativity, keeping nevertheless in mind the restrictions this imposes.

In a stationary homogeneous and isotropic medium of zero electrical conductivity $\sigma \to 0$, Maxwell's curl equations (magneto-quasistatics) read

$$\nabla \times H = 0 \ , \tag{2.2.1}$$

$$\nabla \times E = -\frac{\partial}{\partial t} \mu_0 H \ . \tag{2.2.2}$$

Under the conditions stated, the E and H fields in the so-called "air" are again derived (Sect. 2.1) from a Fitzgerald vector superpotential F; adapting its application to the present problem (Sect. 2.2.8 A) we write

$$E = \nabla \times \frac{\partial F}{\partial t} \ , \tag{2.2.3}$$

$$H = -\frac{1}{\mu_0} \nabla(\nabla \cdot F) \tag{2.2.4}$$

with F satisfying once more, see (2.1.43), the quasistatic differential equation

$$\nabla^2 F = 0 \ . \tag{2.2.5a}$$

The results will prove that, in the present case of planar current flow, the Fitzgerald vector superpotential again reduces to its component along the z-axis only (Sect. 2.1.4), i.e., $F = F_z 1_z$ and hence (2.2.5a) – to

$$\nabla^2 F_z = 0 \ . \tag{2.2.5b}$$

Accordingly, (2.2.3 and 4) yield

$$E_x = \frac{\partial^2 F_z}{\partial y \, \partial t}, \quad E_y = -\frac{\partial^2 F_z}{\partial x \, \partial t}, \quad E_z = 0 \ , \tag{2.2.6}$$

$$H_x = -\frac{1}{\mu_0} \frac{\partial^2 F_z}{\partial x \, \partial z}, \quad H_y = -\frac{1}{\mu_0} \frac{\partial^2 F_z}{\partial y \, \partial z}, \quad H_z = -\frac{1}{\mu_0} \frac{\partial^2 F_z}{\partial z^2} \ . \tag{2.2.7}$$

For sinusoidal steady-state excitation, we introduce – as above – the phasor $\bar{F}_z$ through the relation $F_z = \bar{F}_z \exp(-i\omega t)$; in the same manner, the complex amplitudes $\bar{E}_x$, $\bar{E}_y$, $\bar{E}_z$ of the electric field components and $\bar{H}_x$, $\bar{H}_y$, $\bar{H}_z$ of the magnetic field components are introduced.

2.2.3 Exciting Field

The complex representation of the exciting ("primary") traveling flux density $B_z^{(p)}$, residing in the $z = 0$ plane even in the absence of the conducting sheet, is formulated by means of the phasor $\bar{B}_0$. This exciting field, characterized by a pole pitch τ, advances in the $+x$-direction and its mathematical formulation is taken to be represented by

$$
\begin{aligned}
B_z^{(p)} &= \bar{B}_0 e^{-i\omega t} e^{i(\pi/\tau)x}, & |x| < l, \quad |y| < h, \\
B_z^{(p)} &= 0, & |x| > l, \quad |y| > h.
\end{aligned}
\qquad z = 0 \qquad (2.2.8)
$$

Further, along with the complex amplitude $\bar{B}_0$ pertaining to the imposed magnetic flux density, we also define the complex amplitude of the magnetic flux $\bar{\phi}_0 \equiv (\bar{B}_0 \times 2h \times 2l)$.

We now choose to satisfy the boundary conditions of vanishing normal current flow at the sheet boundaries $y = \pm a$ by an image procedure (Fig. 2.3); hence, the exciting magnetic flux density is taken to alternate (in space) in the y-direction and its analytic representation is given in terms of the continuous wave number m and the discrete, odd wave number $k (= 1, 3, 5, 7 \ldots)$, by

$$
B_z^{(p)} = \frac{\bar{\phi}_0}{\pi^2} \frac{\pi}{2a} e^{-i\omega t} \int\limits_{-\infty}^{\infty} dm
$$

$$
\times \sum_k \frac{\sin\left(k \dfrac{\pi}{2} \dfrac{h}{a}\right)}{\left(k \dfrac{\pi}{2} \dfrac{h}{a}\right)} \frac{\sin\left(\dfrac{\pi}{\tau} - m\right) l}{\left(\dfrac{\pi}{\tau} - m\right) l} e^{imx} \cos\left(k \dfrac{\pi}{2} \dfrac{y}{a}\right), \qquad z = 0 . \tag{2.2.9}
$$

The amplitude $F^{(p)}$ of the primary z-directed Fitzgerald vector superpotential, $F^{(p)}$, from which this field is "derived", satisfies the Laplace equation; this potential is coherent in x and y with $B_z^{(p)}$ as per (2.2.9) and can be expressed in integral form through the (still unknown) weighting functions $\bar{K}_k(m)$:

$$
F^{(p)} = e^{-i\omega t} \int\limits_{-\infty}^{\infty} dm \sum_k \bar{K}_k(m) e^{imx} \cos\left(k \frac{\pi}{2} \frac{y}{a}\right)
$$

$$
\times \cosh\left[\sqrt{\left(k \frac{\pi}{2a}\right)^2 + m^2} \, z\right] . \tag{2.2.10}
$$

For the z-component of the magnetic flux density in the $z = 0$ plane, see (2.2.7), we therefore obtain

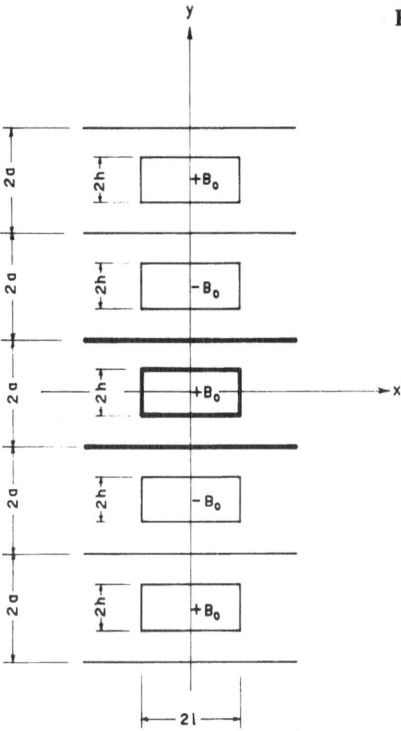

Fig. 2.3. Series of images

$$B_z^{(p)} = -e^{-i\omega t} \int\limits_{-\infty}^{\infty} dm \sum_k \left[\left(k\frac{\pi}{2a}\right)^2 + m^2\right] \pm K_k(m)e^{imx}\cos\left(k\frac{\pi}{2}\frac{y}{a}\right),$$

$$z = 0 \ . \tag{2.2.11}$$

Comparing (2.2.9 and 11), we have

$$\bar{K}_k(m) = -\frac{\bar{\phi}_0}{\pi^2}\frac{\pi}{2a}\frac{\sin\left(k\frac{\pi}{2}\frac{h}{a}\right)}{\left(k\frac{\pi}{2}\frac{h}{a}\right)}$$

$$\times \frac{\sin\left(\frac{\pi}{\tau}-m\right)l}{\left(\frac{\pi}{\tau}-m\right)l}\frac{1}{\left(k\frac{\pi}{2a}\right)^2 + m^2}\ , \tag{2.2.12}$$

and therefore

$$F^{(p)} = -\frac{\bar{\phi}_0}{2\pi a} e^{-i\omega t} \int_{-\infty}^{\infty} dm \sum_{k} \frac{\sin\left(k\frac{\pi}{2}\frac{h}{a}\right)}{\left(k\frac{\pi}{2}\frac{h}{a}\right)} \frac{\sin\left(\frac{\pi}{\tau}-m\right)l}{\left(\frac{\pi}{\tau}-m\right)l}$$

$$\times \frac{e^{imx}\cos\left(k\frac{\pi}{2}\frac{y}{a}\right)}{\left(k\frac{\pi}{2a}\right)^2+m^2}\cosh\left[\sqrt{\left(k\frac{\pi}{2a}\right)^2+m^2}\,z\right]. \qquad (2.2.13)$$

The primary phasor in the primed system of reference $F^{(p)'}$, as obtained now through the Galilean transformation [4], reads

$$\bar{F}^{(p)'} = -\frac{\bar{\phi}_0}{2\pi a} e^{-i\omega t} \int_{-\infty}^{\infty} dm \sum_{k} \frac{\sin\left(k\frac{\pi}{2}\frac{h}{a}\right)}{\left(k\frac{\pi}{2}\frac{h}{a}\right)}$$

$$\times \frac{\sin\left(\frac{\pi}{\tau}-m\right)l}{\left(\frac{\pi}{\tau}-m\right)l} \frac{e^{im(x'+vt')}\cos\left(k\frac{\pi}{2}\frac{y'}{a}\right)}{\left(k\frac{\pi}{2a}\right)^2+m^2}$$

$$\times\cosh\left[\sqrt{\left(k\frac{\pi}{2a}\right)^2+m^2}\,z'\right]. \qquad (2.2.14)$$

Using (2.2.6) adapted to the primed coordinates, we obtain the "primary" electric field components $E_{x'}^{(p)'}$, $E_{y'}^{(p)'}$ in the sheet plane $z = z' = 0$ as follows:

$$E_{x'}^{(p)'} = -\frac{\bar{\phi}_0}{2\pi a} e^{-i\omega t'} i \int_{-\infty}^{\infty} dm \sum_{k} \frac{\sin\left(k\frac{\pi}{2}\frac{h}{a}\right)}{\left(k\frac{\pi}{2}\frac{h}{a}\right)}$$

---

[4] In Sect. 2.1.2 it has been proved that $F^{(p)}$ is invariant for velocities which are negligible with respect to c, see (2.1.27); henceforth, even though Galilean relativity is resorted to, we denote for clarity the time parameter by t' in the primed frame; thus only a kinematic transformation is required here.

$$\times \frac{\sin\left(\dfrac{\pi}{\tau}-m\right)l}{\left(\dfrac{\pi}{\tau}-m\right)l} k \frac{\pi}{2a}(\omega-mv)$$

$$\times \frac{e^{im(x'+vt')}\sin\left(k\dfrac{\pi}{2}\dfrac{y'}{a}\right)}{\left(k\dfrac{\pi}{2a}\right)^2+m^2} \quad . \tag{2.2.15}$$

$$E_{y'}^{(p)'} = +\frac{\bar{\phi}_0}{2\pi a}\, e^{-i\omega t'} \int_{-\infty}^{\infty} dm \sum_k \frac{\sin\left(k\dfrac{\pi}{2}\dfrac{h}{a}\right)}{\left(k\dfrac{\pi}{2}\dfrac{h}{a}\right)}$$

$$\times \frac{\sin\left(\dfrac{\pi}{\tau}-m\right)l}{\left(\dfrac{\pi}{\tau}-m\right)l}\, m(\omega-mv) \frac{e^{im(x'+vt')}\cos\left(k\dfrac{\pi}{2}\dfrac{y'}{a}\right)}{\left(k\dfrac{\pi}{2a}\right)^2+m^2} \quad . \tag{2.2.16}$$

We now turn our attention to the reaction field.

2.2.4 Reaction Field

The electric field in the sheet plane is linked to an electric current density j'. Let A_x', A_y' be the components of the surface current $A' = j'\Delta$. The continuity equation

$$\frac{\partial A_{x'}'}{\partial x'} + \frac{\partial A_{y'}'}{\partial y'} = 0 \tag{2.2.17}$$

is satisfied identically by a stream function $D' = D'(x', y', t')$ defined through

$$\frac{\partial D'}{\partial y'} = -A_{x'}', \qquad \frac{\partial D'}{\partial x'} = +A_{y'}' \quad . \tag{2.2.18}$$

On account of the streamline equation

$$\frac{dy'}{dx'} = \frac{A_{y'}'}{A_{x'}'} \quad ,$$

the family of curves $D' = $ const represents the current flow pattern at the instant $t' = t'_a$.

By means of the still undetermined spectral amplitudes $\bar{f}_k(m)$, we now introduce the stream function D'

$$D' = e^{-i\omega t'} \int_{-\infty}^{\infty} dm \sum_k \bar{f}_k(m) e^{im(x' + vt')} \cos\left(k\frac{\pi}{2}\frac{y'}{a}\right) , \qquad (2.2.19)$$

which in turn yields, via (2.2.18), the surface current distributions

$$A'_{x'} = +\frac{\pi}{2a} e^{-i\omega t'} \int_{-\infty}^{\infty} dm \sum_k \bar{f}_k(m) k e^{im(x' + vt')} \sin\left(k\frac{\pi}{2}\frac{y'}{a}\right) , \qquad (2.2.20)$$

$$A'_{y'} = +ie^{-i\omega t'} \int_{-\infty}^{\infty} dm\, m \sum_k \bar{f}_k(m) e^{im(x' + vt')} \cos\left(k\frac{\pi}{2}\frac{y'}{a}\right) . \qquad (2.2.21)$$

As noted, the original stator has been replaced by iron of infinite horizontal dimensions, but comprising windings so distributed that the exciting magnetic field is confined to the region $|x| < l$; $|y| < h$. This procedure enables us to formulate in the air gap a "secondary" Fitzgerald vector superpotential $F^{(s)}$ (due to the sheet current) comprising again a z-component $F^{(s)}$ only, adapted, in turn, to the boundary conditions of the problem, for which the Laplace equation is solvable by separation of variables.

Now by our assumptions, the stator winding is supplied – in technical terms – by a known imposed voltage totally "compensated" by the primary field; hence the tangential components of the secondary electric field must vanish on the stator at $z = \pm b$. Subject to this boundary condition, the *primed* superpotential $F^{(s)'}$, expressed in the primed system of reference through the still unknown spectral function $\bar{L}_k(m)$, reads

$$F^{(s)'} = e^{-i\omega t'} \int_{-\infty}^{\infty} dm \sum_k \bar{L}_k(m) e^{im(x' + vt')} \cos\left(k\frac{\pi}{2}\frac{y'}{a}\right)$$

$$\times \sinh\left[\sqrt{\left(k\frac{\pi}{2a}\right)^2 + m^2}\,(b \mp z')\right]; \quad z' \gtrless 0 . \qquad (2.2.22)$$

The secondary horizontal magnetic field components are therefore given in the sheet plane, via the primed version of (2.2.7), by

$$H_{x'}^{(s)'} = \pm\frac{1}{\mu_0} ie^{-i\omega t'} \int_{-\infty}^{\infty} dm\, m \sum_k \bar{L}_k(m) e^{im(x' + vt')}$$

$$\times \cos\left(k\frac{\pi}{2}\frac{y'}{a}\right)\left[\sqrt{\left(k\frac{\pi}{2a}\right)^2 + m^2}\right] \cosh\left[\sqrt{\left(k\frac{\pi}{2a}\right)^2 + m^2}\,b\right],$$

$$z' = \pm 0 , \qquad (2.2.23)$$

$$H_{y'}^{(s)'} = \mp \frac{1}{\mu_0} \frac{\pi}{2a} e^{-i\omega t'} \int_{-\infty}^{\infty} dm \sum_k \bar{L}_k(m) k e^{im(x'+vt')}$$

$$\times \sin\left(k\frac{\pi}{2}\frac{y'}{a}\right)\left[\sqrt{\left(k\frac{\pi}{2a}\right)^2 + m^2}\cosh\left[\sqrt{\left(k\frac{\pi}{2a}\right)^2 + m^2}\,b\right]\right],$$

$$z' = \pm 0 . \tag{2.2.24}$$

The integral form of Maxwell's law $\nabla \times H' = j'$ at the interface $z' = 0$ reads

$$H_{y'}'(+0) - H_{y'}'(-0) = -A_{x'}' \tag{2.2.25}$$

$$H_{x'}'(+0) - H_{x'}'(-0) = +A_{y'}' \tag{2.2.26}$$

and hence

$$\bar{L}_k(m) = \frac{\mu_0}{2} \frac{\bar{f}_k(m)}{\sqrt{\left(k\frac{\pi}{2a}\right)^2 + m^2}\cosh\left[\sqrt{\left(k\frac{\pi}{2a}\right)^2 + m^2}\,b\right]} . \tag{2.2.27}$$

The secondary Fitzgerald potential component $F'(s)$ may now be expressed by the spectral coefficients $\bar{f}_k(m)$:

$$F^{(s)'} = \frac{\mu_0}{2} e^{-i\omega t'} \int_{-\infty}^{\infty} dm$$

$$\times \sum_k \frac{\bar{f}_k(m) e^{im(x'+vt')} \cos\left(k\frac{\pi}{2}\frac{y'}{a}\right)}{\sqrt{\left(k\frac{\pi}{2a}\right)^2 + m^2}}$$

$$\times \frac{\sinh\left[\sqrt{\left(k\frac{\pi}{2a}\right)^2 + m^2}\,(b\mp z)\right]}{\cosh\left[\sqrt{\left(k\frac{\pi}{2a}\right)^2 + m^2}\,b\right]} , \quad z \gtrless 0 , \tag{2.2.28}$$

and therefore the secondary electric field components in the sheet plane read

$$E_{x'}^{(s)'} = +\frac{\mu_0}{2}\frac{\pi}{2a} i e^{-i\omega t'} \int_{-\infty}^{\infty} dm \sum_k k \bar{f}_k(m)(\omega - mv)$$

$$\times e^{im(x'+vt')} \sin\left(k\frac{\pi}{2}\frac{y'}{a}\right) \frac{\tanh\left[\sqrt{\left(k\dfrac{\pi}{2a}\right)^2 + m^2 b}\,\right]}{\sqrt{\left(k\dfrac{\pi}{2a}\right)^2 + m^2}} , \qquad (2.2.29)$$

$$E_{y'}^{(s)'} = -\frac{\mu_0}{2} e^{-i\omega t'} \int_{-\infty}^{\infty} dm\, m \sum_k \bar{f}_k(m)(\omega - mv)$$

$$\times e^{im(x'+vt')} \cos\left(k\frac{\pi}{2}\frac{y'}{a}\right) \frac{\tanh\left[\sqrt{\left(k\dfrac{\pi}{2a}\right)^2 + m^2 b}\,\right]}{\sqrt{\left(k\dfrac{\pi}{2a}\right)^2 + m^2}} . \qquad (2.2.30)$$

For the conducting sheet, we now apply Ohm's law

$$\sigma \Delta (\bar{E}_{x'}^{(p)'} + \bar{E}_{x'}^{(s)'}) = \bar{A}_{x'}' , \qquad (2.2.31)$$

$$\sigma \Delta (\bar{E}_{y'}^{(p)'} + \bar{E}_{y'}^{(s)'}) = \bar{A}_{y'}' , \qquad (2.2.32)$$

and obtain, see (2.2.15 and 16; 20 and 21; 29 and 30),

$$\bar{f}_k(m) =$$

$$= \frac{\dfrac{\sigma \Delta \mu_0}{2} \dfrac{2\bar{\phi}_0}{\mu_0} \dfrac{1}{2\pi a} \dfrac{\sin\left(k\dfrac{\pi}{2}\dfrac{h}{a}\right)}{\left(k\dfrac{\pi}{2}\dfrac{h}{a}\right)} \dfrac{\sin\left(\dfrac{\pi}{\tau}-m\right)l}{\left(\dfrac{\pi}{\tau}-m\right)l} (\omega - mv) \dfrac{1}{\left(k\dfrac{\pi}{2a}\right)^2 + m^2}}{i + \dfrac{\sigma \Delta \mu_0}{2} (\omega - mv) \dfrac{\tanh\left[\sqrt{\left(k\dfrac{\pi}{2a}\right)^2 + m^2 b}\,\right]}{\sqrt{\left(k\dfrac{\pi}{2a}\right)^2 + m^2}}} . \qquad (2.2.33)$$

Returning to (2.2.19) and formulating the stream function $D(x, y, t)$ in the *unprimed* system (Sect. 2.2.8 B) we have

$D(x, y, t)$

$$= \frac{2\,\bar{\phi}_0}{\mu_0}\frac{1}{2\pi a}\,\mathrm{e}^{-\mathrm{i}\omega t}\int_{-\infty}^{\infty} dm$$

$$\times \sum_k \frac{\dfrac{\sigma\varDelta\mu_0}{2}\;\dfrac{\sin\left(k\dfrac{\pi}{2}\dfrac{h}{a}\right)}{\left(k\dfrac{\pi}{2}\dfrac{h}{a}\right)}\;\dfrac{\sin\left(\dfrac{\pi}{\tau}-m\right)l}{\left(\dfrac{\pi}{\tau}-m\right)l}\;\dfrac{\omega-mv}{\left(k\dfrac{\pi}{2a}\right)^2+m^2}\,\mathrm{e}^{\mathrm{i}mx}\cos\left(k\dfrac{\pi}{2}\dfrac{y}{a}\right)}{\mathrm{i}+\dfrac{\sigma\varDelta\mu_0}{2}(\omega-mv)\;\dfrac{\tanh\left[\sqrt{\left(k\dfrac{\pi}{2a}\right)^2+m^2}\,b\right]}{\sqrt{\left(k\dfrac{\pi}{2a}\right)^2+m^2}}} \, .$$

$$(2.2.34)$$

Relevant patterns of streamlines are reproduced in Figs. 2.4 to 6. Increasing the ratio $2l/\tau$ reduces the influence of the end effects on the overall shape of the streamlines, and flow patterns resembling those of Fig. 2.7 are obtained. It should be borne in mind, however, that the plotted results are obviously affected by the somewhat *arbitrary* model chosen for analysis.

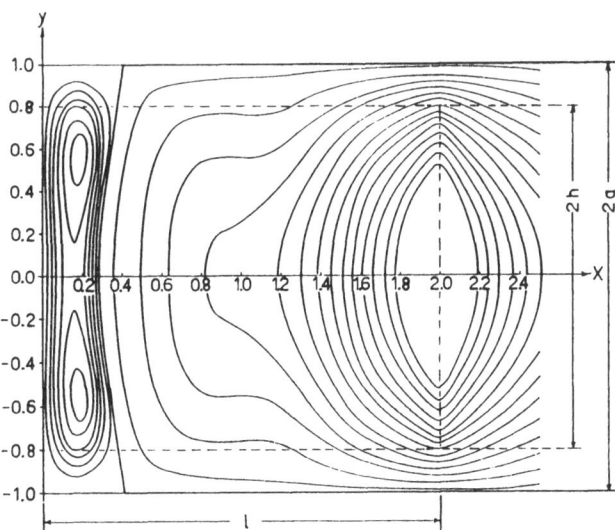

Fig. 2.4. Streamline pattern for sheet of finite width. Ratio of stator length to pole pitch: $2l/\tau = 8$

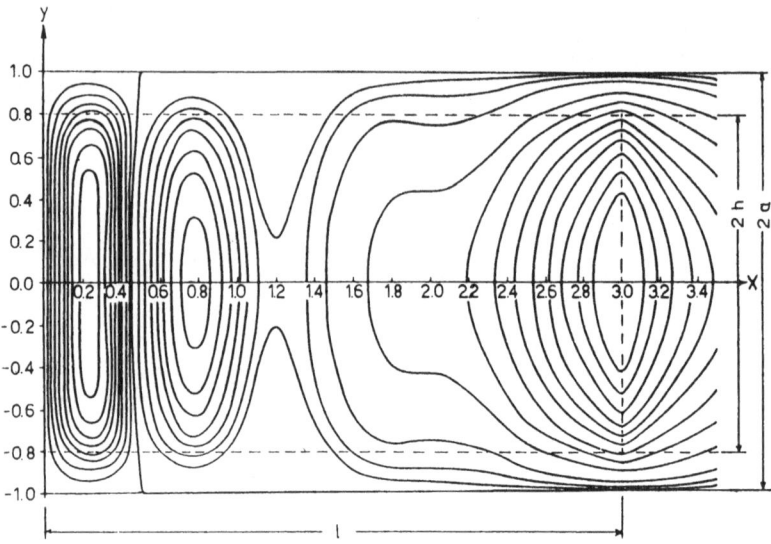

Fig. 2.5. Streamline pattern for sheet of finite width. Ratio of stator length to pole pitch: $2l/\tau = 12$

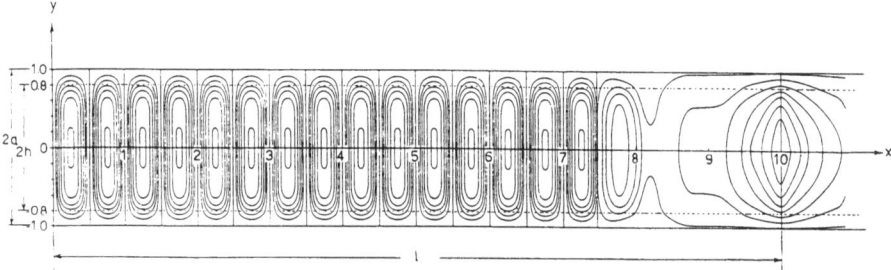

Fig. 2.6. Streamline pattern for sheet of finite width. Ratio of stator length to pole pitch: $2l/\tau = 40$

2.2.5 Two Special Cases

A) Rotor of "Infinite" Extension

If the sheet width $2a$ is much larger than the dimensions $2l$, $2h$, we may use the limit $a \to \infty$. Passing in this case from Fourier series expansion to Fourier integral representation, we obtain for the stream function

$$D(x, y, t) = \frac{2\,\bar{\phi}_0}{\mu_0} \frac{1}{(2\pi)^2} e^{-i\omega t}$$

$$\times \int_{-\infty}^{\infty} \int_{-\infty}^{\infty} \frac{\dfrac{\sigma\Delta\mu_0}{2} \dfrac{\sin nh}{nh} \dfrac{\sin\left(\dfrac{\pi}{\tau}-m\right)l}{\left(\dfrac{\pi}{\tau}-m\right)l} \dfrac{\omega-mv}{n^2+m^2} e^{i(mx+ny)}}{i + \dfrac{\sigma\Delta\mu_0}{2}(\omega-mv)\dfrac{\tanh\left(\sqrt{n^2+m^2}\,b\right)}{\sqrt{n^2+m^2}}}\, dm\, dn.$$

$$(2.2.35)$$

Additionally, if the real stator length obeys the relation $2l \gg 2h$, we assume the limit $l \to \infty$. The x-dependence of the exciting flux density is now represented throughout the x-range, by the monochromatic traveling wave

$$B_z^{(p)} = \bar{B}_0\, e^{-i\omega t} e^{i(\pi/\tau)x}; \qquad -\infty < x < \infty; \qquad |y| < h; \qquad z = 0 \qquad (2.2.36)$$

and (2.2.35) reduces to a simple integral for $m \to \pi/\tau$; thus applying the definition $\bar{\phi}_0 \equiv \bar{B}_0 \times 2h \times 2\pi$, we obtain for the stream function

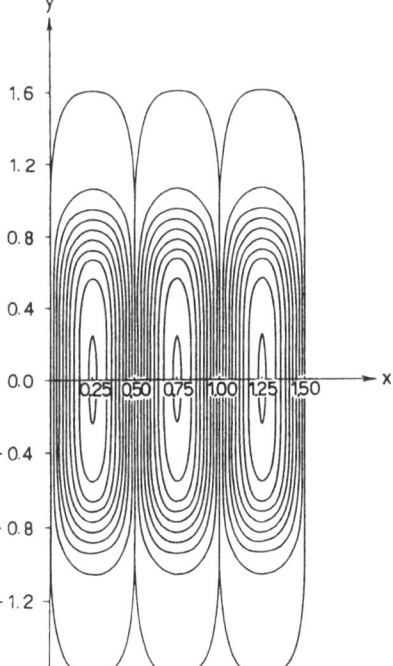

Fig. 2.7. Streamline pattern for sheet of "infinite" width. Ratio of stator length to pole pitch: $2l/\tau \to \infty$

$$D(x, y, t) = \frac{2\bar{\phi}_0}{\mu_0} \frac{1}{(2\pi)^2} e^{-i\omega t}$$

$$\times \int_{-\infty}^{\infty} \frac{\frac{\sigma \Delta \mu_0}{2} \frac{\sin nh}{nh} \frac{\omega - (\pi/\tau)v}{n^2 + \pi^2/\tau^2} e^{i[(\pi/\tau)x + ny]}}{i + \frac{\sigma \Delta \mu_0}{2}\left(\omega - \frac{\pi}{\tau}v\right) \frac{\tanh[\sqrt{n^2 + (\pi/\tau)^2}b]}{\sqrt{n^2 + (\pi/\tau)^2}}} dn \quad . (2.2.37)$$

A streamline pattern for this case is shown in Fig. 2.7.

B) Rotor of "Infinite" Conductivity ($\sigma \to \infty$; Δ = finite)

The case of a rotor of superconducting material involves the concept of "frozen-in fields". For $\sigma \to \infty$, (2.2.37) becomes

$$D(x, y, t) = \frac{2\bar{\phi}_0}{\mu_0} \frac{1}{(2\pi)^2} e^{-i\omega t}$$

$$\times \int_{-\infty}^{\infty} \frac{\sin nh}{nh} \frac{e^{i[(\pi/\tau)x + ny]} dn}{\sqrt{n^2 + (\pi/\tau)^2} \tanh[\sqrt{n^2 + (\pi/\tau)^2}b]} \quad . \quad (2.2.38)$$

Calculating the fields and currents in the sheet plane by the procedure outlined in Sects. 2.2.3 and 4, we obtain for the components of the primary electric field:

$$E_{x'}^{(p)'} = -\bar{B}_0 e^{-i\omega t'} \left(\omega - \frac{\pi}{\tau}v\right) \frac{2h}{2\pi}$$

$$\times \int_{-\infty}^{\infty} n \frac{\sin nh}{nh} \frac{e^{i[(\pi/\tau)(x' + vt')]}e^{iny'}}{\pi^2/\tau^2 + n^2} dn; \quad z' = 0 , \quad (2.2.39)$$

$$E_{y'}^{(p)'} = +\bar{B}_0 e^{-i\omega t'} \left(\omega - \frac{\pi}{\tau}v\right) \frac{\pi}{\tau} \frac{2h}{2\pi}$$

$$\times \int_{-\infty}^{\infty} \frac{\sin nh}{nh} \frac{e^{i[(\pi/\tau)(x' + vt')]}e^{iny'}}{\pi^2/\tau^2 + n^2} dn; \quad z' = 0 . \quad (2.2.40)$$

The secondary field components are now given by:

$$E_{x'}^{(s)'} = +\bar{B}_0 e^{-i\omega t'} \left(\omega - \frac{\pi}{\tau}v\right) \frac{2h}{2\pi}$$

$$\times \int_{-\infty}^{\infty} n \frac{\sin nh}{nh} \frac{e^{i[(\pi/\tau)(x' + vt')]}e^{iny'}}{\pi^2/\tau^2 + n^2} dn; \quad z' = 0 , \quad (2.2.41)$$

$$E_y^{(s)'} = -\bar{B}_0 e^{-i\omega t'} \left(\omega - \frac{\pi}{\tau} v\right) \frac{\pi}{\tau} \frac{2h}{2\pi}$$

$$\times \int_{-\infty}^{\infty} \frac{\sin nh}{nh} \frac{e^{i[(\pi/\tau)(x'+vt')]} e^{iny'}}{\pi^2/\tau^2 + n^2} dn; \quad z' = 0 , \tag{2.2.42}$$

evidently leading towards $E^{(p)'} + E^{(s)'} = 0$, as required by the condition $\sigma \to \infty$; $\Delta =$ finite.

The surface currents are expressed by the integrals

$$A_x = -i \frac{\bar{B}_0}{\mu_0} e^{-i\omega t} \frac{4h}{2\pi}$$

$$\times \int_{-\infty}^{\infty} n \frac{\sin nh}{nh} \frac{e^{i[(\pi/\tau)x + ny]}}{\sqrt{n^2 + (\pi/\tau)^2} \tanh[\sqrt{n^2 + (\pi/\tau)^2} b]} dn , \tag{2.2.43}$$

$$A_y = +i \frac{\bar{B}_0}{\mu_0} e^{-i\omega t} \frac{4h}{2\pi}$$

$$\times \int_{-\infty}^{\infty} \frac{\pi}{\tau} \frac{\sin nh}{nh} \frac{e^{i[(\pi/\tau)x + ny]}}{\sqrt{n^2 + (\pi/\tau)^2} \tanh[\sqrt{n^2 + (\pi/\tau)^2} b]} dn , \tag{2.2.44}$$

and it is seen that their magnitude depends neither on the velocity of motion v nor on the angular frequency ω of the exciting field. The current flow in the present case is governed not by Ohm's law, but by Maxwell's law $j = \nabla \times H$; an infinitesimal displacement of the superconducting strip must generate a field such as to compensate entirely for the exciting magnetic field: the discontinuity in the tangential components of this secondary reaction field fully determines, see (2.2.25, 26), the current distribution in the sheet.

2.2.6 Power and Thrust Considerations

The time-average power $\langle P \rangle$ converted by the gliding armature ("rotor") of width $2a$ and thickness Δ, moving at speed v and carrying the (complex) surface current density components $\bar{A}_x$, $\bar{A}_y$, reads

$$\langle P \rangle = \frac{1}{2} v \tau \operatorname{Re} \left\{ \int_{-a}^{a} \bar{A}_y^* \bar{B}_z^{(p)} dy \right\} , \tag{2.2.45}$$

where $\langle P \rangle$ is here taken per pole pitch τ; the asterisk denotes, as usual, the complex conjugate. Although the induced armature glides in a region pervaded by the primary as well as by the secondary induction, the primary (complex) amplitude $\bar{B}_z^{(p)}$ alone is needed – because of momentum considerations – in order to calculate $\langle P \rangle$.

For brevity, we consider in the sequel only the limiting case of a very wide rotor, i.e. $a \to \infty$, as well as that of a very long stator, i.e. $l \to \infty$.

In the $z = 0$ plane, see (2.2.9), the imposed axial magnetic flux density component is expressed by means of the definite integral

$$B_z^{(p)} = \bar{B}_0 \frac{2h}{2\pi} e^{-i\omega t} e^{i(\pi/\tau)x} \int_{-\infty}^{\infty} \frac{\sin nh}{nh} e^{iny} dn \ . \tag{2.2.46}$$

The secondary superpotential function $F^{(s)}$ is characterized, in turn, by the (complex) spectral density $\bar{f}(n)$, see (2.2.22, 39), which reads

$$\bar{f}(n) = \frac{\dfrac{\sigma\Delta\mu_0}{2} \dfrac{\bar{B}_0}{\mu_0} \dfrac{4h}{2\pi} \left(\omega - \dfrac{\pi}{\tau} v\right) \dfrac{\sin nh}{nh} \dfrac{1}{\pi^2/\tau^2 + n^2}}{i + \dfrac{\sigma\Delta\mu_0}{2}\left(\omega - \dfrac{\pi}{\tau} v\right) \dfrac{\tanh(\sqrt{\pi^2/\tau^2 + n^2} b)}{\sqrt{\pi^2/\tau^2 + n^2}}} \ , \tag{2.2.47}$$

so that

$$\bar{A}_y = i\frac{\pi}{\tau} e^{-i\omega t} e^{i(\pi/\tau)x} \int_{-\infty}^{\infty} \bar{f}(n) e^{iny} dn \ . \tag{2.2.48}$$

Hence

$$\langle P \rangle = \frac{1}{2} v\tau |\bar{B}_0|^2 \left(\frac{h}{\pi}\right)^2 \sigma\Delta \left(\omega - \frac{\pi}{\tau} v\right) \frac{\pi}{\tau} \mathrm{Re} \left\{ \int_{-\infty}^{\infty} dy \right.$$

$$\times \int_{-\infty}^{\infty} \frac{-i \dfrac{\sin nh}{nh} \dfrac{1}{\pi^2/\tau^2 + n^2} e^{-iny}}{-i + \dfrac{\sigma\Delta\mu_0}{2}\left(\omega - \dfrac{\pi}{\tau} v\right) \dfrac{\tanh(\sqrt{\pi^2/\tau^2 + n^2} b}{\sqrt{\pi^2/\tau^2 + n^2}}} dn$$

$$\times \left. \int_{-\infty}^{\infty} \frac{\sin \lambda h}{\lambda h} e^{i\lambda y} d\lambda \right\} \ , \tag{2.2.49}$$

where the dummy variable n in the third integral has been replaced by λ.

We now introduce the phase velocity

$$v_{ph} = \frac{\omega\tau}{\pi} \ , \tag{2.2.50}$$

of the traveling wave of excitation, the slip

$$s = 1 - \frac{v}{v_{ph}} \ , \tag{2.2.51}$$

and furthermore assume a very narrow air gap, i.e., postulate, the validity of the inequality

$$b \ll \tau . \tag{2.2.52}$$

The converted time-average power may therefore be rewritten in the form

$$\langle P \rangle = \frac{1}{2} v | \bar{B}_0 |^2 \left(\frac{h}{\pi} \right)^2 \sigma \Delta s \omega \pi \frac{1}{1 + \left(\dfrac{\sigma \Delta \mu_0 s \omega b}{2} \right)^2}$$

$$\times \int_{-\infty}^{\infty} dy \int_{-\infty}^{\infty} \frac{\sin nh}{nh} \frac{1}{\pi^2/\tau^2 + n^2} e^{-iny} dn \int_{-\infty}^{\infty} \frac{\sin \lambda h}{\lambda h} e^{i\lambda y} d\lambda . \tag{2.2.53}$$

In the integral, see (2.1.48),

$$\int_{-\infty}^{\infty} e^{-iny} e^{i\lambda y} d\lambda = 2 \pi \delta(\lambda - n) , \tag{2.2.54}$$

the Dirac δ-function is defined so that (2.2.53) reduces to the somewhat simpler expression

$$\langle P \rangle = \frac{\dfrac{1}{2} v | \bar{B}_0 |^2 \left(\dfrac{h}{\pi} \right)^2 \sigma \Delta s \omega \pi}{1 + \left(\dfrac{\sigma \Delta \mu_0 s \omega b}{2} \right)^2} 2 \pi \int_{-\infty}^{\infty} \left(\frac{\sin nh}{nh} \right)^2 \frac{dn}{\pi^2/\tau^2 + n^2} . \tag{2.2.55}$$

Resorting to the definite integral

$$\int_0^{\infty} \frac{\sin^2 \alpha x}{x^2 (\beta^2 + x^2)} dx = \frac{\pi}{4\beta^2} \left[2\alpha - \frac{1}{\beta} (1 - e^{-2\alpha\beta}) \right] , \tag{2.2.56}$$

we finally obtain the result

$$\langle P \rangle = \frac{2}{\pi} v \left(\frac{1}{2} \frac{| \bar{B}_0 |^2}{\mu_0} \right) \frac{\dfrac{\sigma \Delta \mu_0 s \omega b}{2}}{1 + \left(\dfrac{\sigma \Delta \mu_0 s \omega b}{2} \right)^2}$$

$$\times \left\{ \frac{h \tau^2}{b} \left[1 - \frac{\tau}{2\pi h} \left(1 - e^{-\frac{2\pi h}{\tau}} \right) \right] \right\} . \tag{2.2.57}$$

Now the time-average thrust

$$\langle F \rangle = \frac{\langle P \rangle}{v} , \tag{2.2.58}$$

attains its extremal values $\pm F_m$ at the critical slip values

$$s = \pm s_c \, , \qquad (2.2.59)$$

respectively, with s_c in turn reading

$$s_c = \frac{2}{\sigma \Delta \mu_0 \omega b} \, . \qquad (2.2.60)$$

The universal thrust-slip characteristic is therefore given by

$$\frac{\langle F \rangle}{\langle F_m \rangle} = \frac{2}{\dfrac{s_c}{s} + \dfrac{s}{s_c}} \, , \qquad (2.2.61)$$

an expression formally identical with that pertaining to rotating induction machines.

2.2.7 Discussion

In the present subsection an approach to the analysis of linear thrust machines is outlined; the value of the proposed approach – based upon the Fitzgerald superpotential – is demonstrated in calculations of performance characteristics and stream-line patterns of some highly idealized versions of the linear motor. Analysis of the streamline pattern, in turn brings out the influence of end effects.

Let us once more stress the fact that our analysis is specifically associated with artificial prescription of an imposed magnetic flux density in the $z = 0$ plane; this is done, on the one hand, in analogy with rotating machinery and – on the other – for the more formal reason of internal unity of the whole chapter (Sect. 2.1.4). In any case, the results yield insight into both the line of attack on, and the energy-exchange of, linear induction devices, and should lead to an improved approach towards analysis and design of such machines.

2.2.8 Appendix

A) The Fitzgerald Vector in Quasistatics

Having adopted from the outset magneto-quasistatics, Maxwell's equation $\nabla \cdot \varepsilon_0 E = \varrho$ becomes an identity, i.e. $0 = 0$; hence, in order to introduce the F vector we resort to an approach somewhat unlike that outlined in Sect. 2.1.2. Accordingly, a conduction current of density j_a is assumed for the air-gap, to which an extremely low conductivity σ_a (relative to copper, σ_c) is now ascribed, i.e.

$$j_a = \sigma_a E; \quad \sigma_a \ll \sigma_c \, . \qquad (2.2.62)$$

In other words,

$$j_a \to 0 \qquad\qquad (2.2.63)$$

for finite values of E.

This assumption is well-founded from the engineering viewpoint, and Maxwell's equation (2.2.1) may now be rewritten as

$$\nabla \times H = j_a , \qquad\qquad (2.2.64)$$

therefore implying

$$\nabla \cdot j_a = \nabla \cdot (\sigma_a E) = 0 . \qquad\qquad (2.2.65)$$

With $\sigma_a \ll \sigma_c$ yet finite, we may now write

$$\nabla \cdot E = 0 , \qquad\qquad (2.2.66)$$

i.e., E may be derived from a vector potential C:

$$E = \nabla \times C . \qquad\qquad (2.2.67)$$

σ_a has therefore been no more than a "play" resorted to in order to introduce the vector C. Returning now to (2.2.1), we also introduce the magnetic scalar potential χ through the equation

$$H = -\nabla \chi . \qquad\qquad (2.2.68)$$

Faraday's law (2.2.2) takes now the form

$$\nabla \times (\nabla \times C) = -\mu_0 \frac{\partial}{\partial t} (-\nabla \chi) \qquad\qquad (2.2.69)$$

and − in Cartesian coordinates −

$$\nabla (\nabla \cdot C) - \nabla^2 C = \mu_0 \nabla \frac{\partial \chi}{\partial t} , \qquad\qquad (2.2.70)$$

where the order of $\partial / \partial t$ and ∇ has been interchanged. We now assume, compare (2.1.9),

$$\nabla \cdot C = \mu_0 \frac{\partial \chi}{\partial t} \qquad\qquad (2.2.71)$$

and obtain the differential equation for C in the form

$$\nabla^2 C = 0 . \qquad\qquad (2.2.72)$$

Further, (2.2.71) becomes an identity once the vector F is introduced by the defining relations

$$C \equiv \frac{\partial F}{\partial t} \tag{2.2.73}$$

$$\chi \equiv \frac{1}{\mu_0} \nabla \cdot F \ . \tag{2.2.74}$$

Hence the expressions

$$E = \nabla \times \frac{\partial F}{\partial t} \tag{2.2.75}$$

$$H = -\frac{1}{\mu_0} \nabla(\nabla \cdot F) \ . \tag{2.2.76}$$

Introducing (2.2.73) into (2.2.72) yields (2.2.5 a).

B) Invariance of the Stream Function D

The expression (2.2.34) was obtained from (2.2.19) by replacing $(x' + v t')$ with x and y' with y, as D is taken to be invariant. In order to prove in general this invariance we resort to restricted relativity and proceed as follows: electric charge conservation yields the equation

$$\nabla \cdot j + \frac{\partial \varrho}{\partial t} = 0 \ , \tag{2.2.77}$$

where j (components: j_x, j_y, j_z) and ϱ stand respectively for the free electric-current and charge densities. To emphasize the four-vector character of these physical parameters, we rewrite the last equation in four-dimensional form

$$\frac{\partial j_x}{\partial x} + \frac{\partial j_y}{\partial y} + \frac{\partial j_z}{\partial z} + \frac{\partial}{\partial i c t}(i\varrho c) = 0 \ . \tag{2.2.78}$$

On the basis of Einstein's restricted relativity, the law of transformation of the four-vector $(j_x, j_y, j_z, i\varrho c)$ into its primed counterpart reads, see (2.1.23),

$$j_x' = \frac{j_x - (v/c)\varrho c}{\sqrt{1 - \beta^2}}; \quad j_y' = j_y; \quad j_z' = j_z; \quad \varrho' c = \frac{\varrho c - (v/c) j_x}{\sqrt{1 - \beta^2}} \ , \tag{2.2.79}$$

and the inverse transformation is

$$j_x = \frac{j_x' + (v/c)\varrho' c}{\sqrt{1 - \beta^2}}; \quad j_y = j_y'; \quad j_z = j_z';$$

$$\varrho c = \frac{\varrho' c + (v/c) j_x'}{\sqrt{1 - \beta^2}} \ . \tag{2.2.80}$$

In the primed frame of reference, for an observer at rest with respect to the sheet, Ohm's law holds in the form $j' = \sigma E'$; now, validity of this law implies – under steady-state conditions – quasineutrality, as no permanent free charge distribution can exist within a region of finite conductivity. Hence $\varrho' = 0$ and, accordingly

$$\varrho = \frac{v}{c^2} j_x \ . \tag{2.2.81}$$

However, replacing c by $\bar{c} = c/n$ with $n \to 0$ ("Cerenkov Electrodynamics"), we also have $\varrho = 0$, so that within the framework of the resulting Galilean approximation we have

$$j_x = j'_x; \quad j_y = j'_y; \quad j_z = j'_z; \quad \varrho = 0 \ . \tag{2.2.82}$$

As the surface current density components are obtained from the defining relations

$$-\frac{\partial D'}{\partial y'} \equiv A'_x = j'_x \Delta; \quad +\frac{\partial D'}{\partial x'} \equiv A'_y = j'_y \Delta \ , \tag{2.2.83}$$

we may write

$$-\frac{\partial D'}{\partial y'} = j_x \Delta; \quad \frac{\partial D'}{\partial x'} = j_y \Delta \ . \tag{2.2.84}$$

However, within the framework of Galilean relativity

$$\frac{\partial}{\partial x'} = \frac{\partial}{\partial x} \ , \tag{2.2.85}$$

and in any case

$$\frac{\partial}{\partial y'} = \frac{\partial}{\partial y} \ , \tag{2.2.86}$$

so that

$$j_x \Delta = -\frac{\partial D'}{\partial y}, \quad j_y \Delta = \frac{\partial D'}{\partial x}, \quad \text{or} \tag{2.2.87}$$

$$A_x = -\frac{\partial D'}{\partial y}, \quad A_y = \frac{\partial D'}{\partial x} \ . \tag{2.2.88}$$

In the unprimed frame, we define from the outset

$$A_x \equiv -\frac{\partial D}{\partial y}, \quad A_y \equiv \frac{\partial D}{\partial x} \ . \tag{2.2.89}$$

Hence,

$$D = D' \ . \tag{2.2.90}$$

2.3 Unipolar Induction – Basic Principles

The phenomenon of unipolar induction was very briefly mentioned in Sect. 2.1.5; it occurs when relative motion arises between the induced medium and the exciting permanent magnet or direct-current electro-magnet.

Early attempts to quantify unipolar induction led to numerous contradictions, and its relativistic character could be appreciated only with the advent of Einstein's theory. The relativistic interplay is illustrated in Sects. 2.1.1 – 5 and may formally be applied by setting the angular supply frequency ω equal to zero in the relevant field expressions. We prefer, however, to adopt a wider approach which has two distinct aspects: (a) a physical explanation of unipolar induction – with an *engineering* point of view in mind; (b) the solution – as an illustration – of *practical problems* involving unipolar induction.

2.3.1 Physical Explanation

Relative motion of the induced strip and the magnetic pole (Fig. 2.1) implies relative motion of the volume magnetization M and the medium; the motion of M with respect to the stationary laboratory gives rise in the latter to an electric polarization P. Now the source density of P is opposed and equal to that of the electric induction counterpart, i.e.

$$\nabla \cdot \varepsilon_0 E = - \nabla \cdot P \ . \tag{2.3.1}$$

Hence the advent of E.

We now introduce the four-dimensional polarization tensor $[\pi]$ of order two

$$[\pi] = \begin{vmatrix} 0 & M_z & -M_y & icP_x \\ -M_z & 0 & M_x & icP_y \\ M_y & -M_x & 0 & icP_z \\ -icP_x & -icP_y & -icP_z & 0 \end{vmatrix} \tag{2.3.2}$$

where the three components M_x, M_y, M_z of M and the three components P_x, P_y, P_z of P are presented in matrix form.

Next, the original assumption that the strip glides with respect to the magnetic pole with constant speed v along the x-axis is replaced by the equivalent assumption that the strip is stationary and the pole recedes from it with constant speed $-v$. The primed frame of reference – stationary with respect to the magnetic pole – is interlinked with the unprimed one through the matrix $[\alpha_{jk}]$ as per (2.1.24), with β now replaced by $-\beta$, i.e.,

$$[\alpha_{jk}] = \begin{vmatrix} \dfrac{1}{\sqrt{1-\beta^2}} & 0 & 0 & \dfrac{-i\beta}{\sqrt{1-\beta^2}} \\[2ex] 0 & 1 & 0 & 0 \\[1ex] 0 & 0 & 1 & 0 \\[1ex] \dfrac{i\beta}{\sqrt{1-\beta^2}} & 0 & 0 & \dfrac{1}{\sqrt{1-\beta^2}} \end{vmatrix} . \tag{2.3.3}$$

The primed source tensor $[\pi']$ has – by definition – the same form as (2.3.2), i.e.,

$$[\pi'] = \begin{vmatrix} 0 & M'_z & -M'_y & icP'_x \\[1ex] -M'_z & 0 & M'_x & icP'_y \\[1ex] M'_y & -M'_x & 0 & icP'_z \\[1ex] -icP'_x & -icP'_y & -icP'_z & 0 \end{vmatrix} . \tag{2.3.4}$$

Transformation of $[\pi']$ back into the laboratory yields the matrix:

$$[\pi] = \begin{vmatrix} 0 & \dfrac{M'_z + vP'_y}{\sqrt{1-\beta^2}} & -\dfrac{M'_y - vP'_z}{\sqrt{1-\beta^2}} & icP'_x \\[3ex] -\dfrac{M'_z + vP'_y}{\sqrt{1-\beta^2}} & 0 & M'_x & \dfrac{i\beta M'_z + icP'_y}{\sqrt{1-\beta^2}} \\[3ex] \dfrac{M'_y - vP'_z}{\sqrt{1-\beta^2}} & -M'_x & 0 & \dfrac{-i\beta M'_y + icP'_z}{\sqrt{1-\beta^2}} \\[3ex] -icP'_x & -\dfrac{i\beta M'_z + icP'_y}{\sqrt{1-\beta^2}} & -\dfrac{-i\beta M'_y + icP'_z}{\sqrt{1-\beta^2}} & 0 \end{vmatrix} . \tag{2.3.5}$$

Hence, identifying (2.3.2) with (2.3.5) we find

$$M_x = M'_x ,$$

$$M_y = \dfrac{M'_y - vP'_z}{\sqrt{1-\beta^2}} , \tag{2.3.6}$$

$$M_z = \dfrac{M'_z + vP'_y}{\sqrt{1-\beta^2}} ,$$

along with

$$P_x = P'_x \; ,$$

$$P_y = \frac{P'_y + (v/c^2) M'_z}{\sqrt{1 - \beta^2}} \; , \qquad (2.3.7)$$

$$P_z = \frac{P'_z - (v/c^2) M'_y}{\sqrt{1 - \beta^2}} \; .$$

Assuming specifically that the magnetic pole comprises in its rest frame no electric polarization, we obtain in the frame of reference linked to the laboratory:

$$M_x = M'_x \; ,$$

$$M_y = \frac{M'_y}{\sqrt{1 - \beta^2}} \; , \qquad (2.3.8)$$

$$M_z = \frac{M'_z}{\sqrt{1 - \beta^2}} \; ,$$

and

$$P_x = 0 \; ,$$

$$P_y = \frac{(v/c^2) M'_z}{\sqrt{1 - \beta^2}} \; , \qquad (2.3.9)$$

$$P_z = -\frac{(v/c^2) M'_y}{\sqrt{1 - \beta^2}} \; .$$

The presence of P is thus directly traced to the motion of the proper magnetization M'; combining (2.3.8 and 9), we have [5]

$$P = \frac{(-v) \times M}{c^2} \; . \qquad (2.3.10)$$

For $\beta^2 \ll 1$, M may obviously be replaced by M'; hence

$$P = \frac{(-v) \times M'}{c^2} \; . \qquad (2.3.11)$$

In spite of the complications entailed by its "mixed" character (both primed and unprimed quantities), (2.3.11) has the advantage of explicitly linking the

---

[5] Here $v = v_x \mathbf{1}_x$ with $v_x = v$ and $\mathbf{1}_x$ the unit vector (Fig. 2.1); in general, $v = v_x \mathbf{1}_x + v_y \mathbf{1}_y + v_z \mathbf{1}_z$ where v_y and v_z are the relevant y and z components, and $\mathbf{1}_y$, $\mathbf{1}_z$ the corresponding unit vectors.

moving magnetization M' to the stationary laboratory. Reverting to (2.3.1), we finally obtain

$$\nabla \cdot E = \nabla \cdot [\mu_0 (v \times M)] \; , \tag{2.3.12}$$

or, for $\beta^2 \ll 1$

$$\nabla \cdot E = \nabla \cdot [\mu_0 (v \times M')] \; . \tag{2.3.13}$$

Insight into the occurrence of unipolar induction has thus been provided[6], and we are now able to consider specific configurations.

2.3.2 Formulation of Problem and Approach to Solution

The first system to be investigated resembles in many respects that described in Sect. 2.1.3 and reproduced in Fig. 2.1. There are, however, also some differences (Fig. 2.8).

The poles are now excited, not by alternating current, but by a direct-current carrying coil (alternatively, they may be regarded as permanent magnetic poles); further, they are not rectangular but circular-cylindrical of diameter $2R$.

The strip – of width $2h > 2R$ and gliding with regulated constant speed v – exhibits an electric conductivity σ; we therefore ascribe to it a finite surface resistance (Sect. 2.2.1) defined by the fictitious limiting process

$$\lim_{\substack{\sigma \to \infty \\ \varDelta \to 0}} (\sigma \varDelta) = \text{finite} \; . \tag{2.3.14}$$

Nevertheless, even though the strip is conducting, we shall waive considerations concerning the secondary magnetic field due to the induced current flow;

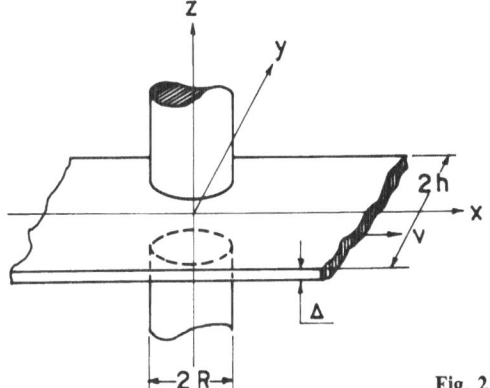

Fig. 2.8. System under investigation

---

[6] An even more formal approach might be based on (2.1.86 – 88) with $E'_x = E'_y = E'_z = 0$, with the primed magnetic field time-independent; now v has to be replaced by $-v$.

it is assumed that this field is negligible compared to the primary (imposed) magnetic field, as is the case in many engineering applications, e.g., eddy-current brakes, measuring instruments, relays, flowmeters, etc. In short, we adopt a so-called low magnetic Reynolds number (R_m) approximation:

$$R_m \equiv \mu_0 \sigma \Delta v \ll 1 \ . \tag{2.3.15}$$

For the proposed model we intend to calculate:

a) the streamline pattern of the induced (surface) current flow;
b) the braking force exerted by the poles (= "stator") on the moving strip (= "glider").

As already pointed out, once the relativistic background has been provided and the presence of E accounted for, we are at liberty to consider a practical problem for which $\beta^2 \ll 1$; moreover, we again specifically refer to a magneto-quasistatic system.

In the primed frame of reference, now attached to the nonmagnetic metallic strip, Faraday's law reads

$$\nabla' \times E' = -\frac{\partial B'}{\partial t'} \ , \qquad B' \equiv \mu_0 H' \ . \tag{2.3.16}$$

In our case (very narrow air gap) the component B'_z of B' in the $z' = 0$ plane is of main interest. Hence, in the primed frame we consider in the sheet plane the equation

$$\frac{\partial E'_y}{\partial x'} - \frac{\partial E'_x}{\partial y'} = -\frac{\partial B'_z}{\partial t'} \ . \tag{2.3.17}$$

Within the framework of the quoted approximations we have [see (1.4.13) with $n = 0$, or, from a different point of view: (1.1.7) with $\beta \to 0$]

$$B'_z = B_z \ , \tag{2.3.18}$$

and further [e.g., (2.1.61), read in reverse, with $\beta \to 0$, or directly from (1.4.26)],

$$\frac{\partial}{\partial t'} = \frac{\partial}{\partial t} + v \frac{\partial}{\partial x} \ . \tag{2.3.19}$$

In more generalized terms

$$\frac{\partial}{\partial t'} = \frac{\partial}{\partial t} + (v \cdot \nabla) \ , \tag{2.3.20}$$

so that in fact we obtain instead of (2.3.17), the relation

$$\frac{\partial E'_y}{\partial x'} - \frac{\partial E'_x}{\partial y'} = -\left[\frac{\partial B_z}{\partial t} + (v \cdot \nabla) B_z \right] \ . \tag{2.3.21}$$

As B_z is time-independent in the unprimed frame, (2.3.21) reduces to the somewhat simpler form

$$\frac{\partial E_y'}{\partial x'} - \frac{\partial E_x'}{\partial y'} = -(\boldsymbol{v} \cdot \nabla) B_z \ . \tag{2.3.22}$$

We next resort again to the generating function D', yielding the surface current densities $A_x' \equiv -\partial D'/\partial y'$; $A_y' \equiv \partial D'/\partial x'$. Applying Ohm's law in the primed frame of reference in component form, i.e.,

$$A_x' = \sigma \Delta E_x' \ , \tag{2.3.23}$$

$$A_y' = \sigma \Delta E_y' \ , \tag{2.3.24}$$

and substituting into (2.3.22), we obtain Poisson's equation

$$\frac{\partial^2 D'}{\partial x'^2} + \frac{\partial^2 D'}{\partial y'^2} = -(\sigma \Delta)(\boldsymbol{v} \cdot \nabla) B_z \ . \tag{2.3.25}$$

At this juncture, the Cartesian systems of coordinates may conveniently be supplemented by circular cylindrical coordinates with axis z, radius $r = \sqrt{x^2 + y^2}$ and angle of azimuth $\alpha = \tan^{-1}(y/x)$. In our case $\partial B_z / \partial r \neq 0$ only in a very narrow circular region at the circumference of the exciting pole; introducing therefore the *radial velocity component* v_r, the last equation is rewritten as

$$\frac{\partial^2 D'}{\partial x'^2} + \frac{\partial^2 D'}{\partial y'^2} = -(\sigma \Delta) v_r \frac{\partial B_z}{\partial r} \ . \tag{2.3.26}$$

Let us now formulate (2.3.26) entirely in the unprimed system. In view of the assumptions of quasistatics as well as $\beta \ll 1$, we once more, see (2.2.85, 86), have

$$\frac{\partial}{\partial x'} = \frac{\partial}{\partial x} \ , \quad \frac{\partial}{\partial y'} = \frac{\partial}{\partial y} \ . \tag{2.3.27}$$

Further, the stream function $D' = D$ is invariant (Sect. 2.2.8 B). Hence, when (2.3.26) is restated in the unprimed frame of reference, it reads:

$$\frac{\partial^2 D}{\partial x^2} + \frac{\partial^2 D}{\partial y^2} = -(\sigma \Delta) v_r \frac{\partial B_z}{\partial r} \ . \tag{2.3.28}$$

The solution of this equation is obtained through the following analogy: a stationary current-density distribution $j = j(x, y) \mathbf{1}_z$ in a homogeneous medium of permeability μ_0 is linked to its vector potential $V = V_z(x, y) \mathbf{1}_z$ by the Poisson equation

$$\frac{\partial^2 V_z}{\partial x^2} + \frac{\partial^2 V_z}{\partial y^2} = -\mu_0 j_z \ . \tag{2.3.29}$$

Expressions (2.3.28, 29) are seen to be identical in form; thus, setting $(\sigma\Delta)v_r(\partial B_z/\partial r)$ equal to $\mu_0 j_z$ and imposing formally identical boundary conditions in both cases, D is obtained once the solution for the vector potential V_z is known. As $\partial B_z/\partial r$ is nonzero only for a very narrow region at $r = R$, j_z is also taken to be confined to a very narrow region with the same functional dependence as $\partial B_z/\partial r$.

Integrating

$$(\sigma\Delta)v_r \int_{0-\zeta}^{\infty} \frac{\partial B_z}{\partial r}\, dr \equiv \mu_0 \int_{0-\zeta}^{\infty} j_z dr \qquad (2.3.30)$$

with $\zeta \ll R$, and recalling that $\int_0^{\infty} j_z dr$ defines a surface current density K_z, we obtain

$$-(\sigma\Delta)v_r B_z = +\mu_0 K_z \ . \qquad (2.3.31)$$

The minus sign on the left is dictated by the fact that the corresponding integral is taken from $r = 0 - \zeta$, inside the magnet where B_z is constant, to $r = \infty$, where $B_z = 0$.

2.3.3 Calculation of Sheet Current

We now consider the induced current in the gliding sheet; it is of necessity confined to the strip $-h < y < h$, hence

$$A_y = \frac{\partial D}{\partial x} = 0 \quad \text{for} \quad y = \pm h \ , \qquad (2.3.32)$$

which implies $D = \text{const}$ for $y = \pm h$.

Using the vector-potential analogy, we have to solve the following problem:

Given the surface-current distribution

$$K_z(\alpha) = -\frac{(\sigma\Delta)v_r B_z}{\mu_0} \qquad (2.3.33)$$

along the circumference of the circle $r = R$, what is the value of the vector potential component V_z? An additional inherent boundary condition, see (2.3.32), on V_z is

$$\frac{\partial V_z}{\partial x} = 0 \quad \text{for} \quad y = \pm h; \quad h > R \ . \qquad (2.3.34)$$

The first step towards a solution is calculation of v_r. Having assumed motion in the x-direction only, we have $v = v\mathbf{1}_x$; now the unit vector $\mathbf{1}_x$, expressed in polar coordinates (unit vectors: $\mathbf{1}_r$, $\mathbf{1}_\alpha$), reads $\mathbf{1}_x = \mathbf{1}_r \cos\alpha - \mathbf{1}_\alpha \sin\alpha$; therefore $v_r = v\cos\alpha$, so that

$$+K_z(\alpha) = -\frac{(\sigma\varDelta)vB_z}{\mu_0}\cos\alpha .$$ (2.3.35)

We thus have the partial differential equation

$$\frac{\partial^2 V_z}{\partial r^2} + \frac{1}{r}\frac{\partial V_z}{\partial r} + \frac{1}{r^2}\frac{\partial^2 V_z}{\partial\alpha^2} = 0 ,$$ (2.3.36)

with the boundary condition (2.3.34) and the excitation as per (2.3.35).

Introducing the – as yet – unspecified constants C_1 and C_2, we preliminarily obtain the solution of (2.3.36) for $h = \infty$ in the form

$$V_z = C_1\frac{r}{R}\cos\alpha, \quad \text{at} \quad r < R ,$$ (2.3.37)

$$V_z = C_2\frac{R}{r}\cos\alpha, \quad \text{at} \quad r > R .$$ (2.3.38)

Prescribing the proper discontinuity of the circular components H_α on the boundary, i.e.

$$H_\alpha\big|_{r=R+0} - H_\alpha\big|_{r=R-0} = K_z(\alpha) ,$$ (2.3.39)

and the continuity of V_z at $r = R$, we find

$$C_1 = C_2 = -\frac{(\sigma\varDelta)vB_z}{2\mu_0}\mu_0 R .$$ (2.3.40)

We now proceed to the solution for a finite width of the strip; to this end we introduce (Fig. 2.9) a series of fictitious exciting currents of alternating sign, distributed over cylinders with radius $r = R$ and infinite along the z-axis. The projected spacing of the circle centers is $2h$.

Further, instead of the vector-potential component V_z we now resort to the *complex* vector potential component $\bar{V}_z$. Thus, setting $w \equiv x + iy = r\exp(i\alpha)$, we have from (2.3.38 and 40)

$$\bar{V}_z = \frac{K_m}{2}\mu_0 R\frac{R}{w}, |w| \equiv R, K_m \equiv -\frac{(\sigma\varDelta)vB_z}{\mu_0} .$$ (2.3.41)

The boundary condition (2.3.34) now reads $\mathrm{Re}\{\partial\bar{V}_z/\partial x\} = 0$ at $y = \pm h$, and is automatically satisfied by the configuration reproduced in Fig. 2.9.

For $|w| > R$, but still inside the original strip, the *resultant* complex vector potential component $\bar{V}_{rz}$ (Subscripts r and z) is obtained from the series

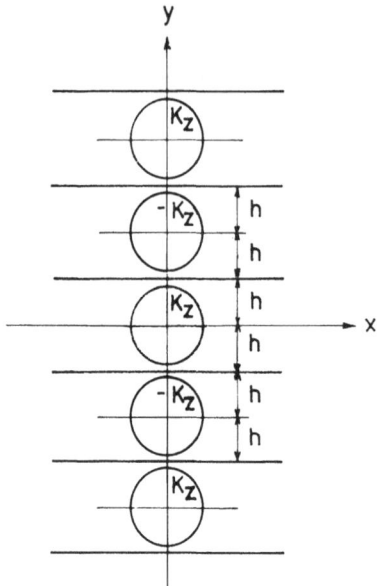

Fig. 2.9. Series of fictitious exciting currents

$$\bar{V}_{rz} = \frac{K_{\mathrm{m}}}{2}\mu_0 R\, \frac{R}{w} - \frac{K_{\mathrm{m}}}{2}\mu_0 R\, \frac{R}{w+2\,ih}$$

$$+ \frac{K_{\mathrm{m}}}{2}\mu_0 R\, \frac{R}{w+4\,ih}$$

$$- \frac{K_{\mathrm{m}}}{2}\mu_0 R\, \frac{R}{w-2\,ih}$$

$$+ \frac{K_{\mathrm{m}}}{2}\mu_0 R\, \frac{R}{w-4\,ih} + \ldots - \ldots \, , \qquad (2.3.42)$$

which yields (Sect. 2.3.5 A):

$$\bar{V}_{rz} = \frac{K_{\mathrm{m}}}{2}\mu_0 R\, \frac{\mathrm{i}(\pi/2)(R/h)}{\sin(\mathrm{i}\pi w/2h)} \, , \qquad |w| > R \, . \qquad (2.3.43)$$

For $|w| < R$, (2.3.37 and 40) yield for the complex vector potential component

$$\bar{V}_z = \frac{K_{\mathrm{m}}}{2}\mu_0 R\, \frac{w}{R} \, . \qquad (2.3.44)$$

Hence, for the strip under consideration, the *resultant* complex vector potential $\bar{V}_{rz}$ in the region $|w| < R$ is given by

$$\bar{V}_{rz} = \frac{K_m}{2} \mu_0 R \frac{w}{r} - \frac{K_m}{2} \mu_0 R \frac{R}{w+2ih}$$

$$+ \frac{K_m}{2} \mu_0 R \frac{R}{w+4ih}$$

$$- \frac{K_m}{2} \mu_0 R \frac{R}{w-2ih}$$

$$+ \frac{K_m}{2} \mu_0 R \frac{R}{w-4ih} + \dots - \dots \ . \tag{2.3.45}$$

Adding and subtracting the expression $(K_m/2)\mu_0 R(R/w)$, we find, see (2.3.42,43),

$$\bar{V}_{rz} = \frac{K_m}{2} \mu_0 R \left(\frac{w}{R} - \frac{R}{w} \right) + \frac{K_m}{2} \mu_0 R \frac{i(\pi/2)(R/h)}{\sin(i\pi w/2h)}; \quad |w| < R . \tag{2.3.46}$$

According to our initial explanation, the function D is given by the real part of $\bar{V}_{rz}$. Thus, for the inner zone $r < R$, we obtain

$$D = \frac{K_m}{2} \mu_0 R \left(\frac{1}{R} - \frac{R}{x^2+y^2} \right) x + \frac{K_m}{2} \mu_0 R \frac{\pi}{2} \frac{R}{h}$$

$$\times \frac{\sinh\left(\dfrac{\pi}{2h}x\right)\cos\left(\dfrac{\pi}{2h}y\right)}{\sinh^2\left(\dfrac{\pi}{2h}x\right)\cos^2\left(\dfrac{\pi}{2h}y\right) + \cosh^2\left(\dfrac{\pi}{2h}x\right)\sin^2\left(\dfrac{\pi}{2h}y\right)} ,$$

$$\tag{2.3.47}$$

whereas, for the outer zone $r > R$,

$$D = \frac{K_m}{2} \mu_0 R \frac{\pi}{2} \frac{R}{h}$$

$$\times \frac{\sinh\left(\dfrac{\pi}{2h}x\right)\cos\left(\dfrac{\pi}{2h}y\right)}{\sinh^2\left(\dfrac{\pi}{2h}x\right)\cos^2\left(\dfrac{\pi}{2h}y\right) + \cosh^2\left(\dfrac{\pi}{2h}x\right)\sin^2\left(\dfrac{\pi}{2h}y\right)} .$$

$$\tag{2.3.48}$$

The components of the surface current density are now readily obtained from the equations $A_x = -\partial D/\partial y$; $A_y = \partial D/\partial x$, and the streamlines are determined from

$$\frac{A_y}{A_x} = \frac{dy}{dx} \ , \quad \text{i.e.,} \tag{2.3.49}$$

$$\frac{\partial D/\partial x}{-\partial D/\partial y} = \frac{dy}{dx} \ , \quad \text{or} \tag{2.3.50}$$

$$D = \text{const} \ . \tag{2.3.51}$$

Thus, from (2.3.47), we obtain the dimensionless current-flow pattern for the inner zone as

$$\frac{2}{\pi} \frac{h}{R} \left(\frac{1}{R} - \frac{R}{x^2+y^2} \right) x$$

$$+ \frac{\sinh\left(\dfrac{\pi}{2h}x\right)\cos\left(\dfrac{\pi}{2h}y\right)}{\sinh^2\left(\dfrac{\pi}{2h}x\right)\cos^2\left(\dfrac{\pi}{2h}y\right)+\cosh^2\left(\dfrac{\pi}{2h}x\right)\sin^2\left(\dfrac{\pi}{2h}y\right)} = \text{const} ,$$

$$\tag{2.3.52}$$

whereas the pattern for the outer zone, see (2.3.48), is given by

$$\frac{\sinh\left(\dfrac{\pi}{2h}x\right)\cos\left(\dfrac{\pi}{2h}y\right)}{\sinh^2\left(\dfrac{\pi}{2h}x\right)\cos^2\left(\dfrac{\pi}{2h}y\right)+\cosh^2\left(\dfrac{\pi}{2h}x\right)\sin^2\left(\dfrac{\pi}{2h}y\right)} = \text{const} .$$

$$\tag{2.3.53}$$

Calculated results are shown in Fig. 2.10.

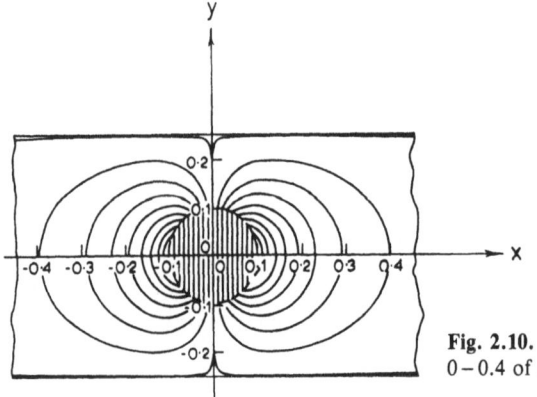

Fig. 2.10. Streamlines for dimensionless values $0-0.4$ of $D = \text{const}$

2.3.4 The Braking Force

The magnetic braking force F_{br} arises from the interaction between the surface current

$$A = A_x 1_x + A_y 1_y \tag{2.3.54}$$

and the magnetic induction B, i.e.,

$$F_{br} = \int_{(a)} (A \times B)\, dx\, dy \ , \tag{2.3.55}$$

where the integral is taken over the surface (a) of the sheet. With B comprising (in the sheet plane) a z-component only – namely B_z – we have

$$F_{br} = 1_x \int_{(a)} \frac{\partial D}{\partial x} B_z dx\, dy + 1_y \int_{(a)} \frac{\partial D}{\partial y} B_z dx\, dy \ . \tag{2.3.56}$$

As the sheet in our problem is constrained to move in the x-direction, the braking force of interest comprises here an x-component F_{br} only, i.e.,

$$F_{br} = \int_{(a)} \frac{\partial D}{\partial x} B_z dx\, dy \ . \tag{2.3.57}$$

Assuming now a closed system, taking into account Green's theorem in the plane for the multiple-connected region defined by the plane $z = 0$ and the contour of the normal pole projection of elementary length dl, and resorting to polar coordinates, we have,

$$F_{br} = B_z \oint D \cos \alpha\, dl \ . \tag{2.3.58}$$

Recalling the analogy $D = \mathrm{Re}\{\bar{V}_{rz}\}$ and setting $dl = R\, d\alpha$, (2.3.43 and 58) yield

$$F_{br} = B_z \mathrm{Re} \left\{ \int_{\alpha=0}^{2\pi} \left[\frac{K_m}{2} \mu_0 R \frac{\mathrm{i}\, \dfrac{\pi}{2} \dfrac{R}{h}}{\sin \dfrac{\mathrm{i}\pi w}{2h}} \right] \cos \alpha\, R\, d\alpha \right\} \ . \tag{2.3.59}$$

The appropriate calculation (Sect. 2.3.5 B) yields the absolute value $|F_{br}|$ of the braking force F_{br}

$$|F_{br}| = \frac{1}{2}(\sigma \Delta) \pi R^2 |v| B_z^2 \left[1 - \frac{\pi^2}{24}\left(\frac{2R}{2h}\right)^2 \right] , \tag{2.3.60}$$

wherein $|v| = +\sqrt{(v)^2} = +\sqrt{v^2}$.

For an "infinitely" wide sheet, i.e. $2h \gg 2R$, the (obviously larger) braking force equals

$$|F_{br}|_\infty = \tfrac{1}{2}(\sigma \Delta) \pi R^2 |v| B_z^2 \ . \tag{2.3.61}$$

The term $(\pi^2/24)(2R/2h)^2$ thus represents the reduction of the braking force due to the finite-width "end effect".

We illustrate the analysis by a numerical example with practical data compatible with the "low-Reynolds number" approximation:

$$\sigma = 3.4 \times 10^7 \, 1/\Omega m \ ,$$

$$\Delta = 2.0 \times 10^{-3} \, m \ ,$$

$$R = 0.010 \, m \ ,$$

$$|v| = 0.05 \, m/s \ ,$$

$$B_z = 0.75 \, Vs/m^2 \ ,$$

$$2R/2h = 2/3 \ .$$

Here $R_m \approx 0.004$ and for the braking force we obtain the value of 0.245 N.

Now the quoted result refers to a narrow metal strip of width 1.5 times the diameter of the magnet. If the sheet is wider, the current paths are less restricted, so that the braking effect is stronger; for an "infinitely" wide strip, the braking force increases in our case to 0.300 N.

The mechanical power consumption is seen to be quite small, namely $0.245 \times 0.05 = 12.25 \, mW$ only.

2.3.5 Appendix

A) Summation of Vector Potential Series

We seek the sum

$$\Sigma = \frac{1}{w} - \frac{1}{w+2ih} - \frac{1}{w-2ih} + \frac{1}{w+4ih} + \frac{1}{w-4ih}$$

$$- \frac{1}{w+6ih} - \frac{1}{w-6ih} + \dots - \dots$$

$$= \frac{1}{w} - \frac{2w}{w^2+4h^2} + \frac{2w}{w^2+16h^2} - \frac{2w}{w^2+36h^2}$$

$$= \frac{1}{w} - 2w \left(\frac{1}{w^2+4h^2} - \frac{1}{w^2+16h^2} + \frac{1}{w^2+36h^2} - \dots \right) . \qquad (2.3.62)$$

This series resembles the Laurent expansion of $\cosec(z)$

$$\cosec(z) = \frac{1}{z} - 2z \left(\frac{1}{z^2-\pi^2} - \frac{1}{z^2-4\pi^2} + \frac{1}{z^2-9\pi^2} - \dots \right) . \qquad (2.3.63)$$

Replacing our original series by

$$\Sigma = \frac{i\,\dfrac{\pi}{2h}}{i\,\dfrac{\pi}{2h}\,w} - \frac{i\,\dfrac{\pi}{2h}}{i\,\dfrac{\pi}{2h}\,w + i\,\dfrac{\pi}{2h}\,2ih} - \frac{i\,\dfrac{\pi}{2h}}{i\,\dfrac{\pi}{2h}\,w - i\,\dfrac{\pi}{2h}\,2ih}$$

$$+ \frac{i\,\dfrac{\pi}{2h}}{i\,\dfrac{\pi}{2h}\,w + i\,\dfrac{\pi}{2h}\,4ih} + \frac{i\,\dfrac{\pi}{2h}}{i\,\dfrac{\pi}{2h}\,w - i\,\dfrac{\pi}{2h}\,4ih} - \ldots + \ldots \qquad (2.3.64)$$

we obtain

$$\Sigma = i\,\frac{\pi}{2h}\left[\frac{1}{i\,\dfrac{\pi}{2h}\,w} - 2i\,\frac{\pi}{2h}\,w\right.$$

$$\times \left.\left(\frac{1}{\left(i\,\dfrac{\pi}{2h}\,w\right) - \pi^2} - \frac{1}{\left(i\,\dfrac{\pi}{2h}\,w\right)^2 - 4\pi^2} \ldots\right)\right], \qquad (2.3.65)$$

and comparing this expression with the expansion of cosec(z), we finally obtain

$$\Sigma = \frac{i\,\dfrac{\pi}{2h}}{\sin\left(i\,\dfrac{\pi}{2h}\,w\right)}\,. \qquad (2.3.66)$$

B) Calculation of Braking Force

We now seek

$$\frac{K_m}{2}\,\mu_0 R^2 B_z R \int_{\alpha=0}^{2\pi}\left[\frac{1}{w} - 2w\right.$$

$$\times \left.\left(\frac{1}{w^2 + 4h^2} - \frac{1}{w^2 + 16h^2} + \frac{1}{w^2 + 36h^2} - \ldots\right)\right]\cos\alpha\,d\alpha\,. \qquad (2.3.67)$$

Here $|w| < h$, as $w = Re^{i\alpha}$; hence we rewrite the original integral in the form

$$\int_{\alpha=0}^{2\pi} \left[\frac{1}{w} - 2w \left(\frac{1}{4h^2\left(1+\dfrac{w^2}{4h^2}\right)} - \frac{1}{16h^2\left(1+\dfrac{w^2}{16h^2}\right)} \right. \right.$$

$$\left. \left. + \frac{1}{36h^2\left(1+\dfrac{w}{16h^2}\right)} \right) \right] \cos\alpha\, d\alpha$$

$$= \int_{\alpha=0}^{\pi} \left\{ \frac{1}{w} - 2w \left[\frac{1}{4h^2}\left(1-\frac{w^2}{4h^2}\cdots\right) - \frac{1}{16h^2}\left(1-\frac{w^2}{16h^2}\cdots\right) \right. \right.$$

$$\left. \left. + \frac{1}{36h^2}\left(1-\frac{w^2}{36h^2}\cdots\right) \right] \right\} \cos\alpha\, d\alpha \ . \qquad (2.3.68)$$

Owing to the orthogonality of the trigonometric functions, we only have now to calculate

$$\int_{\alpha=0}^{2\pi} \left[\frac{1}{w} - 2w\left(\frac{1}{4h^2} - \frac{1}{16h^2} + \frac{1}{36h^2}\cdots \right) \right] \cos\alpha\, d\alpha \ . \qquad (2.3.69)$$

Further, being concerned only with the real part of this integral, the braking force is obtained from

$$F_{\text{br}} = \frac{K_{\text{m}}}{2} \mu_0 R^2 B_z R$$

$$\times \int_{\alpha=0}^{2\pi} \left[\frac{1}{R}\cos\alpha - 2R\cos\alpha\,\frac{1}{(2h)^2}\left(\frac{1}{1^2}-\frac{1}{2^2}+\frac{1}{3^2}\cdots\right) \right] \cos\alpha\, d\alpha$$

$$= \frac{K_{\text{m}}}{2} \mu_0 R^2 B_z \int_0^{2\pi} \left[1 - \frac{1}{2}\left(\frac{2R}{2h}\right)^2 \left(\frac{1}{1^2}-\frac{1}{2^2}+\frac{1}{3^2}\cdots\right) \right] \cos^2\alpha\, d\alpha$$

$$= \frac{K_{\text{m}}}{2} \mu_0 R^2 B_z \left[1 - \frac{\pi^2}{24}\left(\frac{2R}{2h}\right)^2 \right] \pi$$

$$= -\frac{(\sigma\Delta)\,v B_z}{2\mu_0} \mu_0 R^2 \pi B_z \left[1 - \frac{\pi^2}{24}\left(\frac{2R}{2h}\right)^2 \right]$$

$$= -\frac{1}{2}(\sigma\Delta)\,\pi R^2 v B_z^2 \left[1 - \frac{\pi^2}{24}\left(\frac{2R}{2h}\right)^2 \right] , \qquad (2.3.70)$$

where use has been made of the series

$$\frac{1}{1^2} - \frac{1}{2^2} + \frac{1}{3^2} \cdots = \frac{\pi^2}{12} .$$
(2.3.71)

2.4 Unipolar Induction – Extended Array

2.4.1 Formulation of Problem

In Sects. 2.3.1 – 3 we analyzed a very simple example of unipolar induction; having first established the existence of an electric field due to relative motion of the induced medium and the magnetization M, we proceeded, on the basis of the formally established presence of E, to calculate the surface current distribution and the braking force. While this problem is of practical interest in the design and analysis of electromagnetic braking systems, the more complicated case of an extended array of high-strength poles is of interest where traction and propulsion are concerned. In the sequel, we propose to deal with a related problem, namely, the determination of the braking force (per pole pair) exerted on an "infinite" non-magnetic metal sheet gliding as before with constant, externally regulated speed v, in the very narrow air gap of an infinite array of circular cylindrical direct-current electromagnets. Also as before, the sheet exhibits finite dimensions $2h$ and Δ and is characterized by an electric conductivity σ, whereas each pole is of diameter $2R$. For further simplicity and completeness, all physical assumptions of the preceding case are retained here; in particular, a very low magnetic Reynolds number $\mu_0 (\sigma \Delta) v$ is once more postulated.

The braking pole array may be spatially distributed in two characteristics, and the analysis will accordingly concern two distinct cases as reproduced in Figs. 2.11 and 12, respectively:

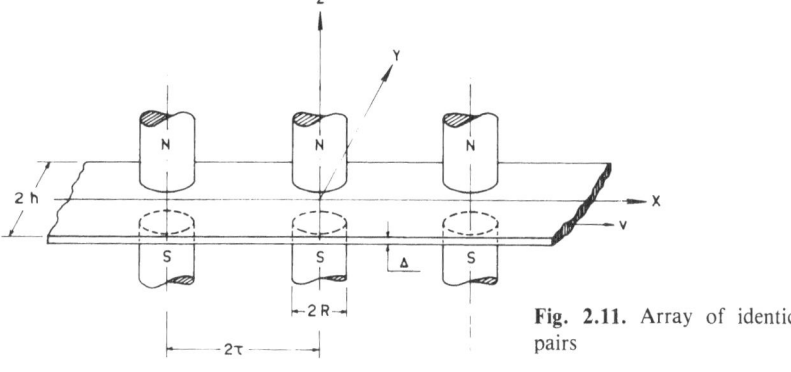

Fig. 2.11. Array of identical pole pairs

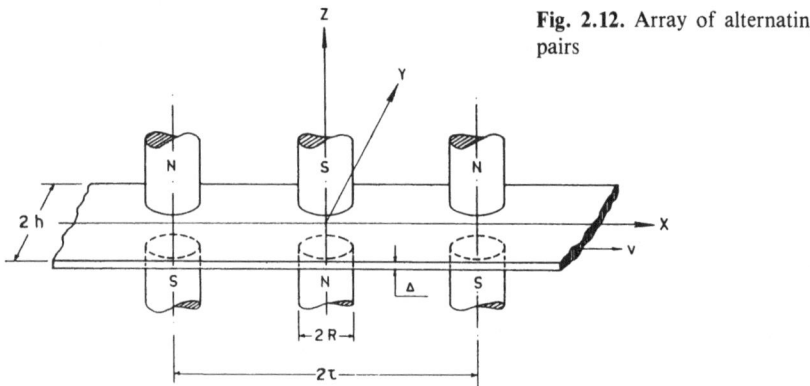

Fig. 2.12. Array of alternating pole pairs

We specifically ask: how does the new absolute value $|F_{br}|$ of the braking force per pole pair compare with that of (2.3.60)?

2.4.2 Solution

We again resort to (2.3.28) which, rewritten in circular polar coordinates, reads

$$\frac{\partial^2 D}{\partial r^2} + \frac{1}{r}\frac{\partial D}{\partial r} + \frac{1}{r^2}\frac{\partial^2 D}{\partial \alpha^2} = -(\sigma \Delta)v_r \frac{\partial B_z}{\partial r} \ . \tag{2.4.1}$$

Once D is known, the braking force is readily obtained from the formerly quoted equation, i.e.

$$F_{br} = B_z \oint D \cos \alpha \, dl \ .$$

Now, the solution of (2.4.1) (Sect. 2.4.3 A, B), subject to the boundary conditions imposed by the problem, yields elliptic functions in the complex plane $w = x + iy$. These functions have real and imaginary periods specified by the complete elliptic integrals K and K′, of parameter k (or m, where $m \equiv k^2$). The interrelationship of the pole-spacing 2τ, the sheet width $2h$, the parameter k and the function K = K(k) is plotted in Fig. 2.13. For given values of 2τ and $2h$, the figure yields the value of k^2 and thus the function K. The absolute value $|F_{br}|$ of the braking force per electromagnet is now obtainable (Sect. 2.4.3) by means of

$$|F_{br}| = \frac{1}{2}(\sigma \Delta)|v|B_z^2 \pi R^2 \left[1 - \frac{1+k^2}{3}\left(\frac{R}{\tau}\right)^2 K^2 \right], \tag{2.4.2}$$

for identical pole pairs (Fig. 2.11), and

$$|F_{br}| = \frac{1}{2}(\sigma \Delta)|v|B_z^2 \pi R^2 \left[1 + \frac{2}{3}(1 - 2k^2)\left(\frac{R}{\tau}\right)^2 K^2 \right], \tag{2.4.3}$$

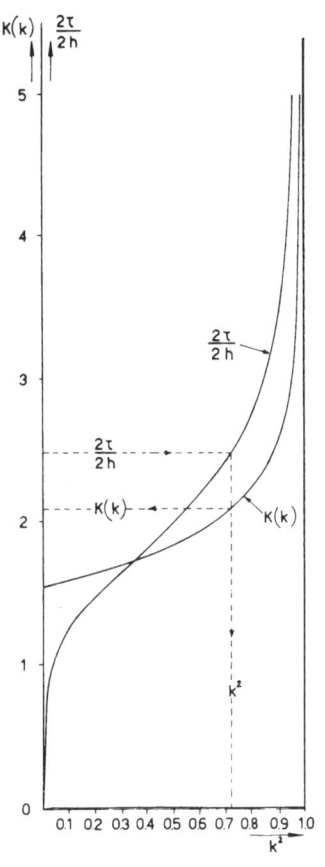

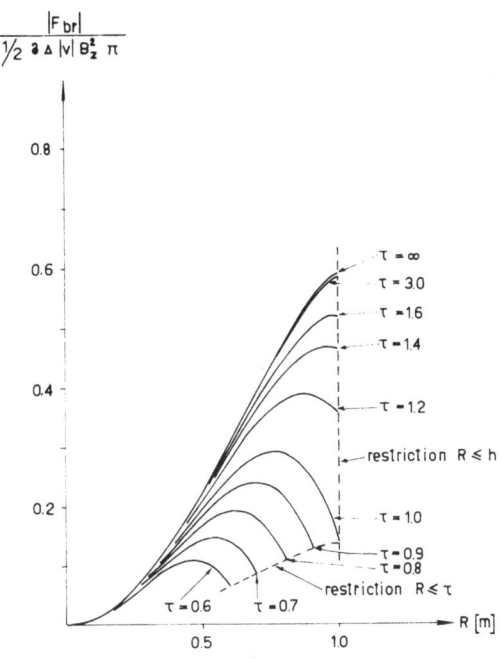

Fig. 2.14. Normalized braking force per pole pair in identical variant, for $2h = 2$; all dimensions in m

◀ **Fig. 2.13.** Relation between dimensionless ratio $2\tau/2h$ and complete elliptic integral $K(k)$

for alternating pole pairs (Fig. 2.12). These equations have been plotted for an arbitrary strip width $2h = 2$ in Figs. 2.14 and 15.

In the case of identical pole pairs, for a chosen value of the pole pitch τ, there exists an "optimal" pole radius $R = R_0$ at which the braking force is maximum (Fig. 2.14); no such maximum exists for the alternating variant (Fig. 2.15). In the limiting case $\tau \to \infty$ (a single electromagnet) both equations reduce [7], as required, to the same form exactly, i.e., to (2.3.60):

$$|F_{\text{br}}| = \frac{1}{2}(\sigma \Delta)\pi R^2 |v| B_z^2 \left[1 - \frac{\pi^2}{24}\left(\frac{R}{h}\right)^2 \right]. \tag{2.4.4}$$

The above results may be used as a guide in optimizing the dimensions of linear braking devices: in practice, the space allocated to the electromagnets is

---

[7] For $\tau \to \infty$ we find, see (2.4.20) and the explanation preceding that equation, $\lim\limits_{\tau \to \infty} (1 + k^2)\dfrac{K^2}{\tau^2}$

$= 2\left(\dfrac{\pi}{4h}\right)^2$; $\lim\limits_{\tau \to \infty} (1 - 2k^2)\dfrac{K^2}{\tau^2} = -\left(\dfrac{\pi}{4h}\right)^2$; hence (2.4.4).

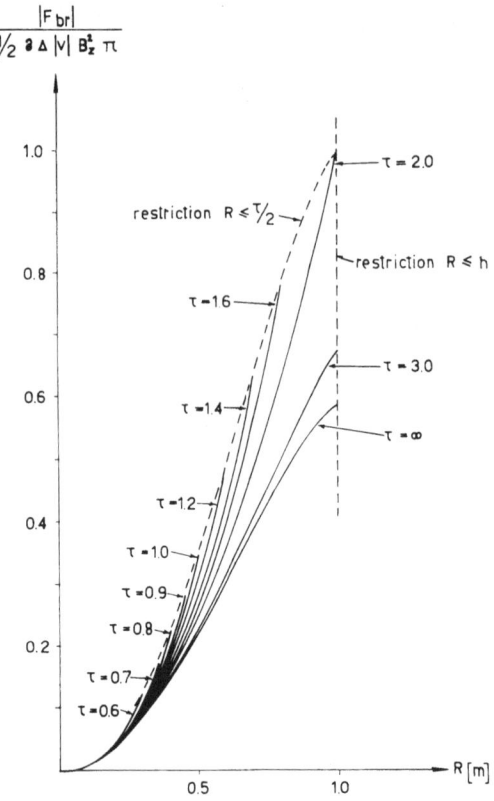

Fig. 2.15. Normalized braking force per pole pair in alternating variant, for $2h = 2$; all dimensions in. m

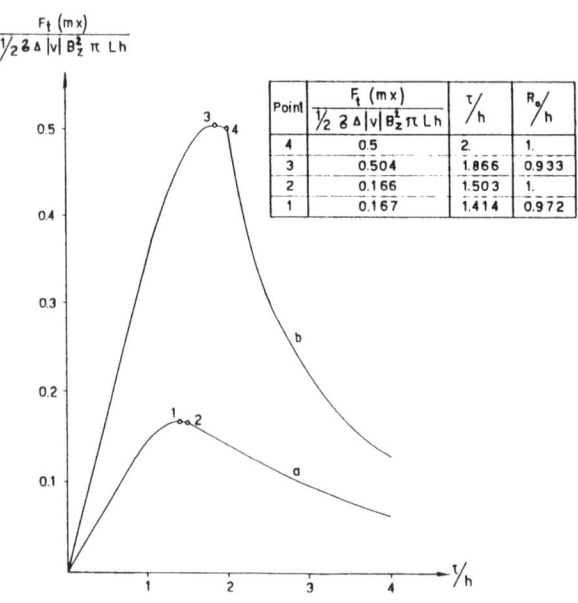

Fig. 2.16. Maximization of normalized interaction force: (*a*) Identical pole pairs; (*b*) Alternating pole pairs. (For explanation of inset table, see Sect. 2.4.3 C)

obviously not infinite in the x-direction, but is restricted to a finite length L. In the case of identical pole pairs, their number accommodated in this finite stretch is $m = L/2\,\tau$. With longitudinal end effects disregarded, the total maximum force of interaction is obtained by multiplying m by the largest braking force per pole pair. The total maximum force of interaction $F_t\,(mx)$, per unit length, divided by the sheet half-width h, is plotted in Fig. 2.16, curve a, in normalized form (Sect. 2.4.3 C).

Similar results (Fig. 2.16, curve b) were obtained for alternating poles.

This concludes the proposed analysis of unipolar induction as extended for alternating poles.

2.4.3 Appendix

A) Identical Pole Pairs

The solution of (2.4.1) is obtained with the aid of the analogy set up in the preceding case (Sect. 2.3.2); the differential equation to be solved is therefore

$$\frac{\partial^2 V_z}{\partial r^2} + \frac{1}{r}\frac{\partial V_z}{\partial r} + \frac{1}{r^2}\frac{\partial^2 V_z}{\partial \alpha^2} = 0; \quad r \geq R \ , \tag{2.4.5}$$

subject once more to

$$\frac{\partial V_z}{\partial x} = 0; \quad y = \pm h \ , \tag{2.4.6}$$

with the excitation on each cylinder again given by

$$K_z = K_m \cos \alpha, \quad K_m = -\frac{(\sigma \Delta)\,v\,B_z}{\mu_0} \ . \tag{2.4.7}$$

The geometrical and electrical conditions of the cylinders are shown in Fig. 2.17.

Now the complete set of solutions of the Laplace equation in polar co-ordinates is given by:

$$V_{z(0)} = (A_0 \ln r + C)(D\alpha + F) \ ,$$

$$V_{z(n)} = r^n (A_n \cos n\alpha + B_n \sin n\alpha) \ , \tag{2.4.8}$$

$$V_{z(-n)} = r^{-n}(A_{-n}\cos n\alpha - B_{-n}\sin n\alpha) \ ,$$

where the constants A, B, etc. have to be adapted to the boundary conditions, and n denotes an integer.

Referring first to the known solution for $h \to \infty$ (infinitely wide sheet) and $\tau \to \infty$ (single exciting pole pair) the vector potential is postulated as

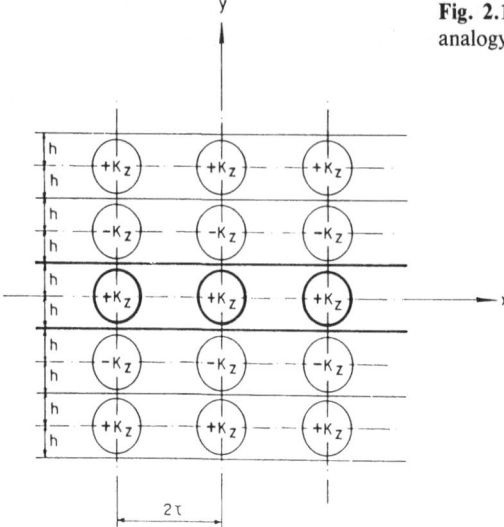

Fig. 2.17. Series of images for electromagnetic analogy of identical pole pairs

$$V_z = \frac{K_m}{2} \mu_0 R \, \frac{r}{R} \cos \alpha, \quad r < R \;, \tag{2.4.9}$$

$$V_z = \frac{K_m}{2} \mu_0 R \, \frac{R}{r} \cos \alpha, \quad r > R \;. \tag{2.4.10}$$

In the case of a finite width $2h$, but with τ still tending to ∞, the complex solution is given by the previous single sequence of images (Fig. 2.9) yielding (2.3.46), i.e.,

$$\bar{V}_z = \frac{K_m}{2} \mu_0 R \left(\frac{w}{R} - \frac{R}{w} \right) + \frac{K_m}{2} \mu_0 R \, \frac{i(\pi/2)(R/h)}{\sin(i\pi w/2h)} \;. \tag{2.4.11}$$

Keeping this result in mind, we now proceed to solve the posed problem; the solution is obviously a double periodic meromorphic function (i.e., with poles as the only singularities) or, in other words, an elliptic function of the generalized solutions (2.4.8) of the Laplace equation in cylindrical coordinates; the logarithmic form [8] is the only one that can be used here in conjunction with an elliptic function: as each cylinder carries equal positive and negative currents, an array of cylindrical *dipoles* has to be considered in determining the potential, at a given point, due to *remote* cylinders.

In the light of these remarks, and based on the given sources and sinks (Fig. 2.17), the solution of the potential problem is formulated by means of the function $\mathrm{sn}(z, k)$ where, in general, z represents the complex variable and

---

[8] The trivial solution proportional to α – see above – is ruled out by the single-valuedness requirement.

k the parameter of the Jacobian elliptic function [Ref. 2.16, Vol. 4, p. 144]. Further, dipoles being formally obtained by differentiation, the sought double periodic solution is generated [Ref. 2.16, Vol. 3, p. 27] by

$$\frac{d}{dz} \ln \operatorname{sn}(z, k) = \frac{\operatorname{cn}(z, k) \operatorname{dn}(z, k)}{\operatorname{sn}(z, k)} . \tag{2.4.12}$$

For the Jacobian functions, we define and use (see below) the argument $v \equiv K w/\tau$; thus, for instance $\operatorname{sn}(z, k) \equiv \operatorname{sn}(K w/\tau) \equiv \operatorname{sn} v$. In specifying the parameter k, we recall that the functions have real and imaginary quarter-periods K and iK' which are complete elliptic integrals of parameter k; (ϕ represents a dummy variable of integration):

$$K = K(k) = \int_0^{\pi/2} (1 - k^2 \sin \phi)^{-1/2} d\phi , \tag{2.4.13}$$

$$iK' = iK'(k) = i \int_0^{\pi/2} [1 - (1 - k^2) \sin^2 \phi]^{-1/2} d\phi . \tag{2.4.14}$$

In the physical system under consideration, the real period along the x-axis is given by the spacing of the magnetic poles of the same sign, whereas the imaginary period along the y-axis equals twice the sheet width (Fig. 2.17).

Hence, k is readily determined through the ratio

$$\frac{2\tau}{4h} = \frac{K(k)}{K'(k)} . \tag{2.4.15}$$

The above considerations thus yield the complex vector potential $\bar{V}_z$, in the double-periodic case, in the form:

$$\bar{V}_z = \frac{K_m}{2} \mu_0 R \left(\frac{w}{R} - \frac{R}{w} \right) + \frac{K_m}{2} \mu_0 R K \frac{R}{\tau} \frac{\operatorname{cn}\left(\dfrac{K}{\tau} w\right) \operatorname{dn}\left(\dfrac{K}{\tau} w\right)}{\operatorname{sn}\left(\dfrac{K}{\tau} w\right)} . \tag{2.4.16}$$

Obviously, for $\tau \to \infty$ (single pole pair), the last equation must reduce to (2.3.46). To verify that such is indeed the case, we proceed as follows:

For $k = 1$: $K \to \infty$ and $K' = \pi/2$; hence, since $\tau \to \infty$ and h is finite, k must equal unity. In addition, we have [Ref. 2.12, p. 106] for $k = 1$

$$\operatorname{sn}(v) = \tanh v , \tag{2.4.17}$$

$$\operatorname{cn}(v) = \frac{1}{\cosh v} , \tag{2.4.18}$$

$$\operatorname{dn}(v) = \frac{1}{\cosh v} . \tag{2.4.19}$$

Now, see again (2.4.15),

$$\lim_{\tau \to \infty} \frac{K}{\tau} = \frac{2}{4h} \frac{\pi}{2} ,$$ (2.4.20)

and thus

$$\lim_{\tau \to \infty} \frac{K_m}{2} \mu_0 R K \frac{R}{\tau} \frac{cn(K w/\tau) dn(K w/\tau)}{sn(K w/\tau)}$$

$$= \frac{K_m}{2} \mu_0 R \frac{1}{2h} \frac{\pi}{2} R \frac{\cosh^{-2}\left(\dfrac{1}{2h} \dfrac{\pi}{2} w\right)}{\tanh\left(\dfrac{1}{2h} \dfrac{\pi}{2} w\right)}$$

$$= \frac{K_m}{2} \mu_0 R \frac{\pi}{2} \frac{R}{h} \frac{1}{\sinh 2\left(\dfrac{\pi}{4h} w\right)}$$

$$= \frac{K_m}{2} \mu_0 R \frac{i \dfrac{\pi}{2} \dfrac{R}{h}}{i \sinh\left(\dfrac{\pi}{2h} w\right)}$$

$$= \frac{K_m}{2} \mu_0 R \frac{i \dfrac{\pi}{2} \dfrac{R}{h}}{\sin \dfrac{i\pi w}{2h}} .$$ (2.4.21)

which, as required, is identical with the second term of (2.3.46).

Although we did not dwell on all the details when deriving (2.4.16), the procedure actually follows from (2.4.21) in the reverse.

Recalling now the analogy $V_z \to D$, we proceed with the determination of braking force F_{br} per pole pair. Using the series expansion [Ref. 2.16, Vol. 3, p. 2 and 24] of the elliptic functions

$$\frac{cn(v) dn(v)}{sn(v)} = \frac{1}{v} - 2(1 + k^2)\frac{v}{3!} \dots ,$$ (2.4.22)

and rewriting (2.4.16) in the form

$$\bar{V}_z = \frac{K_m}{2}\mu_0 R\left\{\frac{w}{R}+R\frac{K}{\tau}\left[\frac{1}{\frac{K}{\tau}w}-2(1+k^2)\frac{\frac{K}{\tau}w}{3!}\cdots\right]-\frac{R}{w}\right\},$$

$$(2.4.23)$$

the integration

$$F_{br} = B_z \mathrm{Re}\{\oint \bar{V}_z \cos\alpha\,dl\} \qquad (2.4.24)$$

is readily effected.

On the pole contour, we once more set $dl = R\,d\alpha$; in view of the orthogonality of the trigonometric functions, we therefore again have only to evaluate (with $w = R[\exp(i\alpha)]$) the expression

$$F_{br} = B_z \mathrm{Re}\left\{\int_{\alpha=0}^{2\pi}\frac{K_m}{2}\mu_0 R\left\{\frac{w}{R}\left[1-\frac{1+k^2}{3}\left(\frac{R}{\tau}\right)^2 K^2\right]\cos\alpha R\,d\alpha\right\}\right.$$

$$= -\frac{1}{2}(\sigma\varDelta)vB_z^2\pi R^2\left[1-\frac{1+k^2}{3}\left(\frac{R}{\tau}\right)^2 K^2\right], \qquad (2.4.25)$$

where the minus sign indicates that the resulting electromagnetic force opposes the external thrust.

Note: Fig. 2.13 was plotted on the basis of expression (2.4.15) and tables of complete elliptic integrals [2.17].

B) Alternating Pole Pairs

For the case of alternating pole pairs, we again have to solve (2.4.5) subject to the boundary condition (2.4.6), but with electrical and geometrical conditions as per Fig. 2.18.

For the positive poles, the solution is given by (2.4.16), i.e.,

$$\bar{V}_{z+} = \frac{K_m}{2}\mu_0 R\left[\frac{w}{R}-\frac{R}{w}+K\frac{R}{\tau}\frac{\mathrm{cn}\left(\frac{K}{\tau}w\right)\mathrm{dn}\left(\frac{K}{\tau}w\right)}{\mathrm{sn}\left(\frac{K}{\tau}w\right)}\right]. \qquad (2.4.26)$$

For the negative poles, the preceding equation yields

$$\bar{V}_{z-} = -\frac{K_m}{2}\mu_0 R\left[\frac{w}{R}-\frac{R}{w}+K\frac{R}{\tau}\frac{\mathrm{cn}\left(\frac{K}{\tau}(w+\tau)\right)\mathrm{dn}\left(\frac{K}{\tau}(w+\tau)\right)}{\mathrm{sn}\left(\frac{K}{\tau}(w+\tau)\right)}\right]. $$

$$(2.4.27)$$

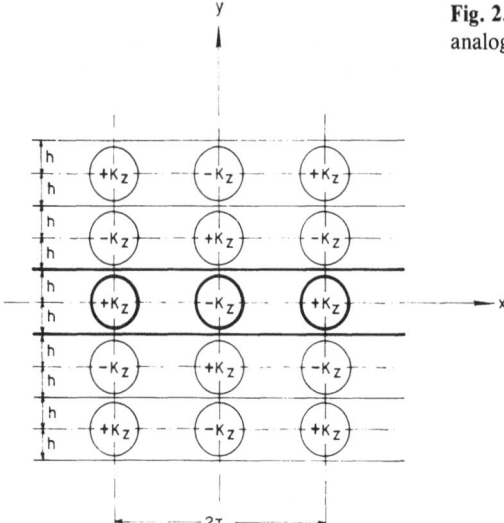

Fig. 2.18. Series of images for electromagnetic analogy of alternating pole pairs

The complex vector potential for the positive-negative variant is obtained by adding the above equations, i.e. $\bar{V}_z = \bar{V}_{z+} + \bar{V}_{z-}$; hence

$$\bar{V}_z = \frac{K_m}{2}\mu_0 R \left[K\frac{R}{\tau} \frac{\mathrm{cn}\left(\dfrac{K}{\tau}w\right)\mathrm{dn}\left(\dfrac{K}{\tau}w\right)}{\mathrm{sn}\left(\dfrac{K}{\tau}w\right)} \right.$$

$$\left. -K\frac{R}{\tau}\frac{\mathrm{cn}\left(\dfrac{K}{\tau}(w+K)\right)\mathrm{dn}\left(\dfrac{K}{\tau}(w+K)\right)}{\mathrm{sn}\left(\dfrac{K}{\tau}(w+K)\right)} \right] . \tag{2.4.28}$$

Introducing the complementary parameter $k'^2 = 1 - k^2$, resorting to the identities [Ref. 2.16, Vol. 3, p. 10]

$$\mathrm{cn}(z+K, k) = -k'\mathrm{sd}(z, k) ,$$

$$\mathrm{dn}(z+K, k) = k'\mathrm{nd}(z, k) , \tag{2.4.29}$$

$$\mathrm{sn}(z+K, k) = \mathrm{cd}(z, k) ,$$

and recalling [Ref. 2.16, Vol. 3, p. 1] that

$$\text{sd} = \frac{\text{sn}}{\text{dn}} ,$$

$$\text{nd} = \frac{1}{\text{dn}} , \tag{2.4.30}$$

$$\text{cd} = \frac{\text{cn}}{\text{dn}} ,$$

so that

$$\frac{\text{cn}(z+\text{K},k)\,\text{dn}(z+\text{K},k)}{\text{sn}(z+\text{K},k)} = \frac{-k'\text{sd}(z,k)k'\dfrac{1}{\text{dn}(z,k)}}{\dfrac{\text{cn}}{\text{dn}}(z,k)}$$

$$= -(1-k^2)\frac{\text{sn}(z,k)}{\text{cn}(z,k)\,\text{dn}(z,k)} , \tag{2.4.31}$$

we finally obtain from (2.4.28)

$$\bar{V}_z = \frac{K_\text{m}}{2}\mu_0 R\text{K}\frac{R}{\tau}\left[\frac{\text{cn}\left(\dfrac{\text{K}}{\tau}w\right)\text{dn}\left(\dfrac{\text{K}}{\tau}w\right)}{\text{sn}\left(\dfrac{\text{K}}{\tau}w\right)} \right.$$

$$\left. +(1-k^2)\frac{\text{sn}\left(\dfrac{\text{K}}{\tau}w\right)}{\text{cn}\left(\dfrac{\text{K}}{\tau}w\right)\text{dn}\left(\dfrac{\text{K}}{\tau}w\right)} \right] . \tag{2.4.32}$$

Resorting once more to (2.4.22), we have

$$\bar{V}_z = \frac{K_\text{m}}{2}\mu_0 R\left[\frac{R}{w} - \frac{1+k^2}{3}\left(\frac{\text{K}}{\tau}\right)^2 R^2\frac{w}{R} - \ldots \right.$$

$$\left. +(1-k^2)\text{K}^2\frac{R^2}{\tau^2}\frac{1}{\dfrac{R}{w} - \dfrac{1+k^2}{3}\left(\dfrac{\text{K}}{\tau}\right)^2 R^2\dfrac{w}{R} - \ldots} \right] . \tag{2.4.33}$$

Again, only the first terms of (2.4.33) are needed for the integration prescribed by (2.4.24). Thus, with $w = R\exp(i\alpha)$ as before, we get

$$
F_{br} = B_z Re \int_{\alpha=0}^{\pi} \frac{K_m}{2} \mu_0 R \left[\frac{R}{w} - \frac{1+k^2}{3} \left(\frac{K}{\tau}\right)^2 R^2 \frac{w}{R} \right.
$$

$$
\left. + (1-k^2)\left(\frac{K}{\tau}\right)^2 R^2 \frac{w}{R} \right] \cos\alpha \, R \, d\alpha
$$

$$
= -\frac{1}{2}(\sigma\varDelta) v B_z^2 \pi R^2 \left[1 + \frac{2}{3}(1-2k^2)K^2 \left(\frac{R}{\tau}\right)^2 \right]. \tag{2.4.34}
$$

C) Optimal Dimensions

a) *Identical Pole Pairs.* The absolute value of the braking force

$$
|F_{br}| = \frac{1}{2}(\sigma\varDelta) |v| B_z^2 \pi R^2 \left[1 - \frac{1+k^2}{3}\left(\frac{R}{\tau}\right)^2 K^2 \right], \tag{2.4.35}
$$

acting on a strip of given width $2h$ attains its maximum $|F_{br}|_{mx}$ (for a prescribed value of τ) at a radius $R = R_0$, according to

$$
R_0^2 = \frac{3}{2} \frac{1}{(1+k^2)K^2} \tau^2 . \tag{2.4.36}
$$

The force

$$
|F_{br}|_{mx} = \frac{1}{2}(\sigma\varDelta) |v| B_z^2 \pi R_0^2 \frac{1}{2} \tag{2.4.37}
$$

multiplied by the pole number m of the poles contained in an array of length L, i.e. $m = L/2\tau$, yields the total maximum force $|F_t|_{mx}$

$$
|F_t|_{mx} = \frac{1}{2}(\sigma\varDelta) |v| B_z^2 \pi R_0^2 \frac{1}{2} \frac{L}{2\tau} , \tag{2.4.38}
$$

or, using normalized units and resorting to (2.4.36) we get

$$
\frac{|F_t|_{mx}}{\frac{1}{2}(\sigma\varDelta) |v| B_z^2 \pi L} = \frac{3}{8} \frac{\tau}{(1+k^2)K^2} . \tag{2.4.39}
$$

Introducing the total maximum force per sheet half width, i.e. $|F_t|_{mx}/h$, and taking into account (2.4.15) we find

$$
\frac{|F_t|_{mx}}{\frac{1}{2}(\sigma\varDelta) |v| B_z^2 \pi L h} = \frac{3}{4} \frac{1}{(1+k^2)KK'} . \tag{2.4.40}
$$

Plotting this expression against the dimensionless pole pitch τ/h (Fig. 2.16a) we obtain a function for the overall largest braking force per unit length, per half width h. Maximization of this function yields, in turn (Fig. 2.16a, point 1) the values τ/h, R_0/h at which the largest braking force per unit length sets in for a given sheet width $2h$ and with longitudinal end effects neglected. The restriction $R \leqslant h$, however, allows (2.4.40) to be plotted only up to a pole pitch τ_h at which, see (2.4.36),

$$\tau_h = \sqrt{\tfrac{2}{3}(1+k^2)}\, h\mathrm{K} \ . \tag{2.4.41}$$

Expressions (2.4.15 and 41) yield $\tau_h = 1.503\,h$ (Fig. 2.16a, point 2) beyond which

$$R_0 = h \tag{2.4.42}$$

remains constant; the largest braking force per electromagnet $|F_{\mathrm{br}}|_{mx}$ is given for this dimensioning, see (2.4.2), by

$$|F_{\mathrm{br}}|_{mx} = \frac{1}{2}\,(\sigma\varDelta)\,|v|B_z^2\pi h^2 \left[1 - \frac{1+k^2}{3}\left(\frac{h}{\tau}\right)^2 \mathrm{K}^2\right] , \tag{2.4.43}$$

and therefore, the overall maximum force $|F_{\mathrm{t}}|_{mx}$ per unit length, per sheet half width h now reads in normalized form

$$\frac{|F_{\mathrm{t}}|_{mx}}{\tfrac{1}{2}\,(\sigma\varDelta)\,|v|B_z^2\pi Lh} = \frac{1}{2}\frac{h}{\tau}\left[1 - \frac{1+k^2}{3}\left(\frac{h}{\tau}\right)^2 \mathrm{K}^2\right] . \tag{2.4.44}$$

Eqs. (2.4.40, 44) (single curve) are plotted in Fig. 2.16a.

b) *Alternating Pole Pairs.* The expression

$$|F_{\mathrm{br}}| = \frac{1}{2}\,(\sigma\varDelta)\,|v|B_z^2\pi R^2 \left[1 + \frac{2}{3}(1-2k^2)\left(\frac{R}{\tau}\right)^2 \mathrm{K}^2\right] ,$$

representing the braking force, attains maximum (for a given pole pitch τ and strip width $2h$) at a radius $R = R_0$ for which

$$R_0^2 = -\frac{3}{4}\frac{1}{(1-2k^2)\mathrm{K}^2}\,\tau^2 \ . \tag{2.4.45}$$

The restriction $R \leqslant \tau/2$ now yields the limiting equation

$$(1-2k^2)\mathrm{K}^2 + 3 = 0 \ , \tag{2.4.46}$$

which in turn, on account of $\tau = 2h(\mathrm{K}/\mathrm{K}')$, yields $\tau/h = 2.7077$, at $R = R_0$; hence, at $R = R_0$ we now have

$$R = R_0 = \frac{\tau}{2} = \frac{1}{2} \times 2.7077\, h > h \ , \tag{2.4.47}$$

contrary to the restriction $R \leqslant h$.

Accordingly, not counting the trivial case $R = 0$, there is no extremum within the interval of interest[9], and the largest value $|F_{br}|_{mx}$ occurs for $R = \tau/2$ (Fig. 2.15).

In the present case, the pole-pair number m is given by $m = L/\tau$ and therefore [substituting $R = \tau/2$ in (2.4.3)]

$$|F_t|_{mx} = \frac{1}{2}(\sigma\varDelta)\,|v|B_z^2\pi\frac{\tau^2}{4}\left[1 + \frac{2}{3}(1 - 2k^2)\frac{\tau^2}{4\tau^2}K^2\right]\frac{L}{\tau} \ . \tag{2.4.48}$$

Dividing by the strip half-width h and resorting to $\tau/h = 2(K/K')$, we have

$$\frac{|F_t|_{mx}}{\frac{1}{2}(\sigma\varDelta)\,|v|B_z^2\pi Lh} = \frac{1}{2}\frac{K}{K'}\left[1 + \frac{2}{3}(1 - 2k^2)\frac{K^2}{4}\right] \ , \tag{2.4.49}$$

which attains maximum at point 3 (Fig. 2.16 b). This equation is useful up to a pole pitch τ_h at which $R = h$; owing to $R \leqslant \tau/2$, which restricts R_0 to $\tau/2$, we have

$$\frac{\tau_h}{h} = \frac{2R_0}{h} = \frac{2h}{h} \ , \tag{2.4.50}$$

i.e. $\tau_h = 2h$ (Fig. 2.16 b, point 4) beyond which $R_0 = h$ and

$$F_t(mx) = \frac{1}{2}(\sigma\varDelta)\,|v|B_z^2\pi h^2\left[1 + \frac{2}{3}(1 - 2k^2)\frac{h^2}{\tau^2}K^2\right]\frac{L}{\tau} \ , \tag{2.4.51}$$

or, again using (2.4.15), we get

$$\frac{F_t(mx)}{\frac{1}{2}(\sigma\varDelta)\,|v|B_z^2\pi Lh} = \frac{1}{2}\frac{K'}{K}\left[1 + \frac{2}{3}(1 - 2k^2)\frac{K'^2}{4}\right] \ . \tag{2.4.52}$$

Eqs. (2.4.49, 52) (single curve) are plotted in Fig. 2.16 b.

2.5 Optimal Dimensions for Braking Electromagnet

The insight gained in the preceding sections may be used in implementing the design of practical systems, for example, that of an electromagnet for eddy-current brakes. This type of brakes is quite frequently resorted to in practice,

---

[9]   All possible stray flux linkages are ruled out.

as the braking force lends itself to continuous control through regulation of the excitation current. Accordingly, we propose to look into the pole design of such an electromagnet with a view to *minimize the material cost* for a specific, required braking force.

To this end, two idealized cases will be considered: (a) a conducting sheet of "infinite" horizontal extent; (b) a conducting sheet "infinite" in the direction of motion, but of finite width.

2.5.1 Two-Way Infinite Sheet

A) Formulation of Problem

The initial braking problem is formulated in analogy to the cases outlined in Sects. 2.2 – 4, i.e., an "infinite" horizontal nonmagnetic metal sheet, of thickness $\varDelta$ and electrical conductivity σ, glides at a constant speed v along the x-axis in the very narrow air gap of a direct-current electromagnet whose pole pieces are rectangular (Fig. 2.19). With the reaction field disregarded as in (2.3.15), what is the braking force acting on the sheet?

B) Solution

The electromagnetic analogy introduced in Sect. 2.3.2 is now somewhat extended so as to cover the specific problem under consideration. The stream function D satisfies here the differential equation, compare (2.3.28),

$$\frac{\partial^2 D}{\partial x^2} + \frac{\partial^2 D}{\partial y^2} = -(\sigma \varDelta) v \frac{\partial B_z(x,\, y)}{\partial x} \ , \tag{2.5.1}$$

again with the induction component $B_z = B_z(x,\, y)$ prescribed in the $z = 0$ plane; the magnetic vector potential of the above analogy once more obeys exactly (2.3.29) with $j_z = j_z(x,\, y)$, i.e.

$$\frac{\partial^2 V_z}{\partial x^2} + \frac{\partial^2 V_z}{\partial y^2} = -\mu_0 j_z(x,\, y) \ . \tag{2.5.2}$$

Writing

$$D \equiv V_z, \quad \text{and} \tag{2.5.3}$$

$$\mu_0 j_z \equiv (\sigma \varDelta) v \frac{\partial B_z}{\partial x} \ , \tag{2.5.4}$$

(2.5.1,2) are not only analogous, but in fact identical.

What is the form of $j_z(x,\, y)$?

In the case of circular pole pieces, our choice of j_z (or rather of its linearly integrated version) was prescribed by spatial coherence requirements following from the $\cos \alpha$ dependence of the radial speed $v_r = v \cos \alpha$; here the

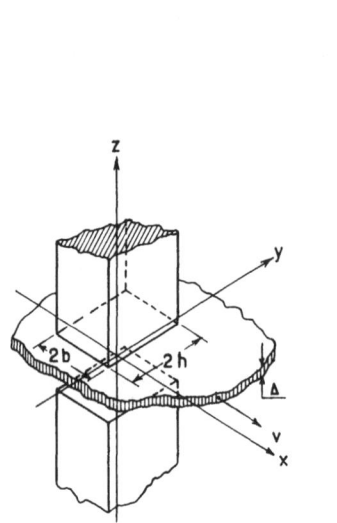

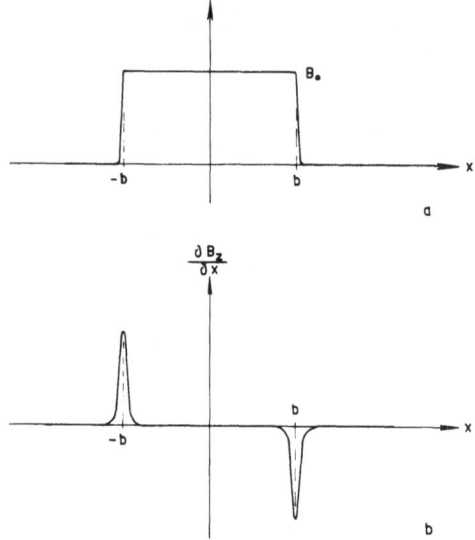

Fig. 2.19. Metal sheet moving in air gap of rectangular electromagnet

Fig. 2.20a,b. Idealized distribution of induction: (a) Imposed flux density; (b) Derivative (arbitrary units) of flux density

choice will rest on a quite similar approach and, with a view to better clarity, we shall consider an idealized version of the prescribed flux density distribution of amplitude B_0 along with its derivative, as reproduced in Fig. 2.20.

As $\partial B_z / \partial x$ differs markedly from zero only at the periphery $x = \pm b$, the required analogy suggests itself in two bus-bars carrying the direct currents ($\pm I$), as shown schematically in Fig. 2.21.

Further, the quite obvious numerical choice, compare Figs. 2.20 and 21, is

$$\pm \mu_0 I = \mp (\sigma \Delta) v B_0 2h \ , \tag{2.5.5}$$

where from now on $B_0 = B_z$.

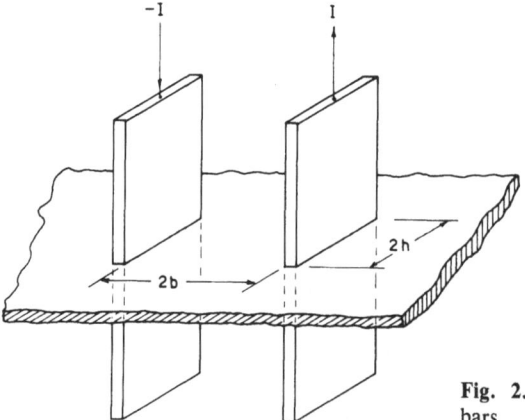

Fig. 2.21. Fictitious current-carrying bus bars

In the $z = 0$ plane, geometrical coincidence is thus obtained between the magnetic "field-lines" $V = \text{const}$ and the electric current "stream-lines" $D = \text{const}$.

Having decided upon the intrinsic form of the analogy, we now proceed to calculate the braking force $F_{br} = F_{br}\mathbf{1}_x$, again via (2.3.57), i.e.,

$$F_{br} = \int\limits_{(a)} \frac{\partial D}{\partial x} B_z dx\, dy \ , \tag{2.5.6}$$

resorting, however, to a somewhat different approach:
Thus,

$$\int\limits_{(a)} \frac{\partial D}{\partial x} B_z dx\, dy = \int\limits_{(a)} \frac{\partial}{\partial x}(DB_z)\, dx\, dy - \int\limits_{(a)} D\frac{\partial B_z}{\partial x}\, dx\, dy \ . \tag{2.5.7}$$

Applying Stokes' theorem, the first surface integral on the right-hand side is replaced by a line integral over the contour l of the sheet; as this contour was assumed to be far removed from the pole, and as B_z comprises only the imposed flux density [10] which, in turn, is clearly localized, we find

$$\int\limits_{(a)} \frac{\partial}{\partial x}(DB_z)\, dx\, dy = \oint\limits_{(l)} DB_z dy = 0 \ . \tag{2.5.8}$$

Hence,

$$F_{br} = -\int\limits_{(a)} D\frac{\partial B_z}{\partial x}\, dx\, dy \ . \tag{2.5.9}$$

Returning to the defining relations (2.5.3,4), the braking force may now be evaluated through the substitutions

$$F_{br} = -\int\limits_{(a)} V_z \frac{\mu_0 j_z}{(\sigma\Delta)v}\, dx\, dy \ , \tag{2.5.10}$$

where (a) obviously refers here only to the current-carrying cross-section of the fictitious bus bars (Fig. 2.21).

Henceforth we consider only the absolute value $|F_{br}|$ of F_{br},

$$|F_{br}| = \frac{\mu_0}{(\sigma\Delta)|v|} \int\limits_{(a)} V_z j_z dx\, dy \ . \tag{2.5.11}$$

Now the (free) magnetic energy W_m per unit length of the bus-bar system may be expressed as

$$W_m = \frac{1}{2} \int\limits_{(a)} V_z j_z dx\, dy \ , \tag{2.5.12}$$

---

[10] See relevant remarks preceding (2.3.15).

so that

$$|F_{br}| = \frac{2\mu_0}{(\sigma\Delta)|v|}\, W_m \ . \qquad (2.5.13)$$

W_m, in turn, is expressed by the self-inductance $\mathscr{L}$ per unit length of the bus bars, i.e.

$$W_m = \tfrac{1}{2}\,\mathscr{L}I^2 \ , \qquad (2.5.14)$$

so that finally, with (2.5.5), we obtain

$$|F_{br}| = \mu_0(\sigma\Delta)|v|\frac{B_0^2}{\mu_0^2}4h^2\mathscr{L} \ . \qquad (2.5.15)$$

The problem is thus reduced to evaluation of $\mathscr{L}$, which yields [2.18]

$$\mathscr{L} = \mu_0\frac{b}{h}\frac{2}{\pi}\left\{\tan^{-1}\frac{h}{b}+\frac{1}{2}\left[\frac{h}{b}\ln\sqrt{1+\left(\frac{b}{h}\right)^2}-\frac{b}{h}\ln\sqrt{1+\left(\frac{h}{b}\right)^2}\right]\right\} \ , \qquad (2.5.16)$$

and therefore

$$|F_{br}| = \mu_0(\sigma\Delta)|v|\cdot\frac{1}{2}\frac{B_0^2}{\mu_0}4bh\cdot\frac{4}{\pi}$$

$$\times\left\{\tan^{-1}\frac{h}{b}+\frac{1}{2}\left[\frac{h}{b}\ln\sqrt{1+\left(\frac{b}{h}\right)^2}-\frac{b}{h}\ln\sqrt{1+\left(\frac{h}{b}\right)^2}\right]\right\} \ . \qquad (2.5.17)$$

The last equation is deliberately stated in a form exhibiting: (a) the dimensionless number $\mu_0(\sigma\Delta)|v|$; (b) the (free) magnetic energy per unit height of the air gap, supposing $B_z = B_0$ throughout; and (c) the dimensionless configuration parameter pertaining to the horizontal $2b$, $2h$ dimensions of the pole pieces.

C) Optimal Pole Shape

The total iron cost C_{Fe} per unit height is a function of its specific gravity γ_{Fe}, the dimension $2b$, $2h$ and the unit material cost (including manufacture) c_{Fe}. Thus

$$C_{Fe} = \gamma_{Fe}\times 2b\times 2h\times c_{Fe} \ , \qquad (2.5.18)$$

where $2b\times 2h\times 1$ is the iron volume per unit length.

The total copper (excitation winding) cost C_{Cu} per unit length is similarly a function of its specific gravity γ_{Cu}, the unit cost c_{Cu} and the geometry; in addition, it comprises two specialized parameters: the space factor k_c and the coil spacing a (Fig. 2.22), namely

$$C_{Cu} = \gamma_{Cu} \times k_c \times 4a \times (a+b+h) \times c_{Cu} \qquad (2.5.19)$$

where $4a \times (a+b+h) \times 1$ is the copper volume per unit length.

The total cost $C = C_{Fe} + C_{Cu}$ was evaluated (in adapted units) on the basis of practical data for c_{Fe}, c_{Cu}, k_c, a; results are reproduced in Fig. 2.23 as a function of b/h for different values of the braking force $|F_{br}|$, measured likewise in adapted units.

It is seen that for a wide range of $|F_{br}|$, a minimum production cost value is obtained within the range $(b/h) \simeq 0.34 \div 0.44$. Another finding of practical significance is that with increasing values of $|F_{br}|$, the minimum cost point is shifted towards increased values of $2h$ which is the pole dimension perpendicular to the direction of motion.

2.5.2 Finite-Width Sheet

Having outlined the general approach to the problem of pole optimization in the case of a sheet of much larger dimensions than those of the pole cross-section, we proceed to the case of a sheet of finite width $2A$ (Fig. 2.24a). The boundary condition precluding current outflow from the sheet at $y = \pm A$ is here again satisfied with the aid of a series of images; the relevant electro-magnetic analogy, comprising a series of infinite current-carrying bus-bars, is reproduced in Fig. 2.24b, with the fictitious surface current K_z obtained by integration of (2.5.4), i.e.,

$$\mu_0 \int_0^\infty j_z dx = (\sigma\Delta) v \int_0^\infty \frac{\partial B_z}{\partial x} dx \;, \qquad (2.5.20)$$

so that

$$\pm K_z = \mp \frac{(\sigma\Delta) v}{\mu_0} B_0 \;. \qquad (2.5.21)$$

Once more the braking force is obtainable from (2.5.6) which, in turn, leads to (2.5.15), so that the problem is actually again reduced in effect to calculation of $\mathscr{L}$. The solution of the boundary value problem defined by Fig. 2.24b yields (Sect. 2.5.3)

$$\mathscr{L} = \frac{\mu_0}{\pi} \sum_{n=0}^{\infty} \frac{\sin^2\left[(2n+1)\frac{\pi}{2}\frac{h}{A}\right]}{\left[(2n+1)\frac{\pi}{2}\frac{h}{A}\right]^2} \cdot \frac{1-\exp\left[-(2n+1)\pi\frac{b}{A}\right]}{(2n+1)} \;. \qquad (2.5.22)$$

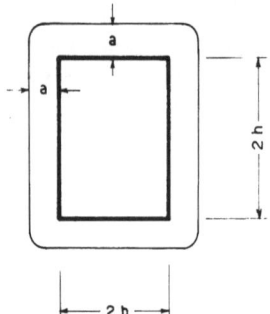

Fig. 2.22. Coil spacing around iron core

Fig. 2.23. Cost C of electromagnet (adapted units) per unit length as function of dimensionless b/h ratio for fixed values of braking force $|F_{br}|$ (adapted units)

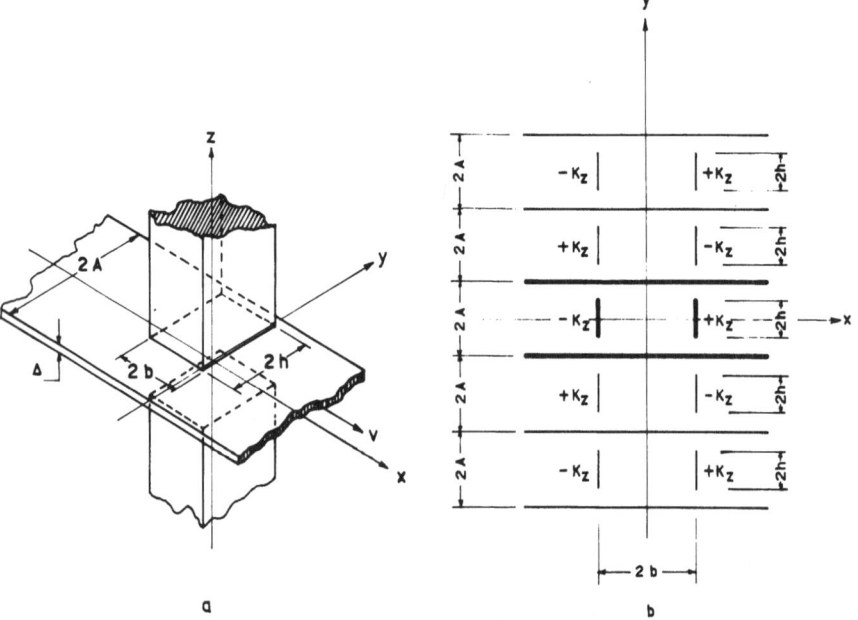

Fig. 2.24a, b. Finite-width sheet: **(a)** Geometry; **(b)** Infinite series of images

As regards the optimal shape, calculations based on the same approach as before show that, for $2A = 1 > 2h$ (adapted units) the minimum cost C now corresponds to a ratio $2b/2h \approx 0.5$, within the braking force range of $1 - 10$ (again – arbitrary units).

It should be emphasized that, in the absence of priorities regarding the relative motion of the magnet and the sheet, the results are evidently valid for a confined magnetic field and linear relative motion. Thus, in spite of the extreme simplification, the model is suitable, in principle, also for linear induction motors (Sect. 2.2); the analogy with a linear motor is more pronounced than with an ordinary induction motor, since the drag force is due to the interaction with eddy currents in a sheet of much larger dimensions than those of the core.

The braking-force calculation presented here indicates an optimum ratio for a single electromagnet only; extension to an array is nevertheless obtainable by means of the main results derived in Sect. 2.4.

2.5.3 Calculation of Self-Inductance per Unit Length

Adapting the notation K_z for one current-carrying bus bar to the present case, i.e. defining, see (2.5.21), $K_z \equiv K_0$, the spatial distribution of the surface current density $K_z(y)$ at $x = +b$ reads (Fig. 2.24 b)

$$K_z(y) = K_0 \frac{4}{\pi} \sum_{n=0}^{\infty} \frac{\sin\left[(2n+1)\frac{\pi}{2}\frac{h}{A}\right]}{(2n+1)} \cos\left[(2n+1)\frac{\pi}{2}\frac{y}{A}\right] . \quad (2.5.23)$$

In the region $x \gtrless b$ the z-component V_z of the vector potential linked to $K_z(y)$ satisfies the Laplace equation

$$\nabla^2 V_z = 0 . \quad (2.5.24)$$

The solution of this equation coherent in y, bounded for $|x| \to \infty$ and satisfying at the interface $z = b$ the condition

$$H_y\big|_{b+0} - H_y\big|_{b-0} = K_z(y) , \quad \text{reads} \quad (2.5.25)$$

$$V_z = \mu_0 K_0 \frac{4A}{\pi^2} \sum_{n=0}^{\infty} \frac{\sin\left[(2n+1)\frac{\pi}{2}\frac{h}{A}\right]}{(2n+1)^2} \cos\left[(2n+1)\frac{\pi}{2}\frac{y}{A}\right]$$

$$\times \exp\left[\mp(2n+1)\frac{\pi}{2}\frac{(x-b)}{A}\right] , \quad x \gtrless b . \quad (2.5.26)$$

In order to calculate the "self" component W_{ms} of the magnetic energy per unit length of *one bus bar* of width $2h$ in the $b = x$ plane, we write

$$W_{ms} = \frac{1}{2} \int_{y=-h}^{h} K_0 V_z dy, \quad x = b, \quad \text{i.e.} \tag{2.5.27}$$

$$W_{ms} = \frac{1}{2} \int_{y=-h}^{h} K_0 \mu_0 K_0 \frac{4A}{\pi^2} \sum_{n=0}^{\infty} \frac{\sin\left[(2n+1)\frac{\pi}{2}\frac{h}{A}\right]}{(2n+1)^2}$$

$$\times \cos\left[(2n+1)\frac{\pi}{2}\frac{y}{A}\right] dy, \tag{2.5.28}$$

so that, with the current $I = K_0 2h$:

$$W_{ms} = \frac{\mu_0}{2\pi} I^2 \sum_{n=0}^{\infty} \frac{\sin^2\left[(2n+1)\frac{\pi}{2}\frac{h}{A}\right]}{\left[(2n+1)\frac{\pi}{2}\frac{h}{A}\right]^2} \cdot \frac{1}{(2n+1)} \tag{2.5.29}$$

Adding to the preceding the "mutual" component W_{mm} due to the surface current at $x = -b$, i.e.

$$W_{mm} = -\frac{\mu_0}{2\pi} I^2 \sum_{n=0}^{\infty} \frac{\sin^2\left[(2n+1)\frac{\pi}{2}\frac{h}{A}\right]}{\left[(2n+1)\frac{\pi}{2}\frac{h}{A}\right]^2} \cdot \frac{1}{(2n+1)}$$

$$\times \exp\left[-(2n+1)\frac{\pi}{2}\frac{2b}{A}\right], \tag{2.5.30}$$

we obtain the total (free) magnetic energy W per unit length of one pair of bus bars as

$$W = \frac{1}{2} I^2 \frac{\mu_0}{\pi} \sum_{n=0}^{\infty} \frac{\sin^2\left[(2n+1)\frac{\pi}{2}\frac{h}{A}\right]}{\left[(2n+1)\frac{\pi}{2}\frac{h}{A}\right]^2} \cdot \frac{1 - \exp\left[-(2n+1)\pi\frac{b}{A}\right]}{(2n+1)}, \tag{2.5.31}$$

and the inductivity $\mathscr{L}$ per unit length therefore reads

$$\mathscr{L} = \frac{\mu_0}{\pi} \sum_{n=0}^{\infty} \frac{\sin^2\left[(2n+1)\frac{\pi}{2}\frac{h}{A}\right]}{\left[(2n+1)\frac{\pi}{2}\frac{h}{A}\right]^2} \cdot \frac{1 - \exp\left[-(2n+1)\pi\frac{b}{A}\right]}{(2n+1)}. \tag{2.5.32}$$

2.6 Unipolar Induction – Circular Motion

The preceding sections dealt exclusively with *rectilinear* motion; in engineering applications, however, *circular* motion is of paramount importance, and we shall try to adapt our analysis accordingly.

In dealing with rectilinear motion in an inertial frame of reference, we had the advantage of the fact that both Maxwell's equations and the relevant constitutive equations had originally been formulated for such a frame, as well as the benefit of Galileo's or Einstein's principle of relativity. Where a rotating frame is involved, however, there are difficulties regarding the valid form of the equations; these difficulties stem from lack of a theory encompassing both the elastic and electrodynamic aspects of the constitutive laws, and they are compounded by the absence of experimental data that might serve as a substitute for such a theory.

In the present context, we shall address ourselves to a relatively simple aspect, namely the formulation and solution of a problem of unipolar induction under circular motion.

2.6.1 Formulation of Problem

In complete analogy to the cases outlined in the preceding sections, we now consider a nonmagnetic circular metal sheet of radius a rotating in the very narrow air gap of a direct-current excited electromagnet (Fig. 2.25).

All the relevant geometric dimensions are reproduced above, with $b \ll R$ and with the sheet thickness Δ of the order of b. Assumption (2.3.14) is again introduced, and the analysis is once more undertaken for very low magnetic

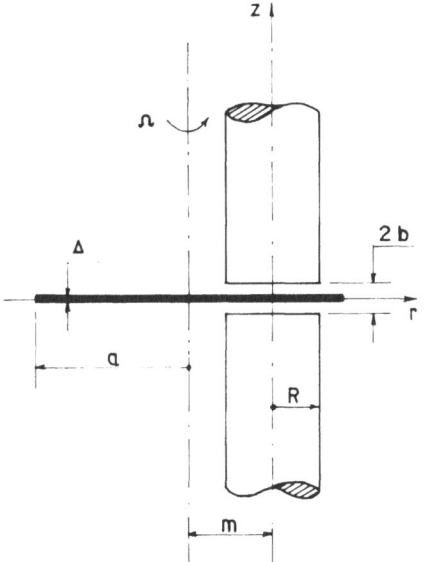

Fig. 2.25. System under investigation

Reynolds numbers R_m; in many electrical instruments (where magnetic damping is frequently resorted to), the driving torque must equal the braking torque at relatively low speeds and, in addition, the radius of the damping disk is limited by weight restrictions. Hence the relevance of the assumption $R_m \ll 1$ which, in turn, enables us to view the magnetic flux density as due to the exciting electromagnet only.

Considering our previous approach, how do we calculate the braking torque exerted on the disk?

2.6.2 Solution

With the electromagnetic analogy outlined in the preceding sections, we rewrite (2.5.13) in the form

$$|F_{br}| \cdot |v| = \frac{2\mu_0}{\sigma \Delta} W_m . \tag{2.6.1}$$

Identifying $|F_{br}| \cdot |v|$ with the absolute value of the mechanical power input (henceforth denoted simply by P), we have

$$P = \frac{2\mu_0}{\sigma \Delta} W_m . \tag{2.6.2}$$

In order to be able to use this relation (from which $|v|$ has now been dropped), the relevant boundary-value problem must first be solved. Now, by virtue of Sect. 2.6.4, the conventional Maxwell equations may be used here even in the primed, non-inertial frame of reference attached to the rotating sheet, along with the defining relations of quasistatics; hence it is possible to adapt our previous solutions to the case of rotation.

According to Fig. 2.26, three systems of coordinates are resorted to: one (r', α') is attached to the circular sheet, which rotates at the uniform mechani-

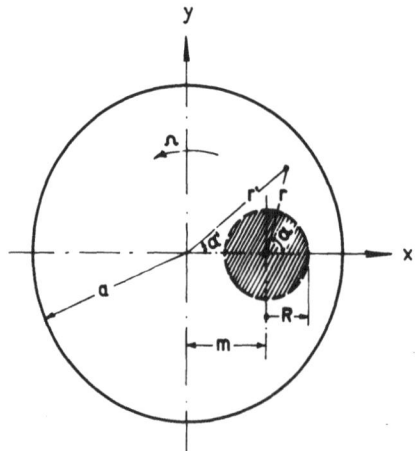

Fig. 2.26. Systems of coordinates

cal circular frequency Ω with respect to the laboratory which is taken to represent an inertial frame of reference; another (r, α) is attached to the normal projection of the magnetic pole; finally, a Cartesian system (x, y) whose $(+x)$-axis coincides with the $\alpha = 0$ angle.

In order to simplify our equations, we use from the outset the complex coordinates

$$w' = r'e^{i\alpha'}, \qquad w = re^{i\alpha} . \tag{2.6.3}$$

As can be seen from Fig. 2.26, these coordinates are obviously interlinked through the relation

$$w' = m + w , \tag{2.6.4}$$

so that

$$r'\cos\alpha' = m + r\cos\alpha , \qquad \text{and} \tag{2.6.5}$$

$$r'\sin\alpha' = r\sin\alpha . \tag{2.6.6}$$

Any control point r' possesses therefore the circular velocity

$$v'_\alpha = \Omega r' , \tag{2.6.7}$$

so that, after decomposition along the x- and y-axes, and using the unit vector relation $1'_\alpha = -1_x \sin\alpha' + 1_y \cos\alpha'$, we find that

$$v_x = -v'_\alpha \sin\alpha' = -\Omega r' \sin\alpha' , \tag{2.6.8}$$

$$v_y = +v'_\alpha \cos\alpha' = +\Omega r' \cos\alpha' . \tag{2.6.9}$$

Along the pole projection, these components attain the values

$$v_x = -\Omega a \sin\alpha , \tag{2.6.10}$$

$$v_y = +\Omega(m + a\cos\alpha) . \tag{2.6.11}$$

The velocity component v_n normal to the pole projection therefore reads

$$v_n = v_x \cos\alpha + v_y \sin\alpha$$

$$= -\Omega a \sin\alpha\cos\alpha + \Omega(m + a\cos\alpha)\sin\alpha$$

$$= \Omega m \sin\alpha . \tag{2.6.12}$$

Having determined v_n, we proceed to the electrical part of the problem through the mode outlined in Sect. 2.3.2; accordingly, (2.3.25) is now rewritten as

$$\frac{\partial^2 D'}{\partial r'^2} + \frac{1}{r'}\frac{\partial D'}{\partial r'} + \frac{1}{r'^2}\frac{\partial^2 D'}{\partial \alpha'^2} = -\sigma \Delta m \Omega \frac{\partial B_z}{\partial r} \sin\alpha . \tag{2.6.13}$$

With the imposed magnetic induction comprising mainly a z-component in the $z = 0$ plane, its derivative $\partial B_z/\partial r$ is once more confined to a very narrow circular region in the vicinity of $r = R$. Resorting to the electromagnetic analogy of sheet currents and assuming along $r = R$ a fictitious surface current distribution $K_z(\alpha)$ in the form

$$K_z(\alpha) = -\frac{\sigma \Delta \Omega m \sin \alpha}{\mu_0} B_z \equiv +K_{z0} \sin \alpha , \qquad (2.6.14)$$

we have to determine the component $V_z^{(1)}$ of the vector potential linked to this current distribution – namely, to solve, instead of the inhomogeneous differential equation (2.6.13), the homogeneous equation

$$\frac{\partial^2 V_z^{(1)}}{\partial r'^2} + \frac{1}{r'} \frac{\partial V_z^{(1)}}{\partial r'} + \frac{1}{r'^2} \frac{\partial^2 V_z^{(1)}}{\partial \alpha'^2} = 0 , \qquad (2.6.15)$$

and to adapt the solutions to the common boundary $r < R$ and $r > R$. Taking advantage of the fact that the Laplace equation preserves its form on transformation from one orthogonal system to another, (2.6.15) may be rewritten as

$$\frac{\partial^2 V_z^{(1)}}{\partial r^2} + \frac{1}{r} \frac{\partial V_z^{(1)}}{\partial r} + \frac{1}{r^2} \frac{\partial^2 V_z^{(1)}}{\partial \alpha^2} = 0 . \qquad (2.6.16)$$

The solution, subject to the condition imposed by the current excitation at $r = R$, is given by

$$V_z^{(1)} = \frac{K_{z0}}{2} \mu_0 R \frac{r}{R} \sin \alpha, \quad r < R ; \qquad (2.6.17)$$

$$V_z^{(1)} = \frac{K_{z0}}{2} \mu_0 R \frac{R}{r} \sin \alpha, \quad r > R . \qquad (2.6.18)$$

We now consider the complex vector potential $\bar{V}_z^{(1)}$; by introducing the complex system of coordinates (2.6.3, 4), the previously found particular solutions read

$$\bar{V}_z^{(1)} = -\mathrm{i} \frac{K_{z0}}{2} \mu_0 R \frac{w' - m}{R} , \quad |w' - m| < R ; \qquad (2.6.19)$$

$$\bar{V}_z^{(1)} = \mathrm{i} \frac{K_{z0}}{2} \mu_0 R \frac{R}{w' - m} , \quad |w' - m| > R , \qquad (2.6.20)$$

where it is understood that the real part is implied.

In the original problem, the stream function D' is constant at the radius $r' = a$; in the analogous problem this leads to the requirement $\mathrm{Re}\{\bar{V}_z^{(1)}\} = \mathrm{const}$ for $|w'| = a$.

Expanding (2.6.20) for $|w'| > m$, we obtain

$$\bar{V}_z^{(1)} = \mathrm{i}\, \frac{K_{z0}}{2}\, \mu_0 R \, \frac{R}{w'} \left[1 + \left(\frac{m}{w'} \right) + \left(\frac{m}{w'} \right)^2 + \dots \right],$$
(2.6.21)

or, for the real component $V_z^{(1)}$ at $|w'| = a$, we get

$$V_z^{(1)} = \frac{K_{z0}}{2}\, \mu_0 \, \frac{R^2}{m} \left[\frac{m}{a} \sin \alpha' + \left(\frac{m}{a} \right)^2 \sin 2\alpha' + \left(\frac{m}{a} \right)^3 \sin 3\alpha' \dots \right].$$
(2.6.22)

This expression is, however, nonzero. The circumference of the circle $|w'| = a$ is thus a source of a so-called "secondary" fictitious potential component $V_z^{(2)}$ which, in conjunction with the "primary" $V_z^{(1)}$, makes the resultant potential at a vanish; in order to offset $V_z^{(1)}$, the potential $V_z^{(2)}$ must have the form

$$V_z^{(2)} = -\frac{K_{z0}}{2}\, \mu_0 \, \frac{R^2}{m} \left[\frac{m}{a} \sin \alpha' + \left(\frac{m}{a} \right)^2 \sin 2\alpha' + \left(\frac{m}{a} \right)^3 \sin 3\alpha' \dots \right].$$
(2.6.23)

The secondary complex potential $\bar{V}_z^{(2)}$ must be an analytic function for $|w'| < a$. Thus, by the Cauchy-Riemann conditions and by (2.6.23), we have

$$\bar{V}_z^{(2)} = \mathrm{i}\, \frac{K_{z0}}{2}\, \mu_0 \, \frac{R^2}{m} \left[\left(\frac{m}{a} \right)^2 \frac{w'}{m} + \left(\frac{m}{a} \right)^4 \left(\frac{w'}{m} \right)^2 \right.$$
$$\left. + \left(\frac{m}{a} \right)^6 \left(\frac{w'}{m} \right)^3 + \dots \right], \quad \text{or}$$
(2.6.24)

$$\bar{V}_z^{(2)} = \mathrm{i}\, \frac{K_{z0}}{2}\, \mu_0 \, \frac{R^2}{m} \, \frac{\left(\dfrac{m}{a} \right)^2 \left(\dfrac{w'}{m} \right)}{1 - \left(\dfrac{m}{a} \right)^2 \left(\dfrac{w'}{m} \right)},$$
(2.6.25)

which leads to

$$\bar{V}_z^{(2)} = \frac{K_{z0}}{2\mathrm{i}}\, \mu_0 R \left[\frac{R}{m} + \left(\frac{a}{m} \right)^2 \frac{R}{w' - \dfrac{a^2}{m}} \right].$$
(2.6.26)

Substitution of $w' = w + m$ and expansion of the second term in the bracketed expression yield

$$\bar{V}_z^{(2)} = -\frac{K_{z0}}{2i}\mu_0 R \left\{ -\frac{R}{m} + \left(\frac{a}{m}\right)^2 \frac{R}{\frac{a^2}{m}-m} \right.$$

$$\left. \times \left[1 + \left(\frac{w}{\frac{a^2}{m}-m}\right) + \left(\frac{w}{\frac{a^2}{m}-m}\right)^2 + \ldots \right] \right\} . \tag{2.6.27}$$

Within the pole projection, we obtain from (2.6.19,27) that the resultant complex vector potential component is given by

$$\bar{V}_z = \bar{V}_z^{(1)} + \bar{V}_z^{(2)} , \tag{2.6.28}$$

$$\bar{V}_z = \frac{K_{z0}}{2i}\mu_0 R \frac{w}{R} - \frac{K_{z0}}{2i}\mu_0 R \left\{ -\frac{R}{m} + \left(\frac{a}{m}\right)^2 \frac{R}{(a^2/m)-m} \right.$$

$$\left. \times \left[1 + \left(\frac{w}{(a^2/m)-m}\right) + \left(\frac{w}{(a^2/m)-m}\right)^2 + \ldots \right] \right\} . \tag{2.6.29}$$

From (2.5.12) − adapted to the surface-current distribution − and owing to the orthogonality of the trigonometric functions, we now obtain by conversion to a line integral over the pole contour l

$$W_m = \frac{1}{2} \oint_{(l)} V_z K_z dl = \frac{1}{2} \mathrm{Re} \oint_{(l)} \frac{K_{z0}}{2i}\mu_0 R \left[\frac{w}{R} - \left(\frac{a}{m}\right)^2 \frac{R^2}{(a^2/m-m)^2} \frac{w}{R}\right]$$

$$\times \left(-\frac{\sigma \Delta \omega m \sin \alpha}{\mu_0} B_z\right) R d\alpha; \qquad w = R e^{i\alpha} ; \tag{2.6.30}$$

i.e.,

$$W_m = \left(\frac{\sigma \Delta \Omega m B_z}{2\mu_0}\right)^2 \mu_0 R^2 \pi \left(1 - \frac{(R/a)^2}{[1-(m/a)^2]^2}\right) . \tag{2.6.31}$$

The power input to the sheet obtained from (2.6.2) therefore reads

$$P = \frac{1}{2}\sigma \Delta (\Omega m B_z)^2 \pi R^2 \left(1 - \frac{(R/a)^2}{[1-(m/a)^2]^2}\right) \tag{2.6.32}$$

so that the braking torque $T_{\mathrm{br}} = P/\Omega$ exerted by the magnet on the rotating sheet is in turn given by

$$T_{\mathrm{br}} = \frac{1}{2}\sigma \Delta \Omega \pi R^2 m^2 B_z^2 \left(1 - \frac{(R/a)^2}{[1-(m/a)^2]^2}\right) . \tag{2.6.33}$$

2.6.3 Numerical Solution

In practice, (2.6.13) may sometimes demand a more sophisticated approach as the spatial distribution $B_z = B_z(r)$ cannot always be approximated by square representations (Fig. 2.27). For such cases, we now outline a numerical approach. Thus, we introduce the notation

$$f(r', \alpha') \equiv -\sigma \Delta \Omega \sin\alpha \, \frac{\partial B_z(r)}{\partial r} \, , \qquad (2.6.34)$$

where $f(r', \alpha')$ is a given function [based on measurements of $B_z(r)$ at points r', α']. When solving

$$\frac{\partial^2 D'}{\partial r'^2} + \frac{1}{r'}\frac{\partial D'}{\partial r'} + \frac{1}{r'^2}\frac{\partial^2 D'}{\partial \alpha'} = f(r', \alpha') \, , \qquad (2.6.35)$$

subject to the boundary condition $D' = $ const for $r' = a$, we consider first the eigenvalue problem

$$\frac{\partial^2 D'}{\partial r'^2} + \frac{1}{r'}\frac{\partial D'}{\partial r'} + \frac{1}{r'^2}\frac{\partial^2 D'}{\partial \alpha'} = -\lambda^2 D' \, , \qquad (2.6.36)$$

with λ^2 (eigenvalue) still undetermined.

For symmetrical conditions with respect to the angle α', we assume a solution of the form

$$D' = R(r') \sin q\alpha' \, , \qquad (2.6.37)$$

where q is an integer and the function $R(r')$ satisfies the Bessel equation

$$\frac{d^2R}{dr'^2} + \frac{1}{r'}\frac{dR}{dr'} + \left(\lambda^2 - \frac{q^2}{r'^2} \right) R = 0 \, . \qquad (2.6.38)$$

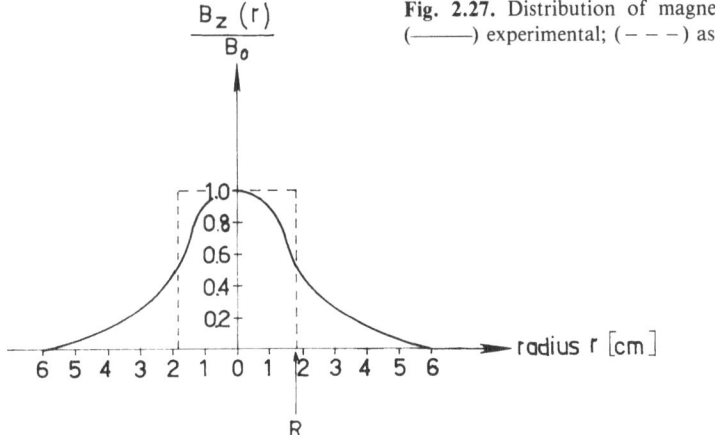

Fig. 2.27. Distribution of magnetic flux density. (———) experimental; (– – –) assumed

The solution of this equation (convergent at $r' = 0$) is given by the Bessel function of first kind, order q:

$$R(r') = D'_{q,\lambda} J_q(\lambda r') \ , \qquad (2.6.39)$$

where the amplitude $D'_{q,\lambda}$ needs still to be determined.

By extending the definition of D' to the circular case, we have $A'_\alpha = -\partial D'/\partial r'$; vanishing of the current density at $r' = a$ therefore imposes the condition

$$\left. \frac{dJ_q(\lambda r')}{dr'} \right|_{r'=a} = 0 \ . \qquad (2.6.40)$$

Denoting by $u_{q,n} = \lambda_{q,n} a$ the nth (positively increasing) solution of this equation, we obtain

$$\lambda_{q,n} = \frac{u_{q,n}}{a} \ . \qquad (2.6.41)$$

On changing the hitherto general notation $D'_{q,\lambda}$ to one adapted to the specific number n, we obtain

$$D' = \sum_{q=0}^{\infty} \sum_{n=1}^{\infty} D'_{q,n} \sin(q\alpha') J_q\left(u_{q,n} \frac{r'}{a}\right) \ . \qquad (2.6.42)$$

The functions in this equation are orthogonal, and we expand $f(r', \alpha')$ in a series of this form by means of the coefficients $C_{q,n}$:

$$f(r', \alpha') = \sum_{q=0}^{\infty} \sum_{n=1}^{\infty} C_{q,n} \sin(q\alpha') J_q\left(u_{q,n} \frac{r'}{a}\right) \ , \qquad (2.6.43)$$

in which

$$C_{q,n} = \frac{\int_{\alpha'=-\pi}^{\pi} \int_{r'=0}^{a} f(r', \alpha') \sin(q\alpha') J_q\left(u_{q,n} \frac{r'}{a}\right) r' \, dr' \, d\alpha'}{\pi \int_{r'=0}^{a} \left[J_q\left(u_{q,n} \frac{r'}{a}\right)\right]^2 r' dr'} \ . \qquad (2.6.44)$$

The coefficients $C_{q,n}$ are obtainable for given values of $\partial B_z/\partial r$. Setting now

$$-(\lambda_{q,n})^2 D' = f(r', \alpha') \ , \qquad (2.6.45)$$

we finally obtain, by means of (2.6.42, 43)

$$D'_{q,n} = -\left(\frac{a}{u_{q,n}}\right)^2 C_{q,n} \ . \qquad (2.6.46)$$

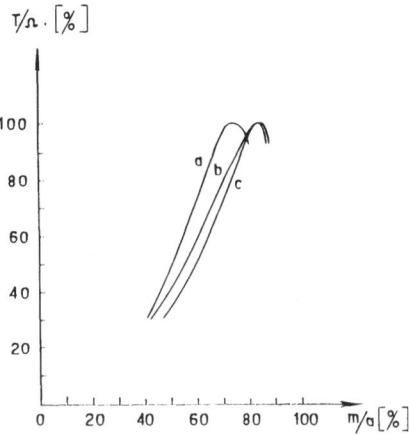

Fig. 2.28. Damping constant T/Ω as function of dimensionless distance m/a between disk axis and pole-face centre. (a) Experimental. (b) Calculated from (2.6.33). (c) Numerical evaluation

Once D' is known, the power P dissipated in the rotating disk is readily calculated, and the braking torque $T = P/\Omega$ is obtained in a straightforward manner.

Some relevant results are reproduced in Fig. 2.28, which is plotted in terms of the braking constant T/Ω.

2.6.4 Appendix

Approach to Maxwell's Equations in the Presence of Rotation

Consider a frame of reference rotating uniformly at the circular frequency Ω. We define its cylindrical coordinates r, α, z, dispensing momentarily with the primes usually affixed to the coordinates of the moving frame; further, the coordinate time is denoted by t. Resorting to the constant phase velocity c of a uniform monochromatic plane wave of light in matter-free space as recorded in an inertial frame of reference, the interval (ds) is given [Ref. 2.6, p. 253] in the rotating frame by

$$(ds)^2 = (c^2 - \Omega^2 r^2)\,dt^2 - 2\,\Omega r^2\,d\alpha\,dt - dz^2 - r^2 d\alpha^2 - dr^2 \ , \qquad (2.6.47)$$

which is a special form of the expression

$$(ds)^2 = g_{\alpha\beta}dx^\alpha dx^\beta + 2\,g_{0\alpha}dx^0 dx^\alpha + g_{00}(dx^0)^2 \ . \qquad (2.6.48)$$

Here x^α, x^β are generalized contravariant space coordinates [11]; x^0 stands for the coordinate $x^0 = ct$ and $x^1 = r$, $x^2 = \alpha$, $x^3 = z$. Moreover, the coefficients $g_{ik} = g_{ki}$ denote the covariant components of the metric tensor (g) of second rank.

---

[11] Departing from the practice of Sect. 2.1.2, we resort in this section to covariant and contravariant notation and accordingly drop the $i = \sqrt{-1}$ convention.

Let us remark at the outset that in accelerated frames of reference, space and time coordinates have no straightforward significance, representing merely a (convenient) form of listing physical occurrences.

In our case, see (2.6.47), (g) is expressed by the array

$$[g] = \begin{vmatrix} g_{00} = 1 - \Omega^2 r^2/c^2 & g_{01} = 0 & g_{02} = -r^2\Omega/c & g_{03} = 0 \\ g_{10} = 0 & g_{11} = -1 & g_{12} = 0 & g_{13} = 0 \\ g_{20} = -r^2\Omega/c & g_{21} = 0 & g_{22} = -r^2 & g_{23} = 0 \\ g_{30} = 0 & g_{31} = 0 & g_{32} = 0 & g_{33} = -1 \end{vmatrix}.$$

$$(2.6.49)$$

The above array defines the four-dimensional time-space metric; the three-dimensional space metric is, in turn, determined by the three-dimensional tensor (γ) whose covariant elements $\gamma_{\alpha\beta}$ are specified by the relation [12]

$$\gamma_{\alpha\beta} = \left(-g_{\alpha\beta} + \frac{g_{0\alpha}g_{0\beta}}{g_{00}} \right). \qquad (2.6.50)$$

Hence, for our rotating system

$$(\gamma) = \begin{vmatrix} \gamma_{11} = 1 & \gamma_{12} = 0 & \gamma_{13} = 0 \\ \gamma_{21} = 0 & \gamma_{22} = \dfrac{r^2}{1 - r^2\Omega^2/c^2} & \gamma_{23} = 0 \\ \gamma_{31} = 0 & \gamma_{32} = 0 & \gamma_{33} = 1 \end{vmatrix}. \qquad (2.6.51)$$

so that the length (dl), expressed in tensor notation by

$$dl^2 = \gamma_{\alpha\beta}dx^\alpha dx^\beta \qquad (2.6.52)$$

is here given by

$$dl^2 = dr^2 + \frac{r^2}{1 - r^2\Omega^2/c^2}d\alpha^2 + dz^2 . \qquad (2.6.53)$$

From now on we shall specifically deal with first-order effects in Ω only, i.e. we shall always assume $(r\Omega)^2 \ll c^2$, while nevertheless retaining terms of the form $r\Omega/c$. At this level of approximation, we have

$$[g] = \begin{vmatrix} g_{00} = 1 & g_{01} = 0 & g_{02} = -r^2\Omega/c & g_{03} = 0 \\ g_{10} = 0 & g_{11} = -1 & g_{12} = 0 & g_{13} = 0 \\ g_{20} = -r^2\Omega/c & g_{21} = 0 & g_{22} = -r^2 & g_{23} = 0 \\ g_{30} = 0 & g_{31} = 0 & g_{32} = 0 & g_{33} = -1 \end{vmatrix}. \qquad (2.6.54)$$

---

[12] The Greek indices run from 1 to 3; obviously, α has here nothing to do with the contravariant coordinate $\alpha = x^2$.

and

$$(\gamma) = \begin{vmatrix} \gamma_{11} = 1 & \gamma_{12} = 0 & \gamma_{13} = 0 \\ \gamma_{21} = 0 & \gamma_{22} = r^2 & \gamma_{23} = 0 \\ \gamma_{31} = 0 & \gamma_{32} = 0 & \gamma_{33} = 1 \end{vmatrix} . \tag{2.6.55}$$

Hence, as opposed to (2.6.53), now

$$dl^2 = dr^2 + r^2 d\alpha^2 + dz^2 , \tag{2.6.56}$$

i.e., the three-dimensional metric is rendered Euclidean.

In order to complete our formalism, we introduce, see below (2.6.60, 61), the three-dimensional vector g in the space whose metric is defined by (γ). The covariant components g_α of g are defined through

$$g_\alpha = -\frac{g_{0\alpha}}{g_{00}} \tag{2.6.57}$$

and its contravariant components g^α read

$$g^\alpha = \gamma^{\alpha\beta} g_\beta , \tag{2.6.58}$$

where $\gamma^{\alpha\beta}$ form the contravariant components of the three-dimensional tensor (γ), expressed by the array $(\bar{\gamma})$:

$$(\bar{\gamma}) = \begin{vmatrix} \gamma^{11} = 1 & \gamma^{12} = 0 & \gamma^{13} = 0 \\ \gamma^{21} = 0 & \gamma^{22} = 1/r^2 & \gamma^{23} = 0 \\ \gamma^{31} = 0 & \gamma^{32} = 0 & \gamma^{33} = 1 \end{vmatrix} . \tag{2.6.59}$$

Hence

$$g_1 = 0; \quad g_2 = r^2 \Omega/c; \quad g_3 = 0 , \quad \text{and} \tag{2.6.60}$$

$$g^1 = 0; \quad g^2 = \Omega/c; \quad g^3 = 0 . \tag{2.6.61}$$

In order to express Maxwell's laws in the rotating frame of reference, we utilize the principle of equivalence phrased as follows: "a noninertial reference system is equivalent to a gravitational field". Accordingly, we first write down Maxwell's laws in the presence of gravitation and then adapt the formulation to rotation.

To this end, we introduce the general (matter-free space included) *three-dimensional* field E, H, D, B with the components chosen so as to coincide in the limit with their counterparts expressed in Lorentzian coordinates.

Here [Ref. 2.6, p. 256]

$$D_\alpha = \frac{\varepsilon_0 E_\alpha}{\sqrt{g_{00}}} + \frac{g^\beta H_{\alpha\beta}}{c} , \tag{2.6.62}$$

$$B^{\alpha\beta} = \frac{\mu_0 H^{\alpha\beta}}{\sqrt{g_{00}}} + \frac{g^\beta E^\alpha - g^\alpha E^\beta}{c} \tag{2.6.63}$$

with $H_{\alpha\beta}$, $B_{\alpha\beta}$ denoting antisymmetric tensors dual to the vectors B, H, i.e.,

$$B^\alpha = -\frac{1}{2\sqrt{\gamma}} e^{\alpha\beta\gamma} B_{\beta\gamma}; \qquad H_\alpha = -\frac{1}{2}\sqrt{\gamma}\, e_{\alpha\beta\gamma} H^{\beta\gamma} . \tag{2.6.64}$$

The term γ (under the square root) stands for the determinant of (γ) and $e^{\alpha\beta\gamma} = e_{\alpha\beta\gamma}$ denotes the unit tensor whose sign changes on transposition of indices: $+1$ if $\alpha\beta\gamma$ are in cyclic order; -1 if they are in countercyclic order; 0 if any two of them are repeated.

Maxwell's equations now read [Ref. 2.6, p. 256]

$$\nabla \times E = -\frac{1}{\sqrt{\gamma}} \frac{\partial}{\partial t}(\sqrt{\gamma} B) , \tag{2.6.65}$$

$$\nabla \cdot B = 0 , \tag{2.6.66}$$

$$\nabla \times H = \frac{1}{\sqrt{\gamma}} \frac{\partial}{\partial t}(\sqrt{\gamma} D) + j , \tag{2.6.67}$$

$$\nabla \cdot D = \varrho . \tag{2.6.68}$$

The contravariant components j^α of j are given by the product of the charge density ϱ with dx^α/dt, i.e. $j^\alpha = \varrho\, dx^\alpha/dt$.

It should be noted that in a three-dimensional system E is a polar vector and H an axial vector; accordingly, their curl expressions differ. Thus, in the case of a polar vector P with covariant components P_1, P_2, P_3, we obtain for the contravariant components of the axial vector $\nabla \times P$ the result

$$(\nabla \times P)^1 = \frac{\partial P_3}{\partial x^2} - \frac{\partial P_2}{\partial x^3} , \tag{2.6.69}$$

$$(\nabla \times P)^2 = \frac{\partial P_1}{\partial x^3} - \frac{\partial P_3}{\partial x^2} , \tag{2.6.70}$$

$$(\nabla \times P)^3 = \frac{\partial P_2}{\partial x^1} - \frac{\partial P_1}{\partial x^3} , \tag{2.6.71}$$

whereas for the axial vector A we have

$$(\nabla \times A)^1 = \frac{1}{\sqrt{\gamma}} \left(\frac{\partial(\sqrt{\gamma} A_3)}{\partial x^2} - \frac{\partial(\sqrt{\gamma} A_2)}{\partial x^3} \right) , \tag{2.6.72}$$

$$(\nabla \times A)^2 = \frac{1}{\sqrt{\gamma}} \left(\frac{\partial(\sqrt{\gamma} A_1)}{\partial x^3} - \frac{\partial(\sqrt{\gamma} A_3)}{\partial x^1} \right) , \tag{2.6.73}$$

$$(\nabla \times A)^3 = \frac{1}{\sqrt{\gamma}} \left(\frac{\partial(\sqrt{\gamma} A_2)}{\partial x^1} - \frac{\partial(\sqrt{\gamma} A_1)}{\partial x^2} \right) . \tag{2.6.74}$$

In ordinary three-dimensional space, however, physical variables are often stated neither in covariant nor in contravariant component form, but – largely because of dimensional considerations – in terms of so-called "physical components" which, it should be noted, are meaningful only in *orthogonal* systems of coordinates. Once these have been introduced, the dissimilarity between $\nabla \times P$ and $\nabla \times A$ vanishes. Thus, if V^k and V_k denote the contravariant and covariant components of a certain vector, respectively, then its physical components $V(k)$ are given by

$$V(k) = \sqrt{V^k V_k} , \tag{2.6.75}$$

the sign of the square root being that of V^k and V_k.

Using contravariant components, we have

$$V^i \equiv \sqrt{(V^i)^2} = \sqrt{\gamma^{ii} V_i V^i} = \sqrt{\gamma^{ii}} V(i) , \tag{2.6.76}$$

and similarly, application of covariant components leads to

$$V_i \equiv \sqrt{(V_i)^2} = \sqrt{\gamma_{ii} V^i V_i} = \sqrt{\gamma_{ii}} V(i) . \tag{2.6.77}$$

We are now ready to turn to the three-dimensional formulation of Maxwell's laws in a rotating frame of reference. In view of the assumption $(r\Omega)^2 \ll c^2$, the cylindrical coordinates r, α, z represent a curvilinear orthogonal system, so that physical components are applicable. Hence the two different curl operations merge into one, yielding

$$(\nabla \times V)(r) = \frac{1}{r} \frac{\partial V(z)}{\partial \alpha} - \frac{\partial V(\alpha)}{\partial z} , \tag{2.6.78}$$

$$(\nabla \times V)(\alpha) = \frac{\partial V(r)}{\partial z} - \frac{\partial V(z)}{\partial r} , \tag{2.6.79}$$

$$(\nabla \times V)(z) = \frac{1}{r} \frac{\partial[r V(\alpha)]}{\partial r} - \frac{1}{r} \frac{\partial V(r)}{\partial \alpha} , \tag{2.6.80}$$

while the scalar divergence of V reads

$$\nabla \cdot V = \frac{1}{r} \frac{\partial[r V(r)]}{\partial r} + \frac{1}{r} \frac{\partial V(\alpha)}{\partial \alpha} + \frac{\partial V(z)}{\partial z} . \tag{2.6.81}$$

Let us now identify the various vector components. From (2.6.64) we have

$$B^1 = -\frac{1}{r}B_{23}, \quad B^2 = -\frac{1}{r}B_{31}, \quad B^3 = -\frac{1}{r}B_{12} , \tag{2.6.82}$$

and

$$H_1 = -rH^{23}, \quad H_2 = -rH^{31}, \quad H_3 = -rH^{12} . \tag{2.6.83}$$

In general, the covariant and contravariant components of a tensor T of rank two are interlinked by $T^{ik} = g^{ij}g^{kl}T_{jl}$; $T_{ik} = g_{ij}g_{kl}T^{jl}$, compare (2.1.26), and those of a vector V by $V^i = g^{ik}V_k$; $V_i = g_{ik}V^k$ with $g_{il}g^{kl} = \delta_i^k$ (δ_i^k denoting the Kronecker delta). Replacing g_{ij} by $\gamma_{\alpha\beta}$, etc., we find

$$B^{23} = \frac{1}{r^2}B_{23} = -\frac{1}{r}B^1 , \tag{2.6.84}$$

$$B^{31} = B_{31} = -rB^2 , \tag{2.6.85}$$

$$B^{12} = \frac{1}{r^2}B_{12} = -\frac{1}{r}B^3 , \tag{2.6.86}$$

along with

$$H_{23} = r^2 H^{23} = -rH_1 , \tag{2.6.87}$$

$$H_{31} = H^{31} = -\frac{H_2}{r} , \tag{2.6.88}$$

$$H_{12} = r^2 H^{12} = -rH_3 . \tag{2.6.89}$$

Reverting to (2.6.62,63), we further find the component equations

$$D_1 = \varepsilon_0 E_1 - \frac{\Omega r}{c^2} H_3 , \tag{2.6.90}$$

$$D_2 = \varepsilon_0 E_2 , \tag{2.6.91}$$

$$D_3 = \varepsilon_0 E_3 + \frac{\Omega r}{c^2} H_1 , \tag{2.6.92}$$

on the one hand, and on the other hand we have

$$-\frac{1}{r}B^3 = -\mu_0 \frac{H_3}{r} + \frac{\Omega}{c} E^1 , \tag{2.6.93}$$

$$rB^2 = \mu_0 \frac{H_2}{r} , \tag{2.6.94}$$

$$-\frac{1}{r}B^1 = -\mu_0 \frac{H_1}{r} - \frac{\Omega}{c^2} E^3 . \tag{2.6.95}$$

Identifying now the indices 1, 2, 3 with r, α, z and introducing the physical components of the vector V, i.e.

$$V(r) = V^r = V_r; \quad V(\alpha) = rV^\alpha = \frac{V_\alpha}{r}; \quad V(z) = V^z = V_z , \quad (2.6.96)$$

we may restate equations (2.6.90 – 95) in the form

$$D(r) = \varepsilon_0 E(r) - \frac{\Omega r}{c^2} H(z) , \quad (2.6.97)$$

$$D(\alpha) = \varepsilon_0 E(\alpha) , \quad (2.6.98)$$

$$D(z) = \varepsilon_0 E(z) + \frac{\Omega r}{c^2} H(r) , \quad (2.6.99)$$

along with

$$B(z) = \mu_0 H(z) - \frac{r\Omega}{c^2} E(r) , \quad (2.6.100)$$

$$B(\alpha) = \mu_0 H(\alpha) , \quad (2.6.101)$$

$$B(r) = \mu_0 H(r) + \frac{r\Omega}{c^2} E(z) . \quad (2.6.102)$$

These equations become even clearer once the velocity v [physical components – see also (2.6.60,61)] is introduced through the definition

$$v = [v(r); \quad v(\alpha); \quad v(z)] , \quad \text{with} \quad (2.6.103)$$

$$v(r) = 0, \quad v(\alpha) = r\Omega, \quad v(z) = 0 , \quad (2.6.104)$$

so that

$$D = \varepsilon_0 (E - v \times \mu_0 H) , \quad (2.6.105)$$

$$B = \mu_0 (H + v \times \varepsilon_0 E) . \quad (2.6.106)$$

It is thus evident that the naive omission, see (2.6.97 – 102), of terms comprising the factor $r\Omega/c$ is untenable in view of the relation $\mu_0 \varepsilon_0 c^2 = 1$. Resorting now to the assumption of magneto-quasistatics – implying $\varepsilon_0 (E)^2 \ll \mu_0 (H)^2$ – we have

$$D = 0 , \quad (2.6.107)$$

$$B = \mu_0 H , \quad (2.6.108)$$

with (2.6.65 – 68) reducing to

$$\nabla \times E = -\frac{\partial B}{\partial t} \ , \tag{2.6.109}$$

$$\nabla \cdot B = 0 \ , \tag{2.6.110}$$

$$\nabla \times H = j \ . \tag{2.6.111}$$

Finally, reintroducing – for the sake of formality – the primes, we have

$$\nabla \times E' = -\frac{\partial B'}{\partial t'} \ , \tag{2.6.112}$$

$$\nabla \cdot B' = 0 \ , \tag{2.6.113}$$

$$\nabla \times H' = j' \ , \tag{2.6.114}$$

with the constitutive relation (Cerenkov electrodynamics extended!) now reduced to a mere definition, i.e.

$$B' = \mu_0 H' \ . \tag{2.6.115}$$

It is thus seen that the approach outlined in Sect. 2.6.2 is based not merely on naive extension to circular motion, but on Maxwell's magneto-quasistatic equations in vacuum, suitably expressed in a rotating frame of reference.

3. Electromagnetic Induction: Steady-State – Stationary Configurations

In this chapter, the first two sections and the last one deal with the circular geometry of induction problems; this simple geometry permits the introduction of well-known functions, so that analytic insight is gained with relative ease. Arbitrary decomposition of the field into "primary" and "secondary" components is resorted to in the calculation of induction and intensity variables. The third section deals with a problem involving linear geometry. The practical device of so-called "internal generators" is applied and a solution is obtained by relatively straightforward means.

3.1 Eddy Currents in Thin Metal Sheets

In many technical devices an ac-supplied "primary" magnetic field induces eddy-current flow in a "secondary" armature. Exact calculation of the overall field distribution is difficult; hence, the secondary field is often neglected in practice and only an estimate of the interaction is obtained. In this section, we intend to outline a general solution of the problem with both fields taken into account; the specific model discussed also shows how to proceed in more complicated cases.

3.1.1 Formulation of Problem

The system is modelled as follows: a very thin homogeneous and isotropic sheet of nonmagnetic metal of thickness Δ and electrical conductivity σ (the armature) is inserted into the narrow air gap ($2b$) of a cylindrical electromagnet (Fig. 3.1). The sheet is assumed to be horizontally "infinite", i.e., much larger than the radius R of the exciting poles; we also recall (Chap. 2) that this sheet is defined mathematically by the limiting process $\lim(\sigma\Delta)$ is finite for $\sigma\to\infty$ and $\Delta\to 0$. The electromagnet is assumed to be made of fictitious "iron" with "infinite" relative permeability μ_r and vanishingly small electrical conductivity. For $\mu_r\to\infty$, the magnetic field intensity H inside the "iron" of the electromagnet tends to zero; hence we may assume that the magnetic circuit of the poles closes at a very large distance from the air gap, where it is connected to an ac voltage source of angular frequency ω by means of fictitious coils of zero resistance. The frequency is again assumed to be sufficiently low so that the time rate of the electric induction may safely be neglected.

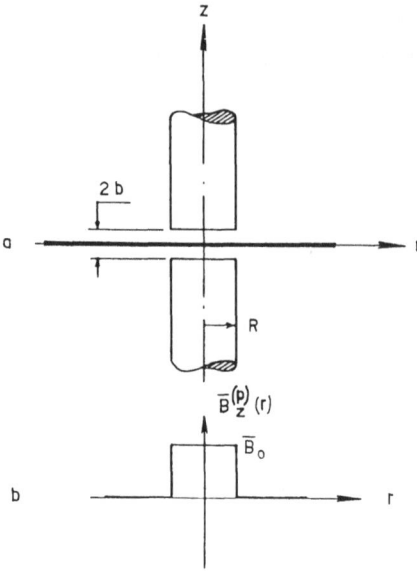

Fig. 3.1a,b. System under investigation: (a) Electromagnet and induced sheet; (b) Assumed spatial dependence of *imposed* (complex) flux density component

Assuming that in the sheet plane the primary, exciting axial magnetic induction component is confined to a region *congruent* with the pole extension, we wish to calculate the current in the sheet and the attendant field distribution.

3.1.2 General Approach to Solution

We introduce for the considered configuration (Fig. 3.1 a), the cylindrical coordinates z (axis), r (radius), α (azimuth). Within the framework of magneto-quasistatics, Maxwell's equations (2.2.1,2) apply.

Further, we are here once more at liberty to resort to a Fitzgerald vector superpotential F comprising a z-component $F_z \equiv F$ only (see also Sect. 2.2.2), which in turn fulfills the Laplace equation

$$\frac{\partial^2 F}{\partial r^2} + \frac{1}{r}\frac{\partial F}{\partial r} + \frac{1}{r^2}\frac{\partial^2 F}{\partial \alpha^2} + \frac{\partial^2 F}{\partial z^2} = 0 \ . \tag{3.1.1}$$

Once F is known, the electric field components E_α, E_r, E_z and their magnetic counterparts H_α, H_r, H_z are immediately obtained by means of partial derivatives, see (2.2.3,4):

$$E_\alpha = -\frac{\partial^2 F}{\partial r\,\partial t}, \quad E_r = \frac{1}{r}\frac{\partial^2 F}{\partial \alpha\,\partial t}, \quad E_z = 0 \ , \tag{3.1.2}$$

$$H_\alpha = -\frac{1}{\mu_0}\frac{1}{r}\frac{\partial^2 F}{\partial \alpha\,\partial z}, \quad H_r = -\frac{1}{\mu_0}\frac{\partial^2 F}{\partial r\,\partial z}, \quad H_z = -\frac{1}{\mu_0}\frac{\partial^2 F}{\partial z^2} \ . \tag{3.1.3}$$

With sinusoidal excitation, the complex amplitude $\bar{F}$ will again be resorted to, i.e., $F = \bar{F}\exp(-i\omega t)$; hence the complex field components $\bar{E}_\alpha$, $\bar{E}_r$, $\bar{E}_z$; $\bar{H}_\alpha$, $\bar{H}_r$, $\bar{H}_z$, under circular symmetry simply read

$$\bar{E}_\alpha = i\omega\frac{\partial \bar{F}}{\partial r}, \quad \bar{E}_r = 0, \quad \bar{E}_z = 0 , \tag{3.1.4}$$

$$\bar{H}_\alpha = 0, \quad \bar{H}_r = -\frac{1}{\mu_0}\frac{\partial^2 \bar{F}}{\partial r \partial z}, \quad \bar{H}_z = -\frac{1}{\mu_0}\frac{\partial^2 \bar{F}}{\partial z^2} . \tag{3.1.5}$$

3.1.3 Primary Field

Adapting (3.1.1) to the complex amplitude of the primary Fitzgerald phasor $\bar{F}^{(p)}$ – defined in the absence of the sheet – and assuming axial symmetry, we obtain the partial differential equation

$$\frac{\partial^2 F^{(p)}}{\partial r^2} + \frac{1}{r}\frac{\partial \bar{F}^{(p)}}{\partial r} + \frac{\partial^2 \bar{F}^{(p)}}{\partial z^2} = 0 . \tag{3.1.6}$$

A solution of this equation – compatible with the symmetry requirements imposed by the magnetic flux density – is obtained in the air gap by means of an as yet undetermined "weighting function" $\bar{f}(\lambda)$ along with a Bessel function of first kind, order zero and argument (λr), i.e.,

$$\bar{F}^{(p)} = \int_{\lambda=0}^{\infty} \bar{f}(\lambda)\,J_0(\lambda r)\cosh(\lambda z)\,d\lambda . \tag{3.1.7}$$

In the $z = 0$ plane, one obtains from (3.1.5,7) the complex amplitude $\bar{B}_z^{(p)} = \mu_0 \bar{H}_z^{(p)}$ of the primary flux density as

$$\bar{B}_z^{(p)} = -\int_{\lambda=0}^{\infty} \bar{f}(\lambda)\lambda^2 J_0(\lambda r)\,d\lambda, \quad z = 0 . \tag{3.1.8}$$

As already pointed out, we assume the complex amplitude of the given, primary quasi-homogeneous magnetic flux density $\bar{B}_z^{(p)}$ to be confined – in the $z = 0$ plane – to the air gap (Fig. 3.1) only, where it has the value $\bar{B}_0$. We may, however, expand this field over the whole $z = 0$ plane by means of the Fourier-Bessel integral in the form

$$\bar{B}_z^{(p)}(r) = \bar{B}_0 R \int_{\lambda=0}^{\infty} J_0(\lambda r)J_1(\lambda R)\,d\lambda; \quad z = 0 . \tag{3.1.9}$$

Equating now (3.1.8 and 9), we have

$$\bar{f}(\lambda) = -\frac{J_1(\lambda R)}{\lambda^2}\bar{B}_0 R \quad \text{whence} \tag{3.1.10}$$

$$\bar{F}^{(p)} = -\bar{B}_0 R \int_{\lambda=0}^{\infty} J_0(\lambda r)\frac{J_1(\lambda R)}{\lambda^2}\cosh(\lambda z)\,d\lambda . \tag{3.1.11}$$

By means of (3.1.4), we obtain the primary electric circular field component

$$\bar{E}_\alpha^{(p)} = i\omega\bar{B}_0 R \int\limits_{\lambda=0}^{\infty} J_1(\lambda r)\frac{J_1(\lambda R)}{\lambda}\cosh(\lambda z)\,d\lambda \ , \tag{3.1.12}$$

which obviously reduces in the $z = 0$ plane to

$$\bar{E}_\alpha^{(p)} = i\omega\bar{B}_0 R \int\limits_{\lambda=0}^{\infty} J_1(\lambda r)\frac{J_1(\lambda R)}{\lambda}\,d\lambda , \quad z = 0 \ . \tag{3.1.13}$$

3.1.4 Secondary Field

Let us now introduce the surface current density A; as implied, symmetry dictates circular current flow only, i.e. $A = A_\alpha I_\alpha$, where, as usual, I_α stands for the circular unit vector. The complex amplitude $\bar{A}_\alpha$ of this current function may be expressed by a spatial integral comprising a weighting function $\bar{h}(\lambda)$

$$\bar{A}_\alpha = \int\limits_{\lambda=0}^{\infty} \bar{h}(\lambda)J_1(\lambda r)\,d\lambda \ . \tag{3.1.14}$$

Assuming momentarily that this is an "imposed" plate current flowing in the sheet *with the exciting poles absent,* we may seek the complex amplitude $\bar{F}^{(s)}$ of the Fitzgerald potential linked to $\bar{A}_\alpha$. *In the absence of the poles,* we are able to obtain a solution of the Laplace equation in free space, for $z \gtrless 0$. Formulating now (3.1.6) for $\bar{F}^{(s)}$ and imposing convergence conditions for $|z| \to \infty$, an integral solution is obtained with the use of an additional spectral density $\bar{k}(\lambda)$:

$$\bar{F}^{(s)} = \int\limits_{\lambda=0}^{\infty} \bar{k}(\lambda)J_0(\lambda r)e^{\mp\lambda z}\,d\lambda, \quad z \gtrless 0 \ . \tag{3.1.15}$$

Hence, with (3.1.5) the radial complex amplitude $\bar{H}_r^{(s)}$ of the secondary magnetic field reads

$$\bar{H}_r^{(s)} = \mp\frac{1}{\mu_0} \int\limits_{\lambda=0}^{\infty} \bar{k}(\lambda)\lambda^2 J_1(\lambda r)e^{\mp\lambda z}\,d\lambda, \quad z \gtrless 0 \ . \tag{3.1.16}$$

Application of Maxwell's first curl law at the sheet boundary yields here the condition

$$\bar{H}_r^{(s)}(z = +0) - \bar{H}_r^{(s)}(z = -0) = \bar{A}_\alpha \ , \tag{3.1.17}$$

which, in turn, by using (3.1.14, 16), leads to

$$-\frac{2}{\mu_0}\bar{k}(\lambda)\lambda^2 = +\bar{h}(\lambda) \ . \tag{3.1.18}$$

Hence, (3.1.15) may be rewritten in the form

$$\bar{F}^{(s)} = -\frac{\mu_0}{2} \int\limits_{\lambda=0}^{\infty} \frac{\bar{h}(\lambda)}{\lambda^2} J_0(\lambda r) e^{\mp \lambda z} d\lambda, \quad z \gtrless 0 . \tag{3.1.19}$$

Thus, the secondary circular electric field amplitude is obtained as

$$\bar{E}_\alpha^{(s)} = \frac{\mu_0}{2} i\omega \int\limits_{\lambda=0}^{\infty} \frac{\bar{h}(\lambda)}{\lambda} J_1(\lambda r) e^{\mp \lambda z} d\lambda, \quad z \gtrless 0 \tag{3.1.20}$$

or, in the sheet plane

$$\bar{E}_\alpha^{(s)} = +\frac{\mu_0}{2} i\omega \int\limits_{\lambda=0}^{\infty} \frac{\bar{h}(\lambda)}{\lambda} J_1(\lambda r) d\lambda; \quad z = 0 . \tag{3.1.21}$$

3.1.5 Calculation of Current Flow

The current will be calculated from Ohm's law expressed in the form $A = \sigma \Delta E$, in which E combines the primary and secondary fields, see also (2.2.31, 32). To this end we mentally divide the sheet in two zones − $0 \leqslant r \leqslant R$ and $R \leqslant r < \infty$ − a division called for by the requirement for solving the Laplace equation subject to the conditions imposed by the exciting poles: this equation is incapable of exact analytic solution in the absence of a three-dimensional coordinate system adapted to the boundaries, in which separation of variables could be applied (Sect. 2.1.4A).

Solution of the problem for the inner and outer zone separately may be justified physically by the fact that the magnetic field in the inner zone is linked both to the exciting coil and to the eddy currents, whereas in the outer zone − assuming fringe effects to be negligible − the magnetic field is linked to the eddy currents only.

We shall now resort to the following approximate procedure: the potential is calculated for a sheet which is not "covered" by the magnetic poles. This solution is valid for the zone $R \leqslant r < \infty$; next, we calculate the current flow in the sheet due to a vortex filament of magnetic flux $\bar{\phi}$ [defined by the limiting process $\lim(\bar{B}_0 \pi R^2) = \bar{\phi}$ for $|\bar{B}_0| \to \infty$ and $R \to 0$]. With the solution of the vortex filament available, we formulate the equivalence of the vortex filament value of $\bar{\phi}$ and the true flux $\bar{\phi}_{tr}$ of the real electromagnet (for which $R \neq 0$); the solution will then be obtained for two different zones.

3.1.6 Outer Zone, $R \leqslant r < \infty$

Combining (3.1.13, 14 and 21) we obtain

$$\bar{h}(\lambda) = \sigma \Delta \left(i\omega \bar{B}_0 R \frac{J_1(\lambda R)}{\lambda} + \frac{\mu_0}{2} i\omega \frac{\bar{h}(\lambda)}{\lambda} \right) \tag{3.1.22}$$

and, passing to the limit $R \rightarrow 0$, we find

$$\bar{h}(\lambda) = \sigma \Delta \left(i\omega \frac{\bar{\phi}}{2\pi} + \frac{\mu_0}{2} i\omega \frac{\bar{h}(\lambda)}{\lambda} \right) \quad \text{i.e.} \tag{3.1.22a}$$

$$\bar{h}(\lambda) = \frac{\bar{\phi}}{\pi \mu_0} \lambda \frac{\sigma \Delta \mu_0 i\omega/2}{\lambda - \sigma \Delta \mu_0 i\omega/2} \ . \tag{3.1.23}$$

Substitution in (3.1.14) leads to

$$\bar{A}_\alpha = \frac{\bar{\phi}}{\pi \mu_0} \int\limits_{\lambda=0}^{\infty} \lambda \frac{\sigma \Delta \mu_0 i\omega/2}{\lambda - \sigma \Delta \mu_0 i\omega/2} J_1(\lambda r) d\lambda \ . \tag{3.1.24}$$

This expression represents the sheet current for all values of r under the assumption of a vortex filament excitation at $r = 0$.

When we attempt in the sequel to calculate the mmf, it will be of some advantage to introduce the stream function $\bar{D}$, see also (2.6.13), from which, in general, the radial and circumferential surface current density components $\bar{A}_r$ and $\bar{A}_\alpha$ follow as $\bar{A}_r = (1/r)(\partial \bar{D}/\alpha)$ and $\bar{A}_\alpha = -\partial \bar{D}/\partial r$. In our case $\bar{A}_r = 0$ and with the notation

$$\frac{1}{r_0} \equiv \frac{1}{2} \sigma \Delta \mu_0 \omega \tag{3.1.25}$$

we readily obtain, see (3.1.24),

$$\bar{D} = \frac{\bar{\phi}}{\pi \mu_0} \int\limits_{\lambda=0}^{\infty} \frac{i(1/r_0)}{\lambda - i(1/r_0)} J_0(\lambda r) d\lambda \ . \tag{3.1.26}$$

Detailed calculations (Sect. 3.1.11) yield

$$\bar{D} = -\frac{\bar{\phi}}{\pi \mu_0} \frac{1}{r_0} \frac{\pi}{2} \left\{ -i \cdot i H_0^{(1)} \left(i\frac{r}{r_0} \right) + \left[I_0 \left(\frac{r}{r_0} \right) - L_0 \left(\frac{r}{r_0} \right) \right] \right\} \tag{3.1.27}$$

and therefore

$$\bar{A}_\alpha = -\frac{\bar{\phi}}{2\pi \mu_0} \frac{1}{r_0^2} \pi \left\{ -i \left[-H_1^{(1)} \left(i\frac{r}{r_0} \right) \right] - \left[I_1 \left(\frac{r}{r_0} \right) - L_1 \left(\frac{r}{r_0} \right) - \frac{2}{\pi} \right] \right\} ;$$

$$R \leqslant r < \infty \ . \tag{3.1.28}$$

The pairs $H_0^{(1)}$ and $H_1^{(1)}$, I_0 and I_1, L_0 and L_1 are Hankel, modified Bessel and modified Struve functions, respectively. The dimensionless expression $\bar{a} = -\bar{A}_\alpha 2\mu_0 r_0^2/\bar{\phi}$ is plotted in Fig. 3.2.

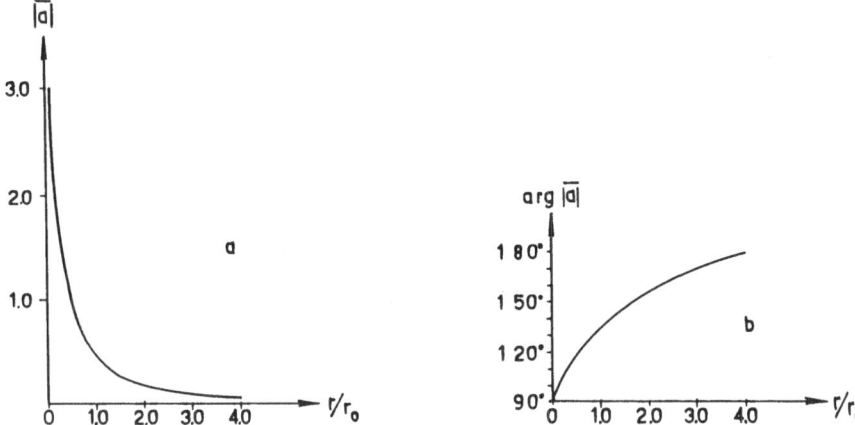

Fig. 3.2a, b. Surface current density in dimensionless units: (a) Amplitude; (b) Angle

3.1.7 Inner Zone, $0 \leqslant r \leqslant R$

Rewriting Maxwell's equations specifically for the volume defined by $0 \leqslant r \leqslant R$, $-b \leqslant z \leqslant b$, we obtain that the complex amplitudes of the field components defined in that region satisfy

$$\frac{\partial \bar{H}_r}{\partial z} - \frac{\partial \bar{H}_z}{\partial r} = \bar{j}_\alpha \quad \text{and} \tag{3.1.29}$$

$$\frac{1}{r} \frac{\partial}{\partial r} (r \bar{E}_\alpha) = \mathrm{i} \omega \mu_0 \bar{H}_z \tag{3.1.30}$$

with $\bar{j}_\alpha$ standing for the circular (volume) current density $\bar{A}_\alpha / \Delta$.

In view of the assumption of an extremely narrow air gap, the field components may be taken as independent of z; integrating (3.1.29) over the z-axis (bearing in mind that $\bar{j}_\alpha$ is confined to the layer Δ and that the magnetic field vanishes on and throughout the fictitious iron poles) and passing to ordinary differentiation, we have

$$-2b(d\bar{H}_z/dr) = \bar{j}_\alpha \Delta = \bar{A}_\alpha . \tag{3.1.31}$$

Combining (3.1.30 and 31) in the sheet plane by means of Ohm's law, and again using ordinary differentiation, we obtain the differential equation of the complex amplitude $\bar{H}_z$ of the z-component of the magnetic field in the $z = 0$ plane

$$\frac{d^2 \bar{H}_z}{dr^2} + \frac{1}{r} \frac{d \bar{H}_z}{dr} + i \omega \mu_0 \frac{\sigma \Delta}{2b} \bar{H}_z = 0 \ . \tag{3.1.32}$$

Hence, due to the requirement of regularity at $r \to 0$

$$\bar{H}_z = \bar{C}_1 J_0 (\sqrt{i \omega \mu_0 \sigma \Delta / 2 b} \, r) \ , \tag{3.1.33}$$

where $\bar{C}_1$ is still undetermined; for the complex amplitude of the electric field component $\bar{E}_\alpha$ in the $z = 0$ plane, we therefore have

$$\bar{E}_\alpha = \frac{2b}{\sigma \Delta} \bar{C}_1 \sqrt{i \omega \mu_0 \frac{\sigma \Delta}{2b}} J_1 \left(\sqrt{i \omega \mu_0 \frac{\sigma \Delta}{2b}} \, r \right) \ , \quad z = 0 \ . \tag{3.1.34}$$

This equation determines the constant $\bar{C}_1$: having assumed exciting coils of vanishingly small electric resistance, the magnetic flux ϕ_{tr} is determined by the external voltage $\bar{V}$ of the supply transformer. Writing $\phi_{tr} = \bar{\phi}_{tr} \exp(-i \omega t)$, Faraday's law reads at $r = R$

$$[\bar{E}_\alpha (r = R)] 2 \pi R = i \omega \bar{\phi}_{tr} \ . \tag{3.1.35}$$

Hence, combining (3.1.34 and 35), we have

$$\bar{C}_1 = \frac{i \omega \bar{\phi}_{tr} / 2 \pi R}{\frac{2b}{\sigma \Delta} \sqrt{i \omega \mu_0 \sigma \Delta / 2 b} \, J_1 (\sqrt{i \omega \mu_0 \sigma \Delta / 2 b} \, R)} \ ; \tag{3.1.36}$$

and therefore,

$$\bar{H}_z = \frac{i \omega \bar{\phi}_{tr}}{2 \pi} \frac{\sigma \Delta}{2b} \frac{J_0 (\sqrt{i \omega \mu_0 \sigma \Delta / 2 b} \, r)}{(\sqrt{i \omega \mu_0 \sigma \Delta / 2 b} \, R) \, J_1 (\sqrt{i \omega \mu_0 \sigma \Delta / 2 b} \, R)} \ , \tag{3.1.37}$$

$$\bar{E}_\alpha = \frac{i \omega \bar{\phi}_{tr}}{2 \pi R} \frac{J_1 (\sqrt{i \omega \mu_0 \sigma \Delta / 2 b} \, r)}{J_1 (\sqrt{i \omega \mu_0 \sigma \Delta / 2 b} \, R)} \ . \tag{3.1.38}$$

3.1.8 Coupling of Zones

In order to couple the zones, we reapply the boundary condition (3.1.35) at $r = R$ and (still retaining the vortex-filament approximation) replace the true exciting flux $\bar{\phi}_{tr}$ with the flux $\bar{\phi}$ mentioned earlier and assumed to be concentrated at the z-axis. The equivalence of $\bar{\phi}_{tr}$ and $\bar{\phi}$ consists in their inducing the same electric field at $r = R$, see (3.1.28), namely

$$-\frac{\bar{\phi}}{2\pi\mu_0}\frac{1}{r_0^2}\pi\left\{-i\left[-H_1^{(1)}\left(i\frac{R}{r_0}\right)\right]-\left[I_1\left(\frac{R}{r_0}\right)-L_1\left(\frac{R}{r_0}\right)-\frac{2}{\pi}\right]\right\}\frac{1}{\sigma\Delta}$$

$$=\frac{i\omega\bar{\phi}_{tr}}{2\pi R} \tag{3.1.39}$$

or

$$\frac{\bar{\phi}}{\bar{\phi}_{tr}}=\left(\frac{\pi}{2}\frac{R}{r_0}\left\{-H_1^{(1)}\left(i\frac{R}{r_0}\right)+\frac{1}{i}\left[I_1\left(\frac{R}{r_0}\right)-L_1\left(\frac{R}{r_0}\right)-\frac{2}{\pi}\right]\right\}\right)^{-1}.$$
$$\tag{3.1.40}$$

Hence, combining the above with (3.1.28), we obtain for the surface current density in the region $R \leqslant r < \infty$

$$\bar{A}_\alpha = -\frac{\bar{\phi}_{tr}}{2\pi\mu_0}\frac{1}{r_0^2}\pi$$

$$\times\frac{\left\{-i\left[-H_1^{(1)}\left(i\frac{r}{r_0}\right)\right]-\left[I_1\left(\frac{r}{r_0}\right)-L_1\left(\frac{r}{r_0}\right)-\frac{2}{\pi}\right]\right\}}{\frac{\pi}{2}\frac{R}{r_0}\left\{-H_1^{(1)}\left(i\frac{R}{r_0}\right)+\frac{1}{i}\left[I_1\left(\frac{R}{r_0}\right)-L_1\left(\frac{R}{r_0}\right)-\frac{2}{\pi}\right]\right\}}, \tag{3.1.41}$$

whereas for $0 \leqslant r \leqslant R$ we obtain from (3.1.38),

$$\bar{A}_\alpha = +i\frac{\bar{\phi}_{tr}}{2\pi\mu_0}\frac{1}{r_0^2}\pi\frac{J_1(\sqrt{i/br_0}\,r)}{\frac{\pi}{2}\frac{R}{r_0}J_1(\sqrt{i/br_0}\,R)}. \tag{3.1.42}$$

3.1.9 Calculation of Ampère-Turns

The above results will be used in order to estimate the complex amplitude of the Ampère-turns (mmf) required for setting up the magnetic flux in the air gap. Now these Ampère-turns are given by the line integral of the overall magnetic field component $\bar{H}_z$ over a closed path, threading the current in the sheet, such a path being chosen so as to traverse the air gap and the iron along the z-axis; the resulting integral, in turn, equals the discontinuity in the complex amplitude $\bar{\chi}$ of the magnetic scalar potential across the surface of the sheet and, in order to calculate it, we resort to our generating function $\bar{D}$: thus, the discontinuity between the secondary radial magnetic field components is exactly compensated, see (3.1.17), by the complex plate current $\bar{A}_\alpha$: further, from the definition of the secondary complex magnetic scalar potential $\bar{\chi}^{(s)}$ we obtain for the radial component of the secondary field $\bar{H}_r^{(s)}$ the expression $\bar{H}_r^{(s)} = - \partial\bar{\chi}^{(s)}/\partial r$, so that

$$\bar{A}_\alpha = \left(-\frac{\partial\bar{\chi}^{(s)}}{\partial r} \right)_{z=+0} - \left(-\frac{\partial\bar{\chi}^{(s)}}{\partial r} \right)_{z=-0} . \tag{3.1.43}$$

However, as $\bar{A}_\alpha = - \partial\bar{D}/\partial r$, we have

$$\bar{D} = \bar{\chi}_{(z=+0)}^{(s)} - \bar{\chi}_{(z=-0)}^{(s)} \tag{3.1.44}$$

i.e., $\bar{D}$ yields in fact the mmf.

Using (3.1.38), we have for the inner zone

$$\bar{A}_\alpha = -\frac{\partial\bar{D}}{\partial r} = \sigma\Delta \frac{i\omega\bar{\phi}_{tr}}{2\pi R} \frac{J_1(\sqrt{i\omega\mu_0\sigma\Delta/2br})}{J_1(\sqrt{i\omega\mu_0\sigma\Delta/2bR})} . \tag{3.1.45}$$

Integrating, resorting to the notation (3.1.25) and substituting $r = R$, we have

$$\bar{D} = \left(\frac{\bar{B}_0}{\mu_0} 2b \right) \frac{(i/r_0bR)J_0(\sqrt{i/r_0bR})}{2J_1(\sqrt{i/r_0bR})} . \tag{3.1.46}$$

Now $D_0 \equiv (\bar{B}_0/\mu_0)2b$ is the mmf required for producing (under conditions of steady-state dc excitation) the space-mean magnetic induction $\bar{B}_0 = B_0$ in the air gap; hence defining by $\xi \equiv R/\sqrt{r_0b}$ the dimensionless pole radius, we find that the dimensionless "inner" complex mmf $\bar{\eta}_i = \bar{D}/D_0$ may succinctly be expressed by the dependence

$$\bar{\eta}_i = i \frac{\xi}{2} \frac{J_0(\sqrt{i}\xi)}{\sqrt{i}J_1(\sqrt{i}\xi)} , \tag{3.1.47}$$

which is reproduced in Fig. 3.3.

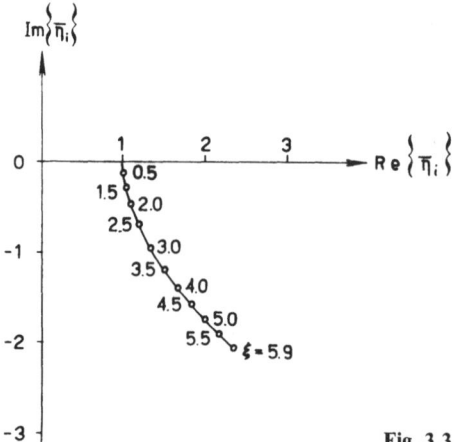

Fig. 3.3. Complex mmf (dimensionless) for inner zone

For the outer zone $r > R$, by resorting to (3.1.27 and 40), and substituting $r = R$, we obtain

$$\bar{D} = -\frac{\bar{\phi}_{tr}}{\pi\mu_0}\frac{1}{r_0}\frac{\pi}{2}\left\{-i\cdot iH_0^{(1)}\left(i\frac{R}{r_0}\right)+\left[I_0\left(\frac{R}{r_0}\right)-L_0\left(\frac{R}{r_0}\right)\right]\right\}$$

$$\times\left(\frac{\pi}{2}\frac{R}{r_0}\left\{-H_1^{(1)}\left(i\frac{R}{r_0}\right)+\frac{1}{i}\left[I_1\left(\frac{R}{r_0}\right)-L_1\left(\frac{R}{r_0}\right)-\frac{2}{\pi}\right]\right\}\right)^{-1}$$

(3.1.48)

Defining now $R/r_0 \equiv \zeta$, the dimensionless complex "outer" mmf $\bar{\eta}_x$ is obtained as

$$\bar{\eta}_x = -\frac{-i\cdot iH_0^{(1)}(i\zeta)+[I_0(\zeta)-L_0(\zeta)]}{-H_1^{(1)}(i\zeta)+\dfrac{1}{i}[I_1(\zeta)-L_1(\zeta)-2/\pi]}\,;$$

(3.1.49)

this expression is reproduced in Fig. 3.4.

Hence, given the values of $\sigma\Delta$, ω, R and b, the total complex mmf in the exciting coil (amplitude and phase) is obtained by adding the "inner" and "outer" mmfs obtainable from (3.1.47,49) in conjunction with Figs. 3.3 and 4.

3.1.10 Discussion

An analytical approach to calculation of the exciting mmf of an electromagnet inducing a plane armature has been provided. Although highly simplifying assumptions have been made, the results obtained enable us to visualize the

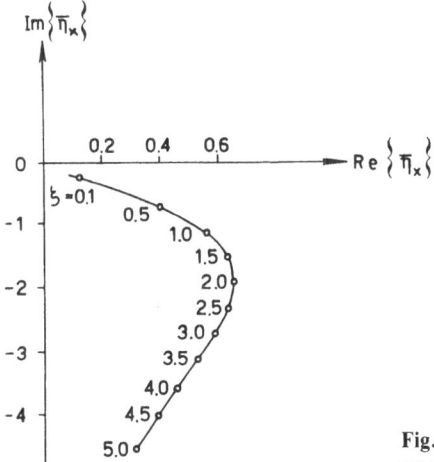

Fig. 3.4. Complex mmf (dimensionless) for outer zone

influence of the parameters of the structure on the final result. In practice, the coil current is often known, while the magnetic induction has to be determined. Under elementary approaches it is assumed that the current and induction are in phase, thereby disregarding the plate reaction. The present analysis proves that such an approximation is valid only for small values of the dimensionless pole radius ξ.

3.1.11 Appendix

Evaluation of Circular Surface Current

Rewriting (3.1.26) in the form

$$\bar{D} = \frac{\bar{\phi}}{\pi\mu_0}\, i\, \frac{1}{r_0} \int\limits_{\lambda=0}^{\infty} \frac{[\lambda + i(1/r_0)]\, J_0(\lambda r)}{\lambda^2 + 1/r_0^2}\, d\lambda \ , \qquad (3.1.50)$$

we obtain the two integrals

$$S_1 = \int\limits_{\lambda=0}^{\infty} \frac{\lambda J_0(\lambda r)}{\lambda^2 + (1/r_0^2)}\, d\lambda; \quad S_2 = \int\limits_{\lambda=0}^{\infty} \frac{J_0(\lambda r)}{\lambda^2 + (1/r_0^2)}\, d\lambda \ . \qquad (3.1.51)$$

which, when evaluated [Ref. 3.1, No. 6.566-2; No. 6.566-5, pp. 687 – 688] yield

$$S_1 = K_0\left(\frac{r}{r_0}\right); \quad S_2 = \frac{\pi}{2(1/r_0)}\left[I_0\left(\frac{r}{r_0}\right) - L_0\left(\frac{r}{r_0}\right)\right] \ . \qquad (3.1.52)$$

Here K_0 stands for the modified Bessel function of second kind and order zero, and may be expressed by means of the Hankel function, i.e.

$$K_0\left(\frac{r}{r_0}\right) = \frac{\pi}{2} i H_0^{(1)}\left(i\frac{r}{r_0}\right) \ . \qquad (3.1.53)$$

Further, I_0 stands for the modified Bessel function of the first kind and order zero, and L_0 for the modified Struve function of order zero. Thus

$$\bar{D} = \frac{\bar{\phi}}{\pi\mu_0}\, i\, \frac{1}{r_0}\left(S_1 + \frac{i}{r_0} S_2\right) \qquad (3.1.54)$$

$$= \frac{\bar{\phi}}{\pi\mu_0}\, i\, \frac{1}{r_0}\left\{\frac{\pi}{2} i H_0^{(1)}\left(i\frac{r}{r_0}\right) + \frac{i}{r_0}\frac{\pi}{2(1/r_0)}\left[I_0\left(\frac{r}{r_0}\right) - L_0\left(\frac{r}{r_0}\right)\right]\right\}$$

$$= -\frac{\bar{\phi}}{\pi\mu_0}\, \frac{1}{r_0}\frac{\pi}{2}\left\{-i\cdot i H_0^{(1)}\left(i\frac{r}{r_0}\right) + \left[I_0\left(\frac{r}{r_0}\right) - L_0\left(\frac{r}{r_0}\right)\right]\right\} \ , \quad (3.1.55)$$

and therefore

$$\bar{A}_\alpha = -\frac{\partial \bar{D}}{\partial r} = +\frac{\bar{\phi}}{\pi \mu_0} \frac{1}{r_0} \frac{\pi}{2} \left\{ -\mathrm{i} \cdot \mathrm{i} \frac{\mathrm{i}}{r_0} \left[-\mathrm{H}_1^{(1)} \left(\mathrm{i} \frac{r}{r_0} \right) \right] \right.$$

$$\left. +\frac{1}{r_0} \left[\mathrm{I}_1 \left(\frac{r}{r_0} \right) - \mathrm{L}_1 \left(\frac{r}{r_0} \right) - \frac{2}{\pi} \right] \right\} . \tag{3.1.56}$$

It should be noted that (3.1.55, 56) are written out in a form which permits direct analytical insight into their properties – even without recourse to a computer.

3.2 Operation of Eddy Current Probe Coil

Non-destructive measurement of electrical conductivity is a well-known tool of modern technology; for this purpose, a magnetic field analyzer comprising, e.g., a Hall detector is quite often used, or, alternatively a follow up of impedance changes in an exciting coil is resorted to. In both instances – in principle at least – an exciting coil is placed near the investigated conductor whose influence is manifested by either field or impedance variations.

Although non-destructive measurement is widespread, closed-form expressions – mainly useful in the preparatory stage of the test, for preliminary estimations – are not always available. It is the aim of the present section to put forward some relevant formulae.

3.2.1 Basic Assumptions and Statement of Problem

A flat, closely-wound coil of mean radius b and a vanishingly small electrical resistance is located (in air) at an elevation h above an "infinite" plane non-magnetic conducting sheet of thickness Δ and electrical conductivity σ (Fig. 3.5). The coil is supplied by an imposed alternating current

$$I(t) = \bar{I} \mathrm{e}^{-\mathrm{i}\omega t} , \tag{3.2.1}$$

where $\bar{I}$ denotes a phasor, as per previous notation.

Assuming magneto-quasistatics, what is the interaction between the coil and the sheet?

As already mentioned, within the framework of magneto-quasistatics, the electric energy density pertaining to the investigated configuration is vanishingly small when compared to its magnetic counterpart; in mathematical terms, this implies that the "air" surrounding the coil has been replaced by a fictitious medium with zero dielectric constant while nevertheless exhibiting a finite magnetic permeability μ_0.

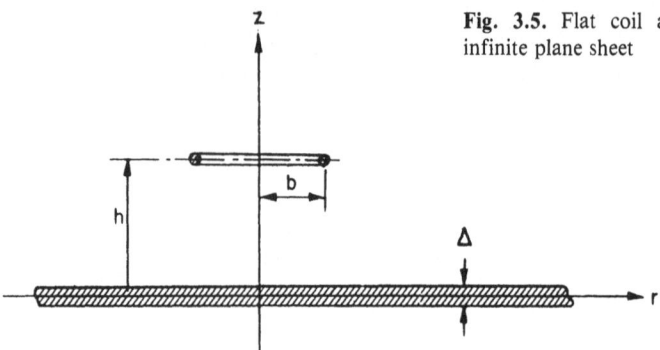

Fig. 3.5. Flat coil at elevation h above infinite plane sheet

The following models are now introduced:

– The *coil* is replaced by an oscillating magnetic dipole of strength $m(t)$:

$$m(t) = \bar{M}e^{-i\omega t} , \qquad (3.2.2)$$

where the phasor $\bar{M}$ is defined by the limiting operation

$$\bar{M} = \lim_{\substack{b \to 0 \\ |\bar{I}| \to \infty}} (\mu_0 \pi b^2 \bar{I}) . \qquad (3.2.3)$$

– The *conductor* is replaced by a mathematical sheet of plane conductivity $\varkappa$, defined by the limiting process to which we have by now quite often resorted, i.e.

$$\varkappa = \lim_{\substack{\sigma \to \infty \\ \Delta \to 0}} (\sigma \Delta) = \text{finite} . \qquad (3.2.4)$$

In the sequel we employ a cylindrical system of coordinates z, r, α, having its axis coincident with that of the coil and its origin at the point of intersection of this axis and the conducting plane: further, in order to facilitate the solution, we again define two electromagnetic fields: the primary (superscript p), due to the dipole alone, and the secondary (superscript s), due to sheet current only.

3.2.2 Primary Field

Within the framework of magneto-quasistatics, the curl of the primary magnetic field H vanishes in air, i.e.

$$\nabla \times H^{(\mathrm{p})} = 0 . \qquad (3.2.5)$$

Accordingly, $H^{(\mathrm{p})}$ is obtainable in this region by means of the relation $H^{(\mathrm{p})} = -\nabla \chi^{(\mathrm{p})}$, where the complex amplitude $\bar{\chi}^{(\mathrm{p})}$ of the primary magnetic scalar potential is in turn given by the equation

$$\bar{\chi}^{(\mathrm{p})} = -\frac{\bar{M}}{4\pi\mu_0} \frac{\partial}{\partial z} \left(\frac{1}{\sqrt{r^2 + (z-h)^2}} \right) . \qquad (3.2.6)$$

The complex amplitude of the radial component of the primary, imposed magnetic field is thus

$$\bar{H}_r^{(p)} = \frac{\bar{M}}{4\pi\mu_0} \frac{3(z-h)r}{[(z-h)^2+r^2]^{5/2}} \quad , \tag{3.2.7}$$

and that of its axial component:

$$\bar{H}_z^{(p)} = \frac{\bar{M}}{4\pi\mu_0} \frac{2(z-h)^2-r^2}{[(z-h)^2+r^2]^{5/2}} \quad . \tag{3.2.8}$$

These equations, in conjunction with the induction law yield the complex amplitude of the primary electric field $E^{(p)}$. In view of the assumed form of the excitation, only its circular component $\bar{E}_\alpha^{(p)}$ differs appreciably from zero; hence from Faraday's law

$$-\frac{\partial \bar{E}_\alpha^{(p)}}{\partial z} = i\omega\mu_0\bar{H}_r^{(p)} \quad , \quad \text{and} \tag{3.2.9}$$

$$\frac{1}{r}\frac{\partial}{\partial r}(r\bar{E}_\alpha^{(p)}) = i\omega\mu_0\bar{H}_z^{(p)} \quad , \tag{3.2.10}$$

so that

$$\bar{E}_\alpha^{(p)} = \frac{\bar{M}}{4\pi} i\omega \frac{r}{[(z-h)^2+r^2]^{3/2}} \quad . \tag{3.2.11}$$

3.2.3 Secondary Field

In the plane sheet, the primary field induces surface currents which, in turn, are linked to the secondary electromagnetic field. Maxwell's equations, adapted to this field in the so-called "air" surrounding the coil, read

$$\nabla \times H^{(s)} = 0 \quad ,$$

$$\nabla \times E^{(s)} = -\frac{\partial}{\partial t}\mu_0 H^{(s)} \quad ,$$

i.e. $\nabla \times (\nabla \times E^{(s)}) = 0$. Temporarily introducing Cartesian coordinates, we may write unambiguously:

$$\nabla(\nabla \cdot E^{(s)}) - \nabla^2 E^{(s)} = 0 \quad . \tag{3.2.12}$$

Having resorted to magneto-quasistatics, our electric field is source-free (Sect. 2.2.8 A), i.e. $\nabla \cdot E^{(s)} = 0$; hence (3.2.12) reduces to

$$\nabla^2 E^{(s)} = 0 \quad . \tag{3.2.13}$$

Transforming now $E^{(s)}$ to the cylindrical coordinate system, with only the circular component $E_\alpha^{(s)}$ again appreciably different from zero, one obtains that its complex amplitude $\bar{E}_\alpha^{(s)}$ satisfies the partial differential equation

$$\frac{\partial^2 \bar{E}_\alpha^{(s)}}{\partial r^2} + \frac{1}{r}\frac{\partial \bar{E}_\alpha^{(s)}}{\partial r} - \frac{\bar{E}_\alpha^{(s)}}{r^2} + \frac{\partial^2 \bar{E}_\alpha^{(s)}}{\partial z^2} = 0 \ . \tag{3.2.14}$$

The solution of this equation – subject to convergence requirements for $r\to 0$, $r\to\infty$ and $|z|\to\infty$ – is expressed by an integral comprising the dummy variable k; this involves a still undetermined (complex) "weighting" function $\bar{f}(k)$ along with a Bessel function of first kind and order one, with argument (kr), i.e.

$$\bar{E}_\alpha^{(s)} = \frac{\bar{M}}{4\pi}\,\mathrm{i}\omega \int\limits_{k=0}^{\infty} \bar{f}(k)\,\mathrm{J}_1(kr)\,\mathrm{e}^{\mp kz}\,k\,dk; \quad z\gtrless 0 \ . \tag{3.2.15}$$

Adapting (3.2.9, 10) to the secondary field, we obtain

$$\bar{H}_r^{(s)} = \pm \frac{\bar{M}}{4\pi\mu_0}\int\limits_{k=0}^{\infty}\bar{f}(k)\,\mathrm{J}_1(kr)\,\mathrm{e}^{\mp kz}k^2\,dk, \quad z\gtrless 0 \ . \tag{3.2.16}$$

$$\bar{H}_z^{(s)} = \frac{\bar{M}}{4\pi\mu_0}\int\limits_{k=0}^{\infty}\bar{f}(k)\,\mathrm{J}_0(kr)\,\mathrm{e}^{\mp kz}k^2\,dk, \quad z\gtrless 0 \ . \tag{3.2.17}$$

The discontinuity in $\bar{H}_r^{(s)}$ at the sheet surface $z = 0$ is compensated by the surface current $\varkappa\bar{E}_\alpha(r,0)$ wherein $\bar{E}_\alpha = \bar{E}_\alpha^{(p)} + \bar{E}_\alpha^{(s)}$. Thus

$$\bar{H}_r^{(s)}(r = +0) - \bar{H}_r^{(s)}(r = -0) = \varkappa\bar{E}_\alpha(r;0) \ . \tag{3.2.18}$$

In order to proceed, (3.2.11) must now be transformed along the lines of (3.2.15). Resorting to

$$\frac{1}{\sqrt{(z-h)^2+r^2}} = \int\limits_{k=0}^{\infty}\mathrm{J}_0(kr)\,\mathrm{e}^{\mp k(z-h)}\,dk; \quad (z-h)\gtrless 0 \ , \tag{3.2.19}$$

and taking advantage of the fact that

$$\frac{\partial}{\partial r}\frac{1}{\sqrt{(z-h)^2+r^2}} = -\frac{1}{[(z-h)^2+r^2]^{3/2}} \ ,$$

we have

$$\bar{E}_\alpha^{(p)} = \frac{\bar{M}}{4\pi}\,\mathrm{i}\omega\int\limits_{k=0}^{\infty}\mathrm{J}_1(kr)\,\mathrm{e}^{\mp k(z-h)}k\,dk; \quad (z-h)\gtrless 0 \ . \tag{3.2.20}$$

Applying now the boundary condition (3.2.18), the expressions (3.2.15, 16 and 20) yield

$$\bar{f}(k) = \mathrm{e}^{-kh}\left(\frac{2k}{\mathrm{i}\omega\varkappa\mu_0} - 1\right)^{-1} \ . \tag{3.2.21}$$

3.2.4 Secondary Axial Field at Coil Center

One of the more practical test instruments initially developed was based upon a small-area Hall detector centrally located inside the exciting coil; the purpose of the Hall probe was to determine the magnetic field variation in the presence of the test specimen. By using (3.2.17,21), the secondary axial field component at $r = 0$, $z = h$ due to this specimen reads

$$\bar{H}_z^{(s)} = \frac{\bar{M}}{4\pi\mu_0} \int_{k=0}^{\infty} \frac{e^{-2kh}}{\dfrac{2k}{i\omega\varkappa\mu_0} - 1} k^2 \, dk, \qquad z = h, \quad r = 0 \ . \qquad (3.2.22)$$

Introducing now the characteristic angular frequency

$$\omega_0 \equiv \frac{1}{\varkappa\mu_0 h} \qquad (3.2.23)$$

along with the dimensionless parameters $\lambda \equiv kh$, $\nu \equiv \omega/\omega_0$, (3.2.22) reduces, on application of (3.2.3), to

$$\bar{H}_z^{(s)} = \frac{b^2\bar{I}}{4h^3} \int_{\lambda=0}^{\infty} \frac{e^{-2\lambda}}{\dfrac{2\lambda}{i\nu} - 1} \lambda^2 \, d\lambda, \qquad z = h, \quad r = 0 \ . \qquad (3.2.24)$$

Accordingly, the dimensionless complex amplitude of the secondary axial field component $\bar{h}_0 = \bar{H}_z^{(s)}/(b^2\bar{I}/4h^3)$ at the coil center is given (Sect. 3.2.7 A) in terms of the dimensionless frequency ν, by

$$\bar{h}_0 = \frac{\nu^3}{8} \left\{ [\text{ci}(\nu)\sin\nu - \text{si}(\nu)\cos\nu] - \frac{1}{\nu} \right\}$$

$$+ i \frac{\nu^3}{8} \left\{ [\text{ci}(\nu)\cos\nu + \text{si}(\nu)\sin\nu] + \frac{1}{\nu^2} \right\} \qquad (3.2.25)$$

with $\text{si}(x) = -\int_x^{\infty} \dfrac{\sin t}{t} \, dt$; $\text{ci}(x) = -\int_x^{\infty} \dfrac{\cos t}{t} \, dt$.

This relation is reproduced in Fig. 3.6.

3.2.5 Change of Coil Impedance

As already pointed out, the impedance variation of a probe coil, on being brought near a conductor, similarly yields information on the test specimen. This variation may be determined as follows:

Eddy-currents in the sheet are linked to a secondary circular electric field component, represented by the phasor $\bar{E}_0^{(s)}$, in the coil plane $z = h$; the

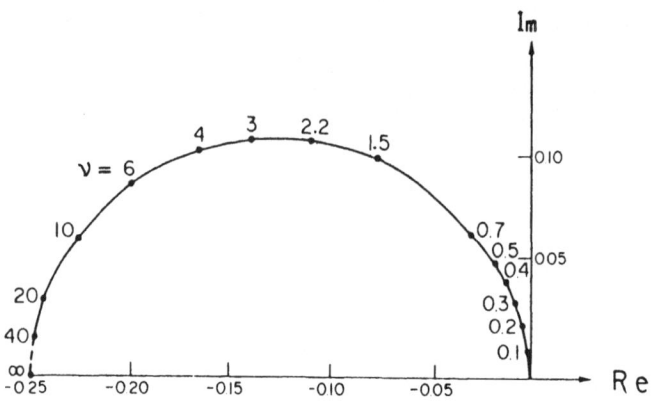

Fig. 3.6. Vector locus of axial magnetic field variation (dimensionless units) at coil center. The dotted part represents very high frequencies, precluded from the magneto-quasistatic approximation

phasor $\bar{V}^{(s)}$ of the secondary circumferential voltage[1] produced in the coil is given by

$$\bar{V}^{(s)} = (2\pi b)\bar{E}_0^{(s)} . \tag{3.2.26}$$

Accordingly, an additional voltage $-\bar{V}^{(s)}$, must be provided by the constant-current generator feeding the excitation coil, at the same current level $\bar{I}$, in order to compensate for the voltage variation at the terminals of the zero-resistance coil.

Defining the impedance variation ΔZ as the ratio

$$\Delta Z = \frac{-\bar{V}^{(s)}}{\bar{I}} , \tag{3.2.27}$$

and using (3.2.15, 21) along with (3.2.26), we have

$$\Delta Z = -\frac{2\pi b}{\bar{I}} \frac{\bar{M}}{4\pi} i\omega \int_{k=0}^{\infty} \frac{e^{-2kh}}{2k/(i\omega\varkappa\mu_0)-1} J_1(kb)k\,dk . \tag{3.2.28}$$

Resorting once more to the dimensionless parameters $\lambda \equiv kh$ and $\nu \equiv \omega/\omega_0$, together with $\beta \equiv b/h$, and introducing now the characteristic resistance

$$R_0 \equiv \frac{\pi b^3}{2\varkappa h^3} , \tag{3.2.29}$$

---

[1] We use here the term "circumferential voltage" as an ad-hoc rendering of the the German term "Umlaufspannung" defined over a closed contour l (differential increment $d\boldsymbol{l}$) by $\bar{V} \equiv \oint \boldsymbol{E} \cdot d\boldsymbol{l}$ $= +i\omega\bar{\phi}$; as usual, $\bar{\phi}$ stands for the complex amplitude of the overall magnetic flux.

we obtain, by (3.2.3), that

$$\frac{\varDelta Z}{R_0} = 2v^2 \int\limits_{\lambda=0}^{\infty} \frac{\lambda^2 e^{-2\lambda} J_1(\beta\lambda)}{4\lambda^2 + v^2}\, d\lambda + i v^3 \int\limits_{\lambda=0}^{\infty} \frac{\lambda e^{-2\lambda} J_1(\beta\lambda)}{4\lambda^2 + v^2}\, d\lambda \ . \quad (3.2.30)$$

For small values of $\beta = b/h$ (dipole approximation!) we retain the first term in the series expansion of the Bessel function $J_1(\beta\lambda)$ and find for the dimensionless impedance increment $\varDelta z = \varDelta Z/R_0$ the expression (Sect. 3.2.7B):

$$\varDelta z = \frac{1}{16}\, v^2 \beta \{v^2[\mathrm{ci}(v)\cos v + \mathrm{si}(v)\sin v] + 1\}$$

$$+ i\frac{1}{16}\, v^3 \beta \{-v[\mathrm{ci}(v)\sin v - \mathrm{si}(v)\cos v] + 1\} \ . \quad (3.2.31)$$

The vector locus of the impedance variation for different values of β and v is shown in Fig. 3.7. As expected, both the real and imaginary of $\varDelta z$ are positive; the net effect on the coil reactance is, however, a decrease, because $\mathrm{Im}\{\varDelta z\}$ must be divided in the present case by $(-i\omega)$.

3.2.6 Discussion

Use of probe coils for non-destructive testing of conducting materials is based on variation of a magnetic field or, alternatively, on that of the coil

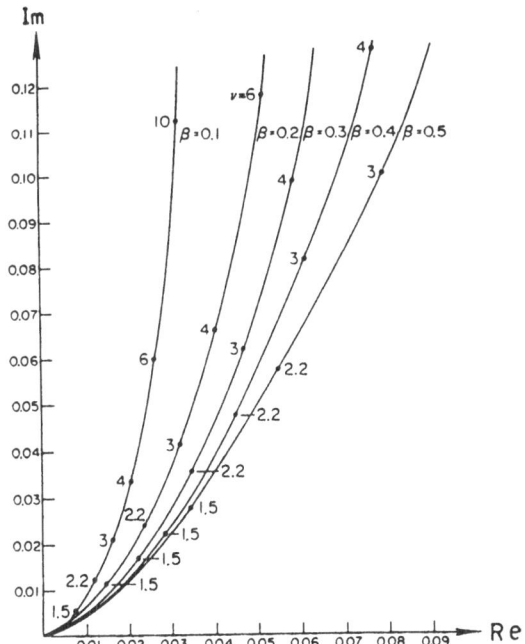

Fig. 3.7. Vector locus of impedance variation (dimensionless units) for different values of the dimensionless radius $\beta = b/h$

impedance; both effects are caused by eddy-currents excited within the test specimen.

In the present section, simple closed-form analytical expressions for these two incremental changes are put forward; it is of some interest to note, compare also Sect. 3.3.8 B, that in (3.2.25) as well as in (3.2.31) the "auxiliary functions" $f(z) \equiv \mathrm{ci}(z)\sin z - \mathrm{si}(z)\cos z$ and $g(z) \equiv -\mathrm{ci}(z)\cos z - \mathrm{si}(z)\sin z$ occur.

3.2.7 Appendix

Performance of Integrations

A) The integral

$$\mathrm{Int}_1 = \int_{\lambda=0}^{\infty} \frac{e^{-2\lambda}}{2\lambda/i\nu - 1}\,\lambda^2 d\lambda \tag{3.2.32}$$

is resolved in two terms: i.e.

$$\mathrm{Int}_1 = -\frac{\nu^2}{4}\int_{\lambda=0}^{\infty}\frac{\lambda^2 e^{-2\lambda}}{\lambda^2+\nu^2/4}\,d\lambda + i2\frac{\nu}{4}\int_{\lambda=0}^{\infty}\frac{\lambda^3 e^{-2\lambda}}{\lambda^2+\nu^2/4}\,d\lambda\ , \tag{3.2.33}$$

which are in turn evaluated by means of the relations [Ref. 3.1, No. 3.356-1; No. 3.356-2, p. 313]

$$\int_{0}^{\infty}\frac{x^{2n+1}e^{-px}}{a^2+x^2}\,dx = (-1)^{n-1}a^{2n}[\mathrm{ci}(ap)\cos(ap)+\mathrm{si}(ap)\sin(ap)]$$

$$+\frac{1}{p^{2n}}\sum_{k=1}^{n}(2n-2k+1)!\,(-a^2 p^2)^{k-1},\quad p>0 \tag{3.2.34}$$

$$\int_{0}^{\infty}\frac{x^{2n}e^{-px}}{a^2+x^2}\,dx = (-1)^{n}a^{2n-1}[\mathrm{ci}(ap)\sin(ap)-\mathrm{si}(ap)\cos(ap)]$$

$$+\frac{1}{p^{2n-1}}\sum_{k=1}^{n}(2n-2k)!\,(-a^2 p^2)^{k-1},\quad p>0\ . \tag{3.2.35}$$

Hence (3.2.25).

B) On substitution of the dimensionless parameters $\lambda = kh$, $\nu = \omega/\omega_0$ and $\beta = b/h$ (3.2.28) reads

$$\Delta Z = -\frac{2\pi b}{\bar{I}}\frac{\mu_0\pi b^2\bar{I}}{4\pi}i\omega\frac{1}{h^2}\int_{\lambda=0}^{\infty}\frac{e^{-2\lambda}}{\dfrac{2\lambda}{i\nu}-1}J_1(\beta\lambda)\lambda\,d\lambda\ . \tag{3.2.36}$$

Now

$$\frac{2\pi b}{\bar{I}}\frac{\mu_0\pi b^2\bar{I}\omega}{4\pi h^2} = \frac{\pi b\mu_0 b^2\omega\varkappa h}{2h^3\varkappa} = \frac{(\varkappa\mu_0 h\omega)\pi b^3}{2h^3\varkappa} = \frac{\nu\pi b^3}{2\varkappa h^3} \equiv \nu R_0 \tag{3.2.37}$$

where

$$R_0 \equiv \frac{1}{\varkappa} \frac{\pi b^3}{2h^3} \tag{3.2.38}$$

is a characteristic resistance.

The dimensionless impedance $\Delta z = \Delta Z / R_0$ is therefore expressed by

$$\Delta z = -iv \int_{\lambda=0}^{\infty} \frac{e^{-2\lambda}}{2\lambda/iv-1} J_1(\beta\lambda) \lambda \, d\lambda = 2v^2 \int_{\lambda=0}^{\infty} \frac{\lambda^2 e^{-2\lambda} J_1(\beta\lambda)}{4\lambda^2 + v^2} \, d\lambda$$

$$+ iv^3 \int_{\lambda=0}^{\infty} \frac{\lambda e^{-2\lambda} J_1(\beta\lambda)}{4\lambda^2 + v^2} \, d\lambda \ . \tag{3.2.39}$$

Approximating $J_1(\beta\lambda)$ by the first term of its series expansion, we obtain

$$\Delta z = \frac{v^2}{4} \beta \int_{\lambda=0}^{\infty} \frac{\lambda^3 e^{-2\lambda}}{\lambda^2 + v^2/4} \, d\lambda + i \frac{v^3}{8} \beta \int_{\lambda=0}^{\infty} \frac{\lambda^2 e^{-2\lambda}}{\lambda^2 + v^2/4} \, d\lambda \tag{3.2.40}$$

or, resorting to (3.2.34, 35) we finally obtain

$$\Delta z = \frac{v^4 \beta}{16} \left[\mathrm{ci}(v) \cos(v) + \mathrm{si}(v) \sin(v) + \frac{1}{v^2} \right]$$

$$+ i \frac{v^4 \beta}{16} \left[-\mathrm{ci}(v) \sin(v) + \mathrm{si}(v) \cos(v) + \frac{1}{v} \right] \ . \tag{3.2.41}$$

3.3 Proximity Effect Between a Plane Metal Screen and a Rectilinear Current Conductor

Rectangular metal cases are often used as enclosures for heavy-current bus bars or connections (a common American practice being the separate ducting for each phase). In addition to the protection they afford against accidental contacts and dust collection, the currents generated in them by electromagnetic induction give rise to magnetic fields, which offset the local exciting field surrounding the phases, thereby weakening the overall interaction between adjacent phases. This relegation of dynamic effects from the phases to the enclosure makes for improved mechanical reliability.

While the design engineer working on the overall layout is concerned with dynamic forces, the electrical engineer dealing with the protective circuitry is also interested in proximity effects; a typical case is the Merz-Price circulating current system, based on current transformers and widely applied in bus-bar protection: even though an equalized current transformer arrangement may by quite satisfactory under steady-state through-fault conditions, significant imbalance sets in under transient surges, with an obvious attendant contribution by a conducting screen in close proximity.

To gain some idea of the field problems encountered in such systems, we shall consider a relevant first-order approximation − namely a plane metal screen in the presence of an extended parallel conductor[2]. The solution of our problem − which we phrase as follow: "What is the proximity effect between a rectilinear current-carrying wire and a thin plane metal sheet?" − is also relevant to the behavior of ground wires and counterpoises [e.g. Ref. 3.2].

3.3.1 Formulation of Problem

A very long ("infinite") wire of circular cross section (diameter $2r_0$) carries an alternating current whose return path is provided by a conducting plane sheet of uniform thickness Δ located at a distance $h \gg r_0$. The sheet exhibits an electric conductivity σ and a magnetic permeability μ_0 and extends infinitely along the $\pm x$ and $\pm y$ axes in a rectangular frame of reference x, y, z (Fig. 3.8) incorporating also the time parameter t. We seek:

a) The per-unit length equivalent impedance change of the wire, due to the finite conductivity of the metal sheet.
b) The flow pattern of the induced current in the extended conductor under an applied current step in the wire.

To simplify the problems posed and their relevant solutions, we shall assume that the wire, characterized by an "infinite" conductivity, comprises a continuous array of internal current generators providing the required excitation (Fig. 3.9).

In practice, the statement that the current-carrying line is "infinite" implies that its length l far exceeds the span h. On the other hand, in power lines, for instance, l is much smaller than the steady-state excitation wavelength λ; specifically, the electrical "shortness" will here be allowed for by disregarding the wire-screen mutual capacitance: the problem is thus relegated to the realm of magneto-quasistatics, allowing us once more to assign to the free space surrounding the wire a fictitious zero permittivity; the conductivity σ_0 of this space is also taken to be vanishingly small compared to that of the return ("ground") plane, so that it exhibits a finite value for its magnetic permeability μ_0 alone.

---

[2] Early studies of the interaction between a plane conductor and an exciting current conductor concerned overhead lines and buried cables [3.3 − 6], or the influence of the adjacent ground on antenna performance and efficiency [3.7 − 11]; similar problems were also dealt with more recently, albeit from somewhat different points of view [3.12 − 15]. These cited research memoirs are of theoretical importance and inherent interest, but only partially applicable to the problem at hand, either because of the complexity of the approaches and results, or because of the underlying assumptions (such as semi-infinite extension of conducting or layered media [3.16] or, alternatively, electrically balanced current meshes).

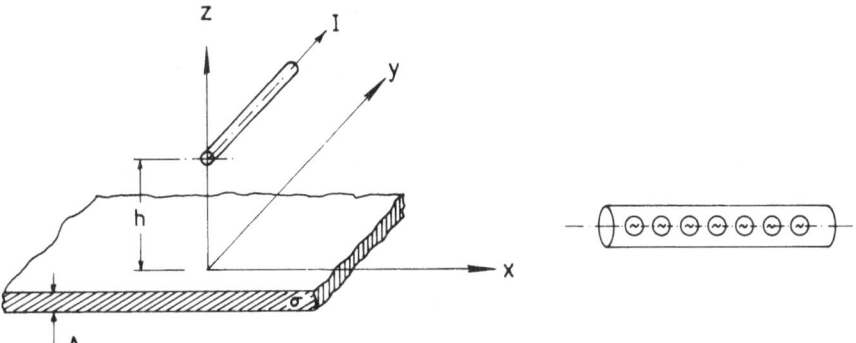

Fig. 3.8. Investigated system

Fig. 3.9. Continuous array of internal generators

Quasistatics will also be postulated for analysis of the transient state: denoting as usual the phase velocity of a plane wave of light in vacuum by c, we assume that the propagation time $\tau_{em} = h/c$ of an electromagnetic disturbance passing from the wire to the screen is negligible relative to some characteristic time of the system T_c. The latter is readily provided by dimensional considerations, namely $T_c \equiv \mu_0 \sigma \Delta h$; our assumption thus implies the inequality $h/c \ll \mu_0 \sigma \Delta h$, or, in terms of the free-space permittivity – $\sqrt{\varepsilon_0/\mu_0} \ll \sigma \Delta$, which is (obviously) satisfied for the fictitious case $\varepsilon_0 = 0$ [3].

3.3.2 Primary Field

The excitation current of complex amplitude $\bar{I}$, alternating at the monochromatic angular frequency ω, is assumed to be confined to a very narrow strip of width, $2\delta = 2r_0$, i.e. the circular cross-section wire is mentally replaced by a flat conductor.

We now choose to represent the complex amplitude $\bar{A}(x)$ of the relevant surface current density $A(x)$ via the Fourier integral

$$\bar{A}(x) = \frac{1}{2\pi} \int_{-\infty}^{\infty} e^{-imx} dm \int_{-\delta}^{\delta} \frac{\bar{I}}{2\delta} e^{im\zeta} d\zeta , \quad \text{i.e.} \tag{3.3.1}$$

$$\bar{A}(x) = \frac{1}{2\pi} \int_{-\infty}^{\infty} e^{-imx} dm \left(\bar{I} \frac{\sin m\delta}{m\delta} \right) . \tag{3.3.2}$$

---

[3] The assumption of quasistatics can also be justified in the context of the transient surge due to lightning, the energy of which has been empirically shown to peak in a frequency region around 10 kHz [3.17]. On the other hand, assuming, e.g. [3.18], $\sigma = 1 \times 10^{-2} (\Omega m)^{-1}$, $\Delta = 1$ m, for a wire span of $h \simeq 40$ m, we find $T_c \simeq 0.5 \times 10^{-6}$ s, which corresponds to a characteristic frequency ($f_c = 1/T_c$) in the range of 1 MHz. Still, it is obvious that the "legality" of such an approach, which in effect reduces the number of the system's degrees of freedom, should be checked in each case against experimental data.

where m and ζ play the role of dummy variables; as $\delta \to 0$ one formally has

$$\bar{A}(x) = \frac{\bar{I}}{\pi} \int_0^\infty \cos mx \, dm \ , \tag{3.3.3}$$

which, although in itself non-integrable (due to the artificial assumption of $\delta \to 0$ for finite $\bar{I}$), enables us to provide closed-form analytical solutions to the problem under consideration.

As already repeatedly stated, within the framework of magneto-quasi-statics Maxwell's first curl equation implies that the magnetic field in the current-free regions is vortex-free, i.e.,

$$\nabla \times \boldsymbol{H} = 0 \ . \tag{3.3.4}$$

Hence, the primary magnetic field $\boldsymbol{H}^{(p)}$, which is linked exclusively to the excitation current, may be derived from a primary magnetic scalar potential $\chi^{(p)}$ via

$$\boldsymbol{H}^{(p)} = -\nabla\chi^{(p)} \ . \tag{3.3.5}$$

Now $\chi^{(p)}$ must satisfy the Laplace equation and, moreover, vanish for $|z| \to \infty$. Hence, the complex amplitude $\bar{\chi}^{(p)}$, compatible on the one hand with the solution $\chi^{(p)} = \bar{\chi}^{(p)}\exp(-i\omega t)$ linked to the excitation (3.3.3), and on the other hand, with the above requirements, may be represented through a still-undetermined spectral density $\bar{f}_p(m)$:

$$\bar{\chi}^{(p)} = \pm\frac{\bar{I}}{\pi} \int_0^\infty \bar{f}_p(m) \sin mx \, e^{\mp m(z-h)} dm \ , \qquad z \gtrless h \ . \tag{3.3.6}$$

Expressing the primary x-directed magnetic field component $H_x^{(p)}$ through its complex amplitude $\bar{H}_x^{(p)}$ we obtain

$$\bar{H}_x^{(p)} = -\frac{\partial\bar{\chi}^{(p)}}{\partial x} = \mp\frac{\bar{I}}{\pi} \int_0^\infty \bar{f}_p(m) \, m \cos mx \, e^{\mp m(z-h)} dm \ , \qquad z \gtrless h \ . \tag{3.3.7}$$

The boundary condition associated with the magnetic field discontinuity reads here

$$\bar{A}_x = \bar{H}_x(z = h+0) - \bar{H}_x(z = h-0) \ , \qquad z = h \ , \tag{3.3.8}$$

so that (3.3.3, 7) yield

$$\bar{f}_p(m) = -\frac{1}{2m} \ . \tag{3.3.9}$$

We therefore obtain

$$\bar{\chi}^{(p)} = \mp\frac{\bar{I}}{2\pi} \int_0^\infty \frac{\sin mx}{m} e^{\mp m(z-h)} dm \ , \qquad z \gtrless h \ . \tag{3.3.10}$$

3.3.3 Secondary Field

In the sequel, we shall again label as "secondary" (superscript s) the field linked to the current in the metal sheet. We assume for its y-directed (complex) surface distribution $\bar{A}^{(s)}(x)$ the dependence

$$\bar{A}^{(s)}(x) = \frac{\bar{I}}{\pi} \int_0^\infty \bar{a}(m) \cos mx \, dm \ , \tag{3.3.11}$$

which comprises the (as yet undetermined) spectral density $\bar{a}(m)$. This current density – coherent in space and synchronous in time with the excitation (3.3.3) – is linked to the secondary (complex) magnetic scalar potential, $\bar{\chi}^{(s)}$, which is represented via the similarly still undetermined spectral amplitude density $\bar{f}^{(s)}(m)$:

$$\bar{\chi}^{(s)} = \pm \frac{\bar{I}}{\pi} \int_0^\infty \bar{f}^{(s)}(m) \sin mx \, e^{\mp mz} dm \ , \qquad z \gtrless 0 \ . \tag{3.3.12}$$

Hence, the secondary field component $\bar{H}_x^{(s)}$ reads

$$\bar{H}_x^{(s)} = \mp \frac{\bar{I}}{\pi} \int_0^\infty \bar{f}^{(s)}(m) m \cos mx \, e^{\mp mz} dm \ , \qquad z \gtrless 0 \ , \tag{3.3.13}$$

and applying once more the boundary condition (3.3.8) – in this case at the interface $z = 0$ and with $\bar{A}(x)$ replaced by $\bar{A}^{(s)}(x)$ – we find

$$\frac{\bar{I}}{\pi} \int_0^\infty \bar{a}(m) \cos mx \, dm = -\frac{2\bar{I}}{\pi} \int_0^\infty \bar{f}^{(s)}(m) m \cos mx \, dm \ , \qquad \text{or} \tag{3.3.14}$$

$$\bar{f}^{(s)}(m) = -\frac{\bar{a}(m)}{2m} \ , \tag{3.3.15}$$

so that finally

$$\bar{\chi}^{(s)} = \mp \frac{\bar{I}}{2\pi} \int_0^\infty \frac{\bar{a}(m)}{m} \sin mx \, e^{\mp mz} dm \ , \qquad z \gtrless 0 \ . \tag{3.3.16}$$

3.3.4 Spectral Density of Sheet Current

Maxwell's second curl equation, relating the electric field E to the magnetic induction B at any instant t, namely

$$\nabla \times E = -\frac{\partial B}{\partial t} \ , \tag{3.3.17}$$

yields a direct relationship between the y-component of E and the z-component of B:

$$\frac{\partial E_y}{\partial x} = -\frac{\partial B_z}{\partial t} \ . \tag{3.3.18}$$

Resorting to (3.3.10, 12), we find – in the sheet plane – for the complex amplitude $\bar{B}_z$

$$\bar{B}_z = -\mu_0 \frac{\partial \bar{\chi}}{\partial z} = -\mu_0 \left[\frac{\bar{I}}{2\pi} \int_0^\infty \sin mx\, e^{-mh} dm + \frac{\bar{I}}{2\pi} \int_0^\infty \bar{a}(m) \sin mx\, dm \right] .$$

(3.3.19)

Here $\bar{\chi} = \bar{\chi}^{(p)} + \bar{\chi}^{(s)}$.

Now Ohm's law relates the complex amplitude $\bar{E}_y$ of the electric field inside the sheet to the surface current density $\bar{A}^{(s)}$ via

$$\bar{E}_y = \frac{1}{\sigma \Delta} \bar{A}^{(s)} .$$

(3.3.20)

Combining (3.3.11, 18 and 20), we have

$$-\frac{1}{\sigma\Delta} \frac{\bar{I}}{\pi} \int_0^\infty \bar{a}(m) m \sin mx\, dm = -i\omega\mu_0 \frac{\bar{I}}{2\pi} \int_0^\infty [e^{-mh} + \bar{a}(m)] \sin mx\, dm ,$$

(3.3.21)

so that finally

$$\bar{a}(m) = \frac{(i\omega\mu_0\sigma\Delta)/2}{m - (i\omega\mu_0\sigma\Delta)/2} e^{-mh} .$$

(3.3.22)

Having determined $\bar{a}(m)$, we are able to express $\bar{\chi}^{(s)}$ explicitly as per (3.3.16):

$$\bar{\chi}^{(s)} = \mp \frac{\bar{I}}{2\pi} \frac{i\omega\mu_0\sigma\Delta}{2} \int_0^\infty \frac{e^{-mh}}{m - (i\omega\mu_0\sigma\Delta)/2} \frac{\sin mx}{m} e^{\mp mz} dm ,$$

$$z \gtrless 0 .$$

(3.3.23)

3.3.5 Reaction on Current-Carrying Conductor

In air, Maxwell's curl equations (3.3.4, 17) imply in the present case the relation, see also (3.2.12),

$$\nabla \times (\nabla \times E) = 0 ,$$

(3.3.24)

or, with Cartesian coordinates used throughout

$$\nabla^2 E = 0 .$$

(3.3.25)

As above, we are primarily interested in the y-component of E; its complex amplitude $\bar{E}_y$ satisfies therefore the Laplace equation

$$\frac{\partial^2 \bar{E}_y}{\partial x^2} + \frac{\partial^2 \bar{E}_y}{\partial z^2} = 0 .$$

(3.3.26)

In view of (3.3.17 and 10), we assume for the primary (complex) component $\bar{E}_y^{(p)}$ in air, a solution in the form

$$\bar{E}_y^{(p)} = \frac{\bar{I}}{\pi} \int_0^\infty \bar{F}_p(m) \cos mx \, e^{\mp m(z-h)} dm \ , \qquad z \gtrless h \ , \tag{3.3.27}$$

$\bar{F}_p(m)$ being as yet arbitrary.

As the induction law, however, requires the equality

$$-\frac{\partial E_y}{\partial z} = -\frac{\partial B_x}{\partial t} \ , \tag{3.3.28}$$

we find

$$\pm \frac{\bar{I}}{\pi} \int_0^\infty m\bar{F}_p(m) \cos mx \, e^{\mp m(z-h)} dm = \pm i \omega\mu_0 \frac{\bar{I}}{2\pi} \int_0^\infty \cos mx \, e^{\mp m(z-h)} dm \ ,$$

$$z \gtrless h \ . \tag{3.3.29}$$

The induction law, however, also implies (see above) the relation

$$\frac{\partial E_y}{\partial x} = -\frac{\partial B_z}{\partial t} \qquad \text{or} \tag{3.3.30}$$

$$-\frac{\bar{I}}{\pi} \int_0^\infty m\bar{F}_p(m) \sin mx \, e^{\mp m(z-h)} dm = -i \omega\mu_0 \frac{\bar{I}}{2\pi} \int_0^\infty \sin mx \, e^{\mp m(z-h)} dm \ ;$$

$$z \gtrless h \ . \tag{3.3.31}$$

Eqs. (3.3.29, 31) are both identically satisfied for

$$\bar{F}_p(m) = \frac{i \omega\mu_0}{2m} \ , \tag{3.3.32}$$

so that

$$\bar{E}_y^{(p)} = \frac{i \omega\mu_0}{2} \bar{I} \int_0^\infty \frac{\cos mx}{m} e^{\mp m(z-h)} dm \ ; \qquad z \gtrless h \ . \tag{3.3.33}$$

With regard to the secondary field component $\bar{E}_y^{(s)}$ linked to the current inside the conducting layer, we introduce, in analogy to (3.3.27), an as yet undetermined spectral density $\bar{F}_s$ so that

$$\bar{E}_y^{(s)} = \frac{\bar{I}}{\pi} \int_0^\infty \bar{F}_s(m) \cos mx \, e^{\mp mz} dm \ ; \qquad z \gtrless 0 \ . \tag{3.3.34}$$

Resorting now to the secondary magnetic scalar potential (3.3.23), we obtain — assisted once more by the induction law as per (3.3.28) and (3.3.30) — that

$$\pm \frac{\bar{I}}{\pi} \int_0^\infty \bar{F}_s(m)\, m \cos mx\, e^{\mp mz}\, dm$$

$$= \pm i \omega \mu_0 \frac{\bar{I}}{2\pi} \frac{i \omega \mu_0 \sigma \Delta}{2} \int_0^\infty \frac{e^{-mh}}{m - (i\omega\mu_0\sigma\Delta)/2} \cos mx\, e^{\mp mz}\, dm \,, \qquad z \gtrless 0 \,,$$

$$\text{(3.3.35)}$$

and

$$-\frac{\bar{I}}{\pi} \int_0^\infty \bar{F}_s(m)\, m \sin mx\, e^{\mp mz}\, dm$$

$$= -i \omega \mu_0 \frac{\bar{I}}{2\pi} \frac{i \omega \mu_0 \sigma \Delta}{2} \int_0^\infty \frac{e^{-mh}}{m - (i\omega\mu_0\sigma\Delta)/2} \sin mx\, e^{\mp mz}\, dm \,, \qquad z \gtrless 0 \,.$$

$$\text{(3.3.36)}$$

The two equations jointly imply

$$\bar{F}_s(m) = \frac{i\omega\mu_0}{2} \frac{(i\omega\mu_0\sigma\Delta)/2}{m - (i\omega\mu_0\sigma\Delta)/2} \frac{e^{-mh}}{m} \,, \qquad \text{(3.3.37)}$$

so that

$$\bar{E}_y^{(s)} = \frac{i\omega\mu_0}{2\pi} \bar{I} \frac{i\omega\mu_0\sigma\Delta}{2} \int_0^\infty \frac{e^{-mh}}{m - (i\omega\mu_0\sigma\Delta)/2} \frac{\cos mx}{m} e^{\mp mz}\, dm \,; \qquad z \gtrless 0 \,.$$

$$\text{(3.3.38)}$$

On the conductor of radius $r_0 \ll h$, we therefore find from (3.3.33)

$$\bar{E}_y^{(p)} = \frac{i\omega\mu_0}{2\pi} \bar{I} \int_0^\infty \frac{e^{-mr_0}}{m}\, dm \,, \qquad \text{(3.3.39)}$$

and, from (3.3.38),

$$\bar{E}_y^{(s)} = \frac{i\omega\mu_0}{2\pi} \bar{I} \int_0^\infty \frac{(i\omega\mu_0\sigma\Delta)/2}{m - (i\omega\mu_0\sigma\Delta)/2} \frac{e^{-2mh}}{m}\, dm \,. \qquad \text{(3.3.40)}$$

Now,

$$\frac{(i\omega\mu_0\sigma\Delta)/2}{m - (i\omega\mu_0\sigma\Delta)/2} \equiv -1 + \frac{m}{m - (i\omega\mu_0\sigma\Delta)/2} \,, \qquad \text{(3.3.41)}$$

so that (3.3.40) becomes

$$\bar{E}_y^{(s)} = -\frac{i\omega\mu_0}{2\pi}\bar{I}\int\limits_0^\infty \frac{e^{-2mh}}{2mh}\,d(2mh)$$

$$+\frac{i\omega\mu_0}{2}\bar{I}\int\limits_0^\infty \frac{e^{-2mh}}{2mh-i\omega\mu_0\sigma\Delta h}\,d(2mh) \ . \tag{3.3.42}$$

Furthermore, denoting

$$2mh \equiv \Lambda \ , \tag{3.3.43}$$

we find that the total (complex) amplitude $\bar{E}_y$ on the conductor is

$$\bar{E}_y = \bar{E}_y^{(p)} + \bar{E}_y^{(s)}$$

$$= \frac{i\omega\mu_0}{2\pi}\bar{I}\left(\int\limits_0^\infty \frac{e^{-\Lambda(r_0/2h)}-e^{-\Lambda}}{\Lambda}\,d\Lambda + \int\limits_0^\infty \frac{e^{-\Lambda}}{\Lambda-i\omega\mu_0\sigma\Delta h}\,d\Lambda\right) \ . \tag{3.3.44}$$

Performing the required integration, we find (Sect. 3.3.8):

$$\bar{E}_y = \frac{i\omega\mu_0}{2\pi}\bar{I}\left[\ln\frac{2h}{r_0} - \sin(\omega\mu_0\sigma\Delta h)\,\mathrm{si}(\omega\mu_0\sigma\Delta h)\right.$$

$$- \cos(\omega\mu_0\sigma\Delta h)\,\mathrm{ci}(\omega\mu_0\sigma\Delta h) - i\cos(\omega\mu_0\sigma\Delta h)\,\mathrm{si}(\omega\mu_0\sigma\Delta h)$$

$$\left. + i\sin(\omega\mu_0\sigma\Delta h)\,\mathrm{ci}(\omega\mu_0\sigma\Delta h)\right] \ , \tag{3.3.45}$$

where once more, see (3.2.25),

$$\mathrm{si}(u) \equiv -\int\limits_u^\infty \frac{\sin v}{v}\,dv \tag{3.3.46}$$

stands for the sine integral, and, similarly

$$\mathrm{ci}(u) \equiv -\int\limits_u^\infty \frac{\cos v}{v}\,dv \ , \tag{3.3.47}$$

for the cosine integral, v being a dummy integration variable, and u an arbitrary argument.

For small values of u, the following approximations hold:

$$\mathrm{si}(u) \simeq u - \frac{\pi}{2} \ , \quad \mathrm{ci}(u) \simeq -\ln\frac{1}{\gamma u} \ , \quad \gamma = 1.781 \ . \tag{3.3.48}$$

Hence, see (3.3.45),

$$-\sin u\,\mathrm{si}(u) - \cos u\,\mathrm{ci}(u) \simeq \ln\frac{1}{\gamma u} \ , \quad u \ll 1 \ , \tag{3.3.49}$$

and further,

$$-\cos u \, \text{si}(u) + \sin u \, \text{ci}(u) \simeq \frac{\pi}{2} - u \ln \frac{1}{\gamma u} \simeq \frac{\pi}{2} \, , \qquad u \ll 1 \, . \tag{3.3.50}$$

In the region $\omega\mu_0\sigma\Delta h \ll 1$, (3.3.45) reduces therefore to the limiting value

$$\bar{E}_y = \frac{i\omega\mu_0}{2\pi} \bar{I} \left(\ln \frac{2h}{r_0} + \ln \frac{1}{\gamma\omega\mu_0\sigma\Delta h} + i \frac{\pi}{2} \right)$$

$$= \frac{i\omega\mu_0}{2\pi} \bar{I} \left(\ln \frac{2}{\gamma\omega\mu_0\sigma\Delta r_0} + i \frac{\pi}{2} \right) \, . \tag{3.3.51}$$

On the other hand, if $u \gg 1$, we have

$$\text{si}(u) \simeq -\frac{\cos u}{u} \, ; \qquad \text{ci}(u) \simeq +\frac{\sin u}{u} \tag{3.3.52}$$

so that

$$-\sin u \, \text{si}(u) - \cos u \, \text{ci}(u) \simeq 0 \, , \qquad u \gg 1 \, , \qquad \text{and} \tag{3.3.53}$$

$$-\cos u \, \text{si}(u) + \sin u \, \text{ci}(u) \simeq \frac{1}{u} \, , \qquad u \gg 1 \, . \tag{3.3.54}$$

Thus, in the case $\omega\mu_0\sigma\Delta h \gg 1$, we therefore obtain

$$\bar{E}_y = \frac{i\omega\mu_0}{2\pi} \bar{I} \left(\ln \frac{2h}{r_0} + i \frac{1}{\omega\mu_0\sigma\Delta h} \right) \simeq \frac{i\omega\mu_0}{2\pi} \bar{I} \ln \frac{2h}{r_0} \, . \tag{3.3.55}$$

Two remarks are in order: (a) the "sense" – or, more accurately, the sign – of the electric current $\bar{I}$ is arbitrary; (b) if the sinusoidal time dependence is assumed as $\exp(-i\omega t)$, any electric circuit comprising a resistance R in series with an inductance L exhibits a complex impedance $R - i\omega L$ (Sect. 3.2.5). Accordingly, in our case, for positive values of the resistance (per unit length) to be obtained, $\bar{E}_y$ must be divided by the value $(-\bar{I})$, i.e. the per unit length impedance z_1 reads in the extremal domains in question

$$\frac{\bar{E}_y}{-\bar{I}} = z_1 = \frac{\omega\mu_0}{2\pi} \frac{\pi}{2} - i \frac{\omega\mu_0}{2\pi} \left(\ln \frac{2h}{r_0} + \ln \frac{1}{\gamma\omega\mu_0\sigma\Delta h} \right) \, ,$$

$$\omega\mu_0\sigma\Delta h \ll 1 \, , \tag{3.3.56}$$

and

$$\frac{\bar{E}_y}{-\bar{I}} = z_1 = \frac{\omega\mu_0}{2\pi} \frac{1}{\omega\mu_0\sigma\Delta h} - i \frac{\omega\mu_0}{2\pi} \left(\ln \frac{2h}{r_0} \right) \, ,$$

$$\omega\mu_0\sigma\Delta h \gg 1 \, . \tag{3.3.57}$$

Evidently, while in the first case the per-unit length resistance

$$R_1 = \frac{\omega\mu_0}{2\pi} \frac{\pi}{2}$$

(3.3.58)

tends to a finite value, in the second case

$$R_1 = \frac{\omega\mu_0}{2\pi} \frac{1}{\omega\mu_0\sigma\Delta h} \to 0 \ .$$

(3.3.59)

Or, with the definition

$$r_1 \equiv \frac{R_1}{(\omega\mu_0)/2\pi} \ ,$$

(3.3.60)

we have

$$r_1 = \begin{cases} \dfrac{\pi}{2}, & \omega\mu_0\sigma\Delta h \ll 1 \ , \\ 0, & \omega\mu_0\sigma\Delta h \gg 1 \ . \end{cases}$$

(3.3.61)

In the intermediate range, we have, see (3.3.45),

$$r_1 = -\cos u \operatorname{si}(u) + \sin u \operatorname{ci}(u) \ ,$$

(3.3.62)

where the notation $u = \omega\mu_0\sigma\Delta h$ was again used, for brevity.

We next consider the reactive term; for the per-unit length inductance L_1, (3.3.56) yields

$$L_1 = \frac{\mu_0}{2\pi}\left(\ln\frac{2h}{r_0} + \ln\frac{1}{\gamma\omega\mu_0\sigma\Delta h}\right) \ , \quad \omega\mu_0\sigma\Delta h \ll 1 \ ,$$

(3.3.63)

whereas (3.3.57) implies

$$L_1 = \frac{\mu_0}{2\pi}\left(\ln\frac{2h}{r_0}\right) \ , \quad \omega\mu_0\sigma\Delta h \gg 1 \ .$$

(3.3.64)

In analogy to the resistance, we introduce the dimensionless incremental inductance

$$l_1 \equiv \frac{L_1}{\dfrac{\mu_0}{2\pi}} - \ln\frac{2h}{r_0} \ ,$$

(3.3.65)

so that

$$l_1 = \begin{cases} \ln \dfrac{1}{\gamma \omega \mu_0 \sigma \Delta h} \,, & \omega \mu_0 \sigma \Delta h \ll 1 \\[4mm] 0\,, & \omega \mu_0 \sigma \Delta h \gg 1 \,. \end{cases} \qquad (3.3.66)$$

Considering once more the intermediate range, (3.3.45) yields

$$l_1 = -\sin u \, \mathrm{si}(u) - \cos u \, \mathrm{ci}(u) \,. \qquad (3.3.67)$$

The dimensionless parameters r_1 and l_1, (3.3.62,67), are reproduced in Fig. 3.10.

It should be noted, that the extremal regions $\omega \mu_0 \sigma \Delta h \gg 1$ and $\omega \mu_0 \sigma \Delta h \ll 1$ are largely of academic interest and are quoted mainly for the sake of completeness: the first case may imply extremely high supply frequencies $\omega \gg (1/\mu_0 \sigma \Delta h)$, so that the line is no longer "short" relative to the wavelength λ, whereas the second case may imply extremely low values of the characteristic time $\mu_0 \sigma \Delta h$, in contrast to the requirement $(h/c) \ll \mu_0 \sigma \Delta h$.

3.3.6 Transient Phenomena

Some of the above results may be used for investigation of line behavior during current surges, for instance, under a sudden current inrush. Furthermore, in the case of overhead lines, the proposed model may yield some information on the effect of lightning strokes.

Let us recall that, in view of (3.3.11 and 22), the sheet (return) surface current density reads

$$\bar{A}^{(s)}(x) = \frac{\bar{I}}{\pi} \int_{m=0}^{\infty} \frac{(\mathrm{i}\,\omega \mu_0 \sigma \Delta)/2}{m - (\mathrm{i}\,\omega \mu_0 \sigma \Delta)/2} \, e^{-mh} \cos mx \, dm \,. \qquad (3.3.68)$$

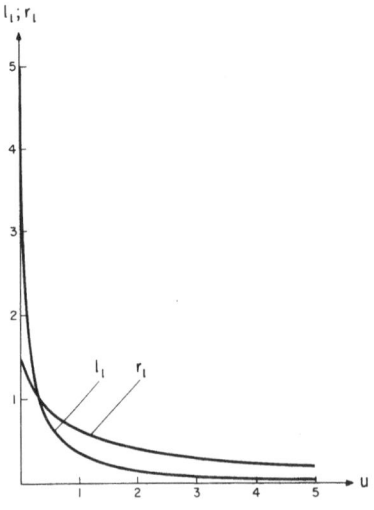

Fig. 3.10. Dimensionless line parameters as dependent on circuit variable $u = \omega \mu_0 \sigma \Delta h$, see (3.3.62,68)

When proceeding from sinusoidal steady-state conditions to transient performance, we must first specify the conditions of excitation. Thus, taking for the line current I (time domain!) a step function of amplitude I_0, we may write, using the complex variable s, the analytical expression

$$I = \frac{I_0}{2\pi i} \int_{P} \frac{e^{st}}{s} \, ds \; . \tag{3.3.69}$$

Replacing now the imaginary variable $(-i\omega)$ by its complex counterpart, s, the sheet current $A^{(s)}$ (time domain!) is obtained as

$$A^{(s)}(x,t) = -\frac{I_0}{\pi} \frac{1}{2\pi i} \int_{P} \frac{e^{st}}{s} \, ds \int_{m=0}^{\infty} \frac{\mu_0 \sigma \Delta s/2}{m + \mu_0 \sigma \Delta s/2} \, e^{-mh} \cos mx \, dm$$

$$= -\frac{I_0}{\pi} \frac{\mu_0 \sigma \Delta}{2} \int_{m=0}^{\infty} \frac{e^{-(2m/\mu_0\sigma\Delta)t}}{\mu_0 \sigma \Delta/2} \, e^{-mh} \cos mx \, dm \; . \tag{3.3.70}$$

Introducing the system time constant T, arbitrarily defined as

$$T \equiv \frac{\mu_0 \sigma \Delta h}{2} \; , \tag{3.3.71}$$

we have

$$A^{(s)}(x,t) = -\frac{I_0}{\pi} \int_0^{\infty} e^{-mh(1 + t/T)} \cos mx \, dm$$

$$= -\frac{I_0}{2\pi} \int_0^{\infty} (e^{-m[h(1 + t/T) - ix]} + e^{-m[h(1 + t/T) + ix]}) \, dm$$

$$= -\frac{I_0}{2\pi} \left(\frac{1}{h(1 + t/T) - ix} + \frac{1}{h(1 + t/T) + ix} \right) \; . \tag{3.3.72}$$

Hence, finally,

$$A^{(s)}(x,t) = -\frac{I_0}{\pi h} \frac{(1 + t/T)}{(1 + t/T)^2 + (x/h)^2} \; . \tag{3.3.73}$$

Significantly, although the system is a distributed one, a single time constant suffices to describe its transients performance; this is due to the simplifying assumption whereby the distributed capacitance is neglected, so that our model exhibits, so to say, only "half a degree of freedom".

The step response as per (3.3.73) is reproduced in Fig. 3.11, wherein

$$A \equiv A^{(s)}(x,t)/(-I_0/\pi h) \; . \tag{3.3.74}$$

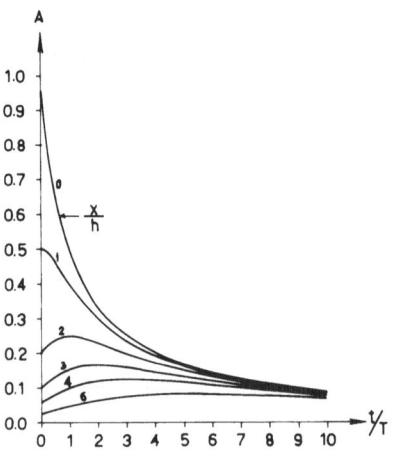

Fig. 3.11a. Secondary current variation during transient operation: Dimensionless surface current vs. dimensionless time, at different locations; see (3.3.73, 74)

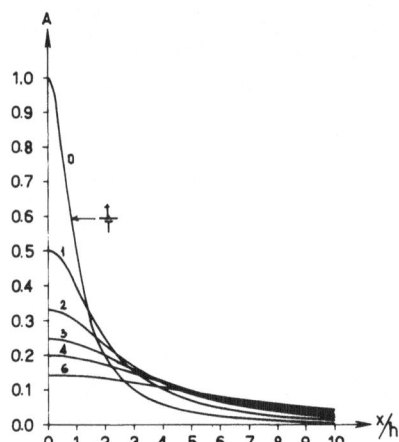

Fig. 3.11b. Secondary current variation during transient operation: Dimensionless surface current vs. dimensionless spacing, at different times; see (3.3.73, 74)

3.3.7 Discussion

Steady-state behavior as well as step response of an infinite plane conducting sheet exposed to a current-carrying conductor were considered. Quoted results are quite simple and enable the practicing engineer to estimate the influence of conducting shields on current-carrying bus-bars, having, moreover, also relevance to earth return problems of lower ground wires, counterpoises, or even overhead lines; nevertheless, these results serve as no more than a guide: the soil usually cannot be regarded as a uniform conducting medium and two-layer formulae often yield, in such cases, better agreement between theory and experiment.

Still, certain qualitative insight into earth currents caused by lightning surges may be provided through the obtained step response function combined with numerical integration of their electronically-recorded excitational time behavior.

3.3.8 Appendix

Performance of Integration of (3.3.44)

A) For the first integral, introducing the real positive number $\varepsilon \ll 1$, we find:

$$I_1 \equiv \int_0^\infty \frac{e^{-\Lambda(r_0/2h)} - e^{-\Lambda}}{\Lambda} \, d\Lambda$$

$$= \int_0^\varepsilon \frac{e^{-\Lambda(r_0/2h)} - e^{-\Lambda}}{\Lambda} \, d\Lambda + \int_\varepsilon^\infty \frac{e^{-\Lambda(r_0/2h)} - e^{-\Lambda}}{\Lambda} \, d\Lambda \,, \qquad (3.3.75)$$

and for the limit at $\varepsilon \to 0$, we have

$$\int_0^\varepsilon \frac{e^{-\Lambda(r_0/2h)} - e^{-\Lambda}}{\Lambda} d\Lambda \to 0 \ , \tag{3.3.76}$$

so that I_1 now reads

$$I_1 \equiv \int_{\Lambda=\varepsilon}^\infty \frac{e^{-\Lambda(r_0/2h)}}{\Lambda} d\Lambda - \int_{\Lambda=\varepsilon}^\infty \frac{e^{-\Lambda}}{\Lambda} d\Lambda \ . \tag{3.3.77}$$

Denoting momentarily

$$\Lambda(r_0/2h) \equiv v, \quad \Lambda \equiv u \ , \tag{3.3.78}$$

we have

$$I_1 \equiv \int_{\varepsilon(r_0/2h)}^\infty \frac{e^{-v}}{v} dv - \int_\varepsilon^\infty \frac{e^{-u}}{u} du \equiv \int_{\varepsilon(r_0/2h)}^\infty \frac{e^{-v}}{v} dv - \int_\varepsilon^\infty \frac{e^{-v}}{v} dv \ . \tag{3.3.79}$$

Now,

$$\int \frac{e^{-v}}{v} dv = e^{-v} \ln v + \int e^{-v} \ln v \, dv = e^{-v} \ln v + v(\ln v - 1) e^{-v} + \ldots \tag{3.3.80}$$

so that

$$I_1 = -\ln \varepsilon(r_0/2h) + \ln \varepsilon \ . \tag{3.3.81}$$

Hence, finally

$$\int_0^\infty \frac{e^{-\Lambda(r_0/2h)} - e^{-\Lambda}}{\Lambda} d\Lambda = \ln(2h/r_0) \ . \tag{3.3.82}$$

B) For the second integral,

$$I_2 \equiv \int_0^\infty \frac{e^{-\Lambda}}{\Lambda - i\omega\mu_0\sigma\Delta h} d\Lambda \ , \tag{3.3.83}$$

integration is performed in the complex plane $\psi = \Lambda + i\Omega$, along the closed contour $0b - ba - a0$ (Fig. 3.12), the result being zero (Cauchy's integral formula), namely,

$$-\int_{\Lambda=0}^\infty \frac{e^{-\Lambda}}{\Lambda - i\omega\mu_0\sigma\Delta h} d\Lambda + \int_{\Omega=0}^{-\infty} \frac{e^{-i\Omega}}{i\Omega - i\omega\mu_0\sigma\Delta h} i \, d\Omega = 0 \ , \tag{3.3.84}$$

i.e.

$$\int_0^\infty \frac{e^{-\Lambda}}{\Lambda - i\omega\mu_0\sigma\Delta h} d\Lambda = \int_0^\infty \frac{e^{i\Omega'}}{\Omega' + \omega\mu_0\sigma\Delta h} d\Omega' \ , \tag{3.3.85}$$

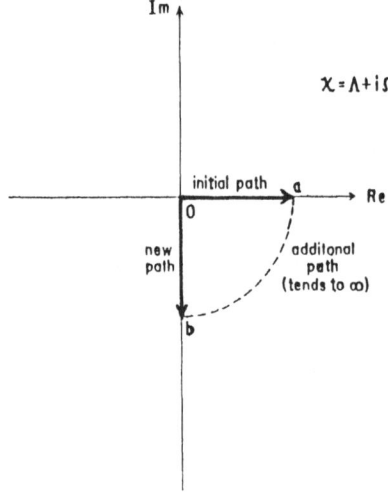

Fig. 3.12. Integration in complex plane

where $\Omega' = -\Omega$. We now undertake an additional change of variables, introducing

$$K \equiv \Omega' + \omega\mu_0\sigma\Delta h \ , \tag{3.3.86}$$

so that

$$\int_{\Omega'=0}^{\infty} \frac{e^{i\Omega'}}{\Omega' + \omega\mu_0\sigma\Delta h} d\Omega' = \int_{\omega\mu_0\sigma\Delta h}^{\infty} \frac{e^{i(K-\omega\mu_0\sigma\Delta h)}}{K} dK$$

$$= e^{-i\omega\mu_0\sigma\Delta h} \int_{\omega\mu_0\sigma\Delta h}^{\infty} \frac{e^{iK}}{K} dK \ . \tag{3.3.87}$$

Now,

$$\int_{\omega\mu_0\sigma\Delta h}^{\infty} \frac{e^{iK}}{K} dK = -\text{ci}(\omega\mu_0\sigma\Delta h) - i\,\text{si}(\omega\mu_0\sigma\Delta h) \ , \tag{3.3.88}$$

and finally,

$$\int_{0}^{\infty} \frac{e^{-\Lambda}}{\Lambda - i\omega\mu_0\sigma\Delta h} d\Lambda$$

$$= e^{-i\omega\mu_0\sigma\Delta h}[-\text{ci}(\omega\mu_0\sigma\Delta h) - i\,\text{si}(\omega\mu_0\sigma\Delta h)]$$

$$= -\sin(\omega\mu_0\sigma\Delta h)\,\text{si}(\omega\mu_0\sigma\Delta h) - \cos(\omega\mu_0\sigma\Delta h)\,\text{ci}(\omega\mu_0\sigma\Delta h)$$

$$-i\cos(\omega\mu_0\sigma\Delta h)\,\text{si}(\omega\mu_0\sigma\Delta h) + i\sin(\omega\mu_0\sigma\Delta h)\,\text{ci}(\omega\mu_0\sigma\Delta h) \ . \tag{3.3.89}$$

Let us remark in passing that this result could in fact have been obtained directly from (3.3.87) with [3.19]

$$\int_0^\infty \frac{\cos t}{t+z}\, dt = g(z) \ , \tag{3.3.90}$$

$$\int_0^\infty \frac{\sin t}{t+z}\, dt = f(z) \ , \tag{3.3.91}$$

where (Sect. 3.2.6)

$$g(z) = -\operatorname{ci}(z)\cos z - \operatorname{si}(z)\sin z \ , \quad \text{and} \tag{3.3.92}$$

$$f(z) = \operatorname{ci}(z)\sin z - \operatorname{si}(z)\cos z \ . \tag{3.3.93}$$

Thus

$$\int_{\Omega'=0}^\infty \frac{e^{i\Omega'}}{\Omega'+\omega\mu_0\sigma\varDelta h}\, d\Omega'$$

$$= \int_{\Omega'=0}^\infty \frac{\cos\Omega'}{\Omega'+\omega\mu_0\sigma\varDelta h}\, d\Omega' + i \int_{\Omega'=0}^\infty \frac{\sin\Omega'}{\Omega'+\omega\mu_0\sigma\varDelta h}\, d\Omega' \ , \tag{3.3.94}$$

so that

$$\int_{\Omega'=0}^\infty \frac{e^{i\Omega'}}{\Omega'+\omega\mu_0\sigma\varDelta h}\, d\Omega' = g(\omega\mu_0\sigma\varDelta h) + if(\omega\mu_0\sigma\varDelta h) \ , \quad \text{i.e.} - \tag{3.3.95}$$

$$\bar{E}_y = \frac{i\omega\mu_0}{2\pi}\, \bar{I}\left(\int_0^\infty \frac{e^{-\varLambda(r_0/2h)}-e^{-\varLambda}}{\varLambda}\, d\varLambda + \int_0^\infty \frac{e^{-\varLambda}}{\varLambda-i\omega\mu_0\sigma\varDelta h}\, d\varLambda \right)$$

$$= \frac{i\omega\mu_0}{2\pi}\, \bar{I}(I_1+I_2)$$

$$= \frac{i\omega\mu_0}{2\pi}\, \bar{I}[\ln(2h/r_0) + g(\omega\mu_0\sigma\varDelta h) + if(\omega\mu_0\sigma\varDelta h)] \ . \tag{3.3.96}$$

3.4 On the Inductance of Printed Spiral Coils

Advances in microwave integrated circuits have stimulated a trend towards replacement of distributed elements by lumped ones [3.20–25], thereby achieving considerable economy in size – a feature in heavy demand in sophisticated electronic packages. Recently, monolithic GaAs FET active devices have been described, in which printed coils serve as feedback inductances [3.26, 27] as peaking or resonating elements [3.28] or alternatively as rf chokes [3.29]. Among the diverse forms tested in practice, the printed circular spiral coil – or a variant thereof – seems to have acquired considerable popularity (e.g., [Ref. 3.20, Fig. 4], [Ref. 3.23, Fig. 29], [Ref. 3.26, Fig. 6a]).

This coil poses quite a few problems for the design engineer, namely, inherently low inductance, high magnetic stray fields and hence – coupling effects. As the characteristic dimension of the lumped element is much smaller than the signal wavelength, the magneto-quasistatic approach may be invoked, and design is usually based on classical low-frequency formulas, e.g. [3.30 – 33]. However, comparison of commonly used designs shows discrepancies, and empirical approaches are preferred in many cases. In the present section, the (quasi-) static self-inductance of a thin-film spiral coil is investigated along purely basic principles.

3.4.1 Formulation of Problem

A flat ("two-dimensional") printed current-carrying spiral of outer radius R_0 comprising N turns is centered (Fig. 3.13) in a circular cylindrical system of coordinates. The current I alternates monochromatically at an angular frequency ω, such that the signal wavelength is much larger than R_0. What is the self-inductance of the coil under the conditions stated?

As noted above, thanks to the inequality (wave length) ≫ (mean coil radius), the problem can be treated as though it were static, see (3.4.1, 2) below. Thus, instead of resorting in the following to phasor notation, we envisage a situation where $\omega \to 0$, which is readily obtained if the above inequality is sharpened to the requirement of an "infinite" wavelength. Ascribing now a finite value to the (free) magnetic energy density – which is here much higher than its electrical counterpart – we are once more entitled

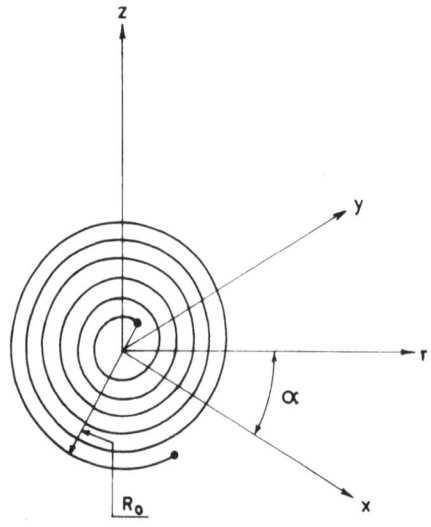

Fig. 3.13. Flat circular spiral coil

to consider the coil as surrounded by a fictitious medium of vanishing dielectric constant, zero electrical conductivity, and finite free-space magnetic permeability μ_0.

3.4.2 Solution

As already stated several times, within the realm of (quasi-) statics, the magnetic field H is curl-free throughout the entire space, except for the region occupied by the flat coil. Thus, in the current-free surroundings

$$\nabla \times H = 0, \quad z \gtrless 0 . \tag{3.4.1}$$

Further, in the absence of magnetic charge, the magnetic field is everywhere source-free, i.e.,

$$\nabla \cdot \mu_0 H = 0 , \tag{3.4.2}$$

so that the magnetic vector potential V is introduced through the usual defining relation

$$\mu_0 H = \nabla \times V , \tag{3.4.3}$$

and therefore, by the curl equation (3.4.1) we have

$$\nabla \times (\nabla \times V) = 0 . \tag{3.4.4}$$

In view of the coil geometry, it is further assumed that the magnetic vector potential comprises only a circular component V_α, so that, in fact,

$$V = V_\alpha 1_\alpha \tag{3.4.5}$$

(1_α: unit vector). Hence the vector differential equation (3.4.4) reduces to the scalar partial differential equation

$$\frac{\partial^2 V_\alpha}{\partial r^2} + \frac{1}{r} \frac{\partial V_\alpha}{\partial r} - \frac{V_\alpha}{r^2} + \frac{\partial^2 V_\alpha}{\partial z^2} = 0 , \tag{3.4.6}$$

and the field components H_r, H_α and H_z are, therefore, in turn expressed by the partial derivatives

$$\mu_0 H_r = -\frac{\partial V_\alpha}{\partial z}, \quad \mu_0 H_\alpha = 0, \quad \mu_0 H_z = \frac{1}{r} \frac{\partial}{\partial r} (r V_\alpha) . \tag{3.4.7}$$

For the solution of (3.4.6) to be compatible with the posed problem, we impose the following requirements:

– The potential function V_α vanishes at $z \to \pm \infty$, i.e.,

$$\lim_{|z| \to \infty} V_\alpha = 0 . \tag{3.4.8}$$

− In the $z = 0$ plane, the axial derivative of V_α is discontinuous by an amount prescribed by the surface current density $A(r) = A_\alpha(r)I_\alpha$:

$$\frac{\partial V_\alpha}{\partial z}\bigg|_{z=+0} - \frac{\partial V_\alpha}{\partial z}\bigg|_{z=-0} = -\mu_0 A_\alpha(r) \ . \tag{3.4.9}$$

Introducing the separation constant λ^2 and assuming for V_α the product

$$V_\alpha = R(r) \cdot Z(z) \ , \tag{3.4.10}$$

wherein R and Z are functions of r and z, respectively, we arrive at the ordinary differential equations

$$r^2 \frac{d^2 R}{dr^2} + r \frac{dR}{dr} + (\lambda^2 r^2 - 1)R = 0 \quad \text{and} \tag{3.4.11}$$

$$\frac{d^2 Z}{dz^2} - \lambda^2 Z = 0 \ . \tag{3.4.12}$$

The solution of (3.4.12) compatible with (3.4.8) is obtained as

$$Z = e^{\mp \lambda z}, \quad z \gtrless 0 \ , \tag{3.4.13}$$

whereas (3.4.11) is directly solved − subject to regular behavior at $r = 0$ − by a Bessel function of the first kind of order one with argument (λr), notation $J_1(\lambda r)$. In addition, the still unknown spectral density − or "weighting function" − $a(\lambda)$ is introduced, such that

$$V_\alpha = \int_{\lambda=0}^{\infty} a(\lambda) J_1(\lambda r) e^{\mp \lambda z} d\lambda; \quad z \gtrless 0 \ . \tag{3.4.14}$$

In order to determine $a(\lambda)$, we resort to the Fourier-Bessel integral; thus, choosing for $A_\alpha(r)$ an expression compatible with (3.4.14) and introducing the dummy variable of integration ϱ, we have

$$A_\alpha(r) = \int_{\lambda=0}^{\infty} \lambda \, d\lambda J_1(\lambda r) \int_{\varrho=0}^{\infty} A_\alpha(\varrho) J_1(\lambda \varrho) \varrho \, d\varrho \ . \tag{3.4.15}$$

Returning to (3.4.9), we find

$$-2 \int_0^\infty \lambda a(\lambda) J_1(\lambda r) d\lambda$$

$$= -\mu_0 \int_0^\infty J_1(\lambda r) \lambda \, d\lambda \int_0^\infty A_\alpha(\varrho) J_1(\lambda \varrho) \varrho \, d\varrho \ , \quad \text{or} \tag{3.4.16}$$

$$2a(\lambda) = \mu_0 \int_0^\infty A_\alpha(\varrho) J_1(\lambda \varrho) \varrho \, d\varrho \ , \tag{3.4.17}$$

so that finally

$$V_\alpha(r, z) = \frac{\mu_0}{2} \int_0^\infty J_1(\lambda r) e^{\mp \lambda z} d\lambda \int_0^\infty A_\alpha(\varrho) J_1(\lambda \varrho) \varrho \, d\varrho, \quad z \gtreqless 0 . \quad (3.4.18)$$

Clearly, once the functional dependence $A_\alpha(\varrho)$ of the surface current is known, the vector potential $V = V_\alpha I_\alpha$ is fully determined: thus, assuming, as already mentioned, for the coil an outer radius R_0 (with vanishingly small inner radius) and, in addition, an even distribution of N turns carrying the excitation current I, we obtain

$$A_\alpha = \frac{NI}{R_0} . \quad (3.4.19)$$

Introducing further, for brevity, the defining relations

$$\xi \equiv \lambda R_0; \quad \gamma \equiv \frac{\varrho}{R_0} , \quad (3.4.20)$$

the potential function V_α may be rewritten as

$$V_\alpha(r, z) = \frac{\mu_0}{2} NI \int_{\xi=0}^\infty J_1(\xi(r/R_0)) e^{\mp \xi(z/R_0)} d\xi \int_{\gamma=0}^1 J_1(\xi \gamma) \gamma \, d\gamma,$$

$$z \gtreqless 0 . \quad (3.4.21)$$

Now in (quasi-) statics, the (free) magnetic energy W_m residing in a field – linked to a current density j and to its vector potential function V – reads

$$W_m = \frac{1}{2} \int_{(v)} j \cdot V dv , \quad (3.4.22)$$

where the integration extends over all incremental volumes dv which make up the volume v of the current carrying conductor. In the present case the volume integral reduces to a surface integral limited to the coil region. Thus

$$W_m = \frac{1}{2} \int_{r=0}^{R_0} A_\alpha V_\alpha 2 \pi r \, dr, \quad z = 0 , \quad (3.4.23)$$

i.e. (Sect. 3.4.4)

$$W_m = \frac{1}{2} \mu_0 I^2 N^2 \pi R_0 \int_0^\infty d\xi \left[\int_{\gamma=0}^1 J_1(\xi \gamma) \gamma \, d\gamma \right]^2 . \quad (3.4.24)$$

On the other hand, the (quasi-) static self-inductance L of a coil is defined through

$$W_{\mathrm{m}} = \tfrac{1}{2} L I^2 \, , \tag{3.4.25}$$

so that finally

$$L = \mu_0 N^2 \pi R_0 \int_0^\infty d\xi \left[\int_0^1 J_1(\xi \gamma) \, \gamma \, d\gamma \right]^2 . \tag{3.4.26}$$

Before turning to numerical integration we perform the following straightforward calculations: for the definite integral in the brackets we have

$$\int_0^1 J_1(\xi \gamma) \, \gamma \, d\gamma = \frac{1}{\xi^2} \int_0^\xi J_1(\xi \gamma)(\xi \gamma) \, d(\xi \gamma) \, . \tag{3.4.27}$$

Denoting for simplicity

$$\xi \gamma \equiv \eta \tag{3.4.28}$$

and integrating by parts we have

$$\frac{1}{\xi^2} \int_0^\xi J_1(\eta) \eta \, d\eta = \frac{1}{\xi^2} \left[-\eta J_0(\eta) \big|_0^\xi + \int_0^\xi J_0(\eta) \, d\eta \right] . \tag{3.4.29}$$

Introducing

$$\int_0^\xi J_0(\eta) \, d\eta \equiv \Sigma_0(\xi) \tag{3.4.30}$$

there results

$$\frac{1}{\xi^2} \int_0^\xi J_1(\eta) \eta \, d\eta = \frac{1}{\xi^2} [\Sigma_0(\xi) - \xi J_0(\xi)] = \frac{1}{\xi} \left[\frac{\Sigma_0(\xi)}{\xi} - J_0(\xi) \right] , \tag{3.4.31}$$

so that

$$\int_0^1 J_1(\xi \gamma) \, \gamma \, d\gamma = \frac{1}{\xi} \left[\frac{\Sigma_0(\xi)}{\xi} - J_0(\xi) \right] . \tag{3.4.32}$$

From (3.4.26) we obtain

$$L = \mu_0 N^2 \pi R_0 \int_0^\infty \frac{1}{\xi^2} \left[\frac{\Sigma_0(\xi)}{\xi} - J_0(\xi) \right]^2 d\xi \tag{3.4.33}$$

and resorting now to numerical integration we find

$$\int_0^\infty \frac{1}{\xi^2} \left[\frac{\Sigma_0(\xi)}{\xi} - J_0(\xi) \right]^2 d\xi = 0.175 \, .$$

Hence, finally

$$L = 0.175\,\mu_0\,\pi\,R_0 N^2 \;. \tag{3.4.34}$$

The inductance, which in practice is relatively quite small, may be increased if two coils are printed, one on each side of the insulating board of, say, thickness h. The overall self-inductance L_{ov} of a pair of similar coils in series aiding is

$$L_{ov} = L + 2M + L \;, \tag{3.4.35}$$

where M stands for the mutual inductance. The latter is computed as follows: Assuming two identical coils at a spacing h, their coupled (index c), free magnetic exchange energy, $W_{c,m}$, reads

$$W_{c,m} = \int_{r=0}^{R_0} A_\alpha V_\alpha 2\,\pi r\,dr, \quad z = h \;. \tag{3.4.36}$$

Equating now

$$W_{c,m} = I^2 M \tag{3.4.37}$$

we find (Sect. 3.4.4)

$$I^2 \mu_0 N^2 \pi R_0 \int_0^\infty e^{-(h/R_0)\,\xi}d\xi \left[\int_0^1 J_1(\xi\gamma)\,\gamma\,d\gamma\right]^2 = I^2 M \tag{3.4.38}$$

so that, referring to (3.4.32),

$$M = \mu_0 N^2 \pi R_0 \int_0^\infty e^{-(h/R_0)\,\xi}\left\{\frac{1}{\xi^2}\left[\frac{\Sigma_0(\xi)}{\xi} - J_0(\xi)\right]^2\right\}d\xi \;. \tag{3.4.39}$$

Now

$$\frac{M}{L} = \frac{1}{0.175}\int_0^\infty e^{-(h/R_0)\,\xi}\left\{\frac{1}{\xi^2}\left[\frac{\Sigma_0(\xi)}{\xi} - J_0(\xi)\right]^2\right\}d\xi \;, \tag{3.4.40}$$

i.e., see (3.4.35),

$$\frac{L_{ov}}{L} \equiv l = 2\left(1 + \frac{M}{L}\right) \;. \tag{3.4.41}$$

The dimensionless overall self-inductance l is reproduced in Fig. 3.14 as a function of the dimensionless spacing h/R_0.

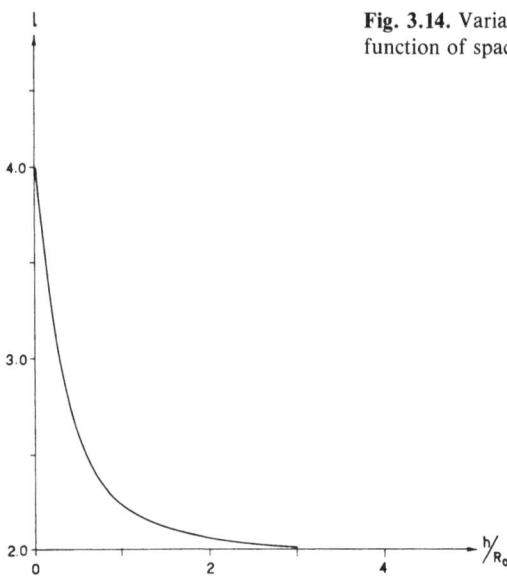

Fig. 3.14. Variation of dimensionless coil inductance as function of spacing

3.4.3 Discussion

The outlined approach to the inductance of spiral coils is based on a solution of the field problem and dispenses with empirical results; it may be extended – by means of numerical integration – to the analysis and design of printed spirals of more complex configurations, such as squares, rectangles, incomplete circles, etc.

It should be borne in mind that, due to the assumption of quasistatics (obviously justified per definition in the case of lumped elements), any capacitive effects of the substrate evidently have – within the context of the proposed solution – only secondary importance.

3.4.4 Appendix

A) Equation (3.4.24)

Substituting (3.4.19, 21) into (3.4.23) we have

$$W_m = \frac{1}{2} \int_{r=0}^{R_0} \frac{NI}{R_0} 2\pi r\, dr \left[\frac{\mu_0}{2} NI \int_{\xi=0}^{\infty} J_1 \left(\frac{\xi}{R_0} r \right) d\xi \int_{\gamma=0}^{1} J_1(\xi\gamma)\gamma\, d\gamma \right]$$

$$= \frac{1}{2} \mu_0 \frac{N^2 I^2}{R_0} \pi \int_{r=0}^{R_0} r\, dr \int_{\xi=0}^{\infty} J_1 \left(\xi \frac{r}{R_0} \right) d\xi \int_{\gamma=0}^{1} J_1(\xi\gamma)\gamma\, d\gamma$$

$$= \frac{1}{2} \mu_0 \frac{N^2 I^2}{R_0} \pi \int_{r/R_0=0}^{1} \frac{r}{R_0} R_0 d \left(\frac{r}{R_0} \right) R_0 \int_{\xi=0}^{\infty} J_1 \left(\xi \frac{r}{R_0} \right) d\xi \int_{\gamma=0}^{1} J_1(\xi\gamma)\gamma\, d\gamma$$

$$= \frac{1}{2}\mu_0 \frac{N^2 I^2}{R_0} \pi R_0^2 \int\limits_{r/R_0=0}^{1} d\left(\frac{r}{R_0}\right) \int\limits_{\xi=0}^{\infty} \frac{r}{R_0} J_1\left(\xi \frac{r}{R_0}\right) d\xi \int\limits_{\gamma=0}^{1} J_1(\xi\gamma)\gamma \, d\gamma$$

$$= \frac{1}{2}\mu_0 N^2 I^2 \pi R_0 \int\limits_{\xi=0}^{\infty} d\xi \int\limits_{r/R_0=0}^{1} J_1\left(\xi \frac{r}{R_0}\right) \frac{r}{R_0} d\left(\frac{r}{R_0}\right) \int\limits_{\gamma=0}^{1} J_1(\xi\gamma)\gamma \, d\gamma.$$

$$(3.4.42)$$

With the dummy variable

$$\frac{r}{R_0} \equiv v \qquad\qquad\qquad (3.4.43)$$

we have

$$W_m = \frac{1}{2}\mu_0 N^2 I^2 \pi R_0 \int\limits_{\xi=0}^{\infty} d\xi \int\limits_{v=0}^{1} J_1(\xi v) v \, dv \int\limits_{\gamma=0}^{1} J_1(\xi\gamma)\gamma \, d\gamma, \qquad (3.4.44)$$

i.e.,

$$W_m = \frac{1}{2}\mu_0 N^2 I^2 \pi R_0 \int\limits_{\xi=0}^{\infty} d\xi \left[\int\limits_{\gamma=0}^{1} J_1(\xi\gamma)\gamma \, d\gamma\right]^2. \qquad (3.4.45)$$

B) Equation (3.4.38)

Equation (3.4.36) is similar to (3.4.23), with two differences:

1. It does not comprise the factor $1/2$;
2. In the expression for V_α, (3.4.21), we have now to retain the factor $\exp[-\xi(h/R_0)]$, as integration is in this case performed not in the plane $z = 0$, but in the plane of elevation $z = h$. Hence (3.4.40).

4. Electromagnetic Induction: Transient Phenomena – Stationary Configuration

Four examples are analysed in this chapter. The first two deal with the field build up in an idealized ground and an idealized iron core, respectively. Although modeling is heavily resorted to, insight into the relevant processes underlying the field build-up is gained. Another section deals with the field switching in superconducting materials. Even though such a problem can be tackled only through quantum mechanics, a classical approach originally put forward by London and later extended by von Laue is applied. The interrelation between the London and Maxwell equations at field inception is accentuated in this section.

The final part of the chapter deals with electromechanical oscillations in mercury. Obviously, two aspects had to be tackled in this context, namely the electrical one, on the one hand, and the mechanical one, on the other hand. Results are compared with experimental data.

4.1 Surge Impedance of Extended Grounding Rods

Lightning protection is often required in order to prevent equipment failure, and adequate ground provision is essential for minimizing damage under surge conditions; in many practical circumstances, multiple grounds and mesh-grounding systems have to be used – mainly in high soil-resistance areas.

Experiment has shown that equipment insulation is frequently punctured and transformer and/or circuit breaker bushings tend to flash over under current surges; it is therefore evident that resistances under surge conditions are high compared with their static counterparts.

In the present section, an *estimate* of the surge resistance of ground rods is sought for the case of a lightning stroke. The outlined approach is of interest also for more complicated ground systems as well as for arbitrarily imposed transient current inputs.

4.1.1 Mathematical Model and Attendant Field Problem

The circular ground rod (or pipe) is assumed to have a radius ϱ_0 and to be buried at a depth $h \gg \varrho_0$ beneath the soil surface (Fig. 4.1). The soil is model-

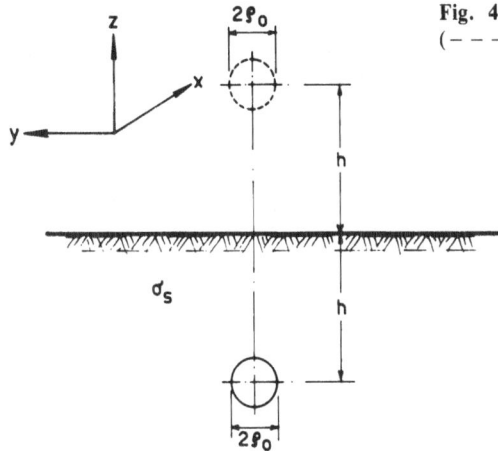

Fig. 4.1. Assumed configuration with image
$(---)$

led by an isotropic, homogeneous semi-infinite region, exhibiting an electric conductivity σ_s, vacuum magnetic permeability μ_0 and vanishingly small dielectric coefficient ε_r [4.1, 2]: this implies a fictitious configuration in which the magnetic diffusion time constant τ_d outweighs its electric relaxation counterpart τ_r, i.e.

$$\tau_r \ll \tau_d \ . \tag{4.1.1}$$

Hence, according to this point of view, we regard the system — immediately after the stroke — as described by a magneto-quasistatic approximation.

As a further practical simplification we assume the magnetic field H within the soil mainly to surround the rod; hence, in a plane $x =$ const, the electric field E is characterized by a vanishingly small emf, i.e.,

$$\oint_{(l)} E \cdot dl = 0, \quad x = \text{const} , \tag{4.1.2}$$

where dl stands for the linear vector element. Accordingly, in such a plane we have by Stokes' theorem

$$\nabla \times E = 0, \quad x = \text{const} , \tag{4.1.3}$$

and therefore an electric scalar potential ϕ may be introduced through the usual defining relation

$$E = -\nabla\phi, \quad x = \text{const} . \tag{4.1.4}$$

Now the normal derivative (direction n) of this potential must vanish at the soil surface:

$$\frac{\partial \phi}{\partial n} = 0, \quad z = h \ ; \tag{4.1.5}$$

this requirement imposes, in turn, the condition specifying the sign of the fictitious image current: if the rod current is taken as $+I$, the image is likewise given by $+I$ [4.2]. We are therefore dealing with an unbalanced transmission line immersed in a *homogeneous, infinite medium* characterized by the soil conductivity σ_s.

The rods are now mentally replaced by fictitious pipes of negligible resistance, their specific conductivity σ_r being taken to be many orders of magnitude higher than that of the soil. Thus, while an ordinary transmission line is characterized by the four distributed parameters per unit length L (inductance), C (capacitance), R (resistance) and G (conductance), our fictitious, rather specialized line, comprises only L and G.

Having defined the model, we seek answers to the following questions:
a) How does a current surge propagate along the rod?
b) How does a current-surge *field* propagate in the soil concentrically with the "injecting" grounding conductor AB specified in Fig. 4.2?

A reply to these queries will lead us, in turn, to an estimate of suitably defined surge impedances.

4.1.2 Solution of Field Problem

a) Denoting the (horizontal) rod potential and current by the symbols V and I, respectively, the propagation equations read

$$\frac{\partial V}{\partial x} = -L \frac{\partial I}{\partial t} , \tag{4.1.6}$$

$$\frac{\partial I}{\partial x} = -GV . \tag{4.1.7}$$

The constants L, G are estimated at hand of the approximations [4.2]

$$L = \frac{\mu_0}{2\pi} \ln \frac{2h}{\varrho_0}; \quad G = 2\pi\sigma_s \frac{1}{\ln(2h/\varrho_0)} , \tag{4.1.8}$$

and the current surge is modelled by a step function

$$1(t) = \begin{cases} 0, & t < 0 , \\ 1, & t > 0 \end{cases} \tag{4.1.9}$$

of strength I_0. Our aim is therefore to solve (4.1.6, 7) simultaneously, subject to the initial conditions

$$V = 0, \quad I = 0, \quad t < 0 , \tag{4.1.10}$$

and to the imposed forcing function

$$I = I_0 \cdot 1(t) \tag{4.1.11}$$

applied at an arbitrarily selected input plane labelled $x = 0$.

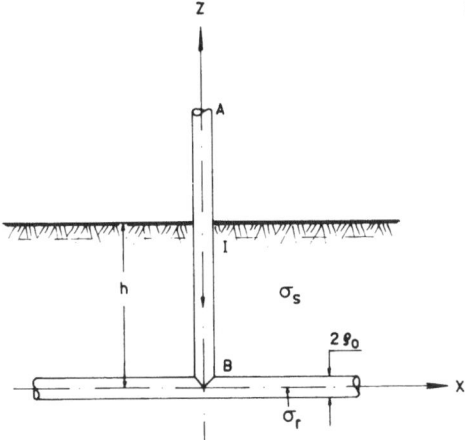

Fig. 4.2. Grounding system

The solutions are expressed through the error integral Φ of argument α defined by the equation

$$\Phi(\alpha) \equiv \frac{2}{\sqrt{\pi}} \int_0^\alpha e^{-\xi^2} d\xi \ .$$ (4.1.12)

Thus (Sect. 4.1.5)

$$I = \frac{1}{2} I_0 \left[\pm 1 - \Phi \left(\sqrt{\frac{GL}{4t}} x \right) \right], \quad x \gtrless 0, \quad t > 0 , \quad \text{and}$$ (4.1.13)

$$V = I_0 \sqrt{\frac{L}{G4\pi t}} e^{-(GLx^2/4t)}, \quad x \gtrless 0, \quad t > 0 \ .$$ (4.1.14)

These expressions provide the answer to our question (a) – see above.

A lightning stroke starts with a relatively weak predischarge which proceeds from a cloud to ground, and which is followed closely by a very luminous "return stroke" in the opposite direction. This predischarge (also known as the "stepped leader") apparently begins with a local electric breakdown and typically comprises a negative charge of about 5 A · s near the ground, propagating at velocities of the order of 1.0×10^5 m/s. Although photographic measurements show the "stepped leader" to have a discharge diameter of the order of one meter, the current is assumed to be confined to a very narrow core, whose radius we identify here with the *half-width* x_0 of the surge. Thus, with

$$\frac{GLx_0^2}{4t} = \frac{1}{2} ,$$ (4.1.15)

we obtain the time instant t_0 at which V attains its highest value [see remark after (4.1.19) below] for $x = x_0$, namely

$$t \equiv t_0 = \frac{GLx_0^2}{2} \; . \tag{4.1.16}$$

Assuming arbitrarily $x_0 = 0.05$ and propagation in a soil characterized by $\sigma_s = 10^{-2} (\Omega m)^{-1}$, we obtain approximately $t_0 \simeq 1.6 \times 10^{-11}$ s, a value small even when compared to the soil relaxation time $\tau_r = \varepsilon_r \varepsilon_0 / \sigma_s$: for the considered time resolution V instantaneously reaches its highest value.

In order to estimate surge impedances, we now proceed as follows: rewriting (4.1.14) in the form

$$V = \frac{I_0}{Gx_0\sqrt{\pi}} \sqrt{\frac{GLx_0^2}{4t}} e^{-(GLx_0^2/4t)} \; , \tag{4.1.17}$$

and introducing the dimensionless time parameter τ defined as

$$\tau \equiv \frac{4t}{GLx_0^2} \; , \tag{4.1.18}$$

we are able to express the dimensionless line voltage $v(\tau)$ by the equation

$$v(\tau) \equiv \frac{V}{I_0/(Gx_0\sqrt{\pi})} = \frac{1}{\sqrt{\tau}} e^{-(1/\tau)} \; . \tag{4.1.19}$$

The function $v(\tau)$ (reproduced in Fig. 4.3) attains attains its highest value of $\exp(-1/2)/\sqrt{2} = 0.429$ at $\tau = 2$; the ratio between the maximum voltage V_{max} and I_0, i.e., the horizontal (subscript h) surge (subscript s) impedance $R_{s,h}$ defined at $x = x_0$ for $t = t_0$, thus equals

$$\frac{V_{max}}{I_0} \equiv R_{s,h} = \frac{0.429}{Gx_0\sqrt{\pi}} \; . \tag{4.1.20}$$

For an estimate let us assume some practical values, e.g., $\sigma_s \simeq 1 \times 10^{-2} (\Omega m)^{-1}$; $h = 1$ m; $\varrho_0 = 0.05$ m; $x_0 = 0.05$ m, obtaining $R_{s,h} \simeq 285 \; \Omega$. We consider $R_{s,h}$ as an estimate of the surge resistance, at the onset of the stroke.

b) In the second part of the problem, we consider the field in the vicinity of the vertical injection rod, in the absence of any disturbance due to the grounding pipe.

In the fictitious soil ($\varepsilon_r = 0$) of conductivity σ_s and permeability μ_0, Maxwell's first curl equation once more simplifies to

$$\nabla \times H = \sigma_s E \; , \tag{4.1.21}$$

whereas the induction law retains its usual form

$$\nabla \times E = -\frac{\partial}{\partial t} \mu_0 H \; . \tag{4.1.22}$$

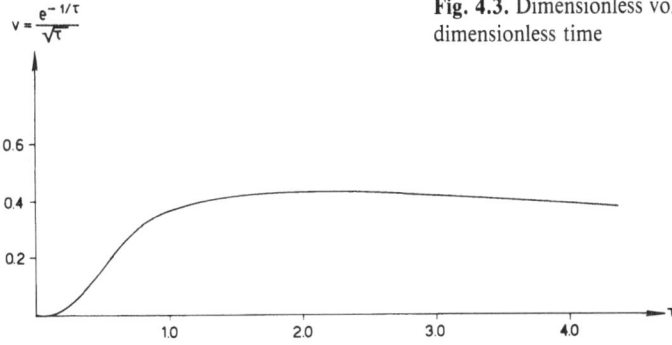

Fig. 4.3. Dimensionless voltage as dependent on dimensionless time

We now assume in the nonmagnetic soil the existence of a vector potential V from which the field H is derived by the usual curl operation

$$\mu_0 H = \nabla \times V \ . \tag{4.1.23}$$

Hence (4.1.22) may be integrated to yield

$$E = -\frac{\partial V}{\partial t} - \nabla \phi \ , \tag{4.1.24}$$

wherein ϕ denotes once more the electric scalar potential.

Multiplying both sides of (4.1.21) by μ_0 and substituting (4.1.23,24), we obtain

$$\nabla \times (\nabla \times V) = -\sigma_s \mu_0 \left(\frac{\partial V}{\partial t} + \nabla \phi \right) \quad \text{or} \tag{4.1.25}$$

$$\nabla(\nabla \cdot V) - \nabla^2 V = -\sigma_s \mu_0 \frac{\partial V}{\partial t} - \sigma_s \mu_0 \nabla \phi \ . \tag{4.1.26}$$

We simplify this equation by arbitrarily imposing upon the (as yet undefined) quantity $\nabla \cdot V$ the condition

$$\nabla \cdot V = -\sigma_s \mu_0 \phi \ , \tag{4.1.27}$$

and thus obtain for V the three dimensional vector diffusion equation

$$\nabla^2 V = \sigma_s \mu_0 \frac{\partial V}{\partial t} \ . \tag{4.1.28}$$

Now (4.1.27) is converted into an identity through introduction of the modified Hertz vector potential π, defined in turn by the equations

$$V \equiv -\sigma_s \mu_0 \pi \ , \tag{4.1.29}$$

$$\phi \equiv \nabla \cdot \pi \ . \tag{4.1.30}$$

Thus, π itself satisfies the diffusion equation

$$\nabla^2 \pi = \sigma_s \mu_0 \frac{\partial \pi}{\partial t} \ . \tag{4.1.31}$$

We now resort to a procedure introduced by *Ollendorff* in his classic book "Erdströme" (Ground Currents) ([Ref. 4.1, p. 319)]. In the infinite homogeneous and isotropic medium which models the fictitious soil (after the Kelvin reflection idea has here been extended heuristically), we provisionally replace the vertical rod AB by a dipole of infinitesimal extension l and moment $p(t)$ defined through the equality

$$p(t) = I_0 l \cdot 1(t) \ . \tag{4.1.32}$$

Now, with the aid of the complex variable s, the unit step $1(t)$ is represented, using Cauchy's integral and the Bromwich theorem, by

$$1(t) = \frac{1}{2\pi i} \int_{\overset{\curvearrowright}{\uparrow}} \frac{e^{st}}{s} \, ds \ , \tag{4.1.33}$$

where the symbol beneath the integral indicates that the pole $s = 0$ has been excluded from the integration contour. Hence

$$p(t) = I_0 l \frac{1}{2\pi i} \int_{\overset{\curvearrowright}{\uparrow}} \frac{e^{st}}{s} \, ds \ . \tag{4.1.34}$$

In a spherical system of coordinates (r, θ, α), with origin at the centroid of the dipole, the *z-directed* Hertzian component π_z satisfies the spherical symmetric differential equation

$$\frac{1}{r} \frac{\partial^2 (r \pi_z)}{\partial r^2} = \sigma_s \mu_0 \frac{\partial \pi_z}{\partial t} \ . \tag{4.1.35}$$

For the dipole, we assume the solution of (4.1.35) to be represented by the time-coherent expression

$$\pi_z = I_0 l \frac{1}{2\pi i} \int_{\overset{\curvearrowright}{\uparrow}} \frac{e^{st}}{s} f(r, s) \, ds \ , \tag{4.1.36}$$

where the function $f(r, s)$ is as yet unknown; when substituted in (4.1.35), the above yields the ordinary differential equation

$$\frac{d^2 (r f)}{dr^2} = \sigma_s \mu_0 s (r f) \ . \tag{4.1.37}$$

Convergence being dictated by physical considerations, the integral yields

$$f = A \frac{e^{-r\sqrt{\sigma_s \mu_0 s}}}{r} \ , \tag{4.1.38}$$

where A is yet to be specified, and the root is subject to the requirement

$$\text{Re}\{\sqrt{\sigma_s\mu_0 s}\} > 0 \ . \tag{4.1.39}$$

The scalar potential ϕ now reads, see (4.1.30),

$$\phi = \frac{\partial \pi_z}{\partial z} = -I_0 l \frac{A \cos\theta}{2\pi i} \int_{\mathcal{P}} \frac{e^{st}}{s} \left(\frac{1}{r^2} + \frac{\sqrt{\sigma_s\mu_0 s}}{r}\right) e^{-r\sqrt{\sigma_s\mu_0 s}} ds \ . \tag{4.1.40}$$

With $r\to 0$, the potential

$$\phi = -I_0 l \frac{A \cos\theta}{r^2} \frac{1}{2\pi i} \int_{\mathcal{P}} \frac{e^{st}}{s} ds \ , \tag{4.1.41}$$

is obtained, which is that of a dipole of intensity

$$p(t) = -(I_0 l A)4\pi\sigma_s 1(t) \ . \tag{4.1.42}$$

Comparison with (4.1.34) immediately yields

$$A = -\frac{1}{4\pi\sigma_s} \ , \tag{4.1.43}$$

so that, see (4.1.36, 38),

$$\pi_z = -I_0 l \frac{1}{4\pi\sigma_s r} \frac{1}{2\pi i} \int_{\mathcal{P}} \frac{e^{st - \sqrt{\sigma_s\mu_0 s}\, r}}{s} ds \ . \tag{4.1.44}$$

The branch cut is still, as shown in Sect. 4.1.5 (Fig. 4.6), and for all $t > 0$ integration may once more be performed along it; with the notation

$$\sqrt{\sigma_s\mu_0 s} \equiv i\lambda \ , \tag{4.1.45}$$

we thus have

$$\pi_z = -I_0 l \frac{1}{4\pi\sigma_s r} \frac{2}{2\pi i} \int_{\lambda = -\infty}^{\infty} \frac{e^{-[(\lambda^2/\sigma_s\mu_0)t + i\lambda r]}}{\lambda} d\lambda \ . \tag{4.1.46}$$

At this stage, we are ready to resume consideration of the injecting rod: having dispensed with the horizontal pipe — while adhering to the "infinite" soil model — our rod is taken to extend over $z \to \pm\infty$. We now replace l by the extension Δz and resort to the notation

$$r^2 = (x^2 + y^2) + z^2 \equiv \varrho^2 + z^2 \ . \tag{4.1.47}$$

Each rod element is represented by a dipole for which the *elementary* Hertzian potential $\Delta\pi_z$ now reads

$$\Delta \pi_z = -I_0 \Delta z \, \frac{1}{4\pi\sigma_s\sqrt{\varrho^2+z^2}} \, \frac{1}{\pi i} \int_{\lambda=-\infty}^{\infty} \frac{e^{-[(\lambda^2/\sigma_s\mu_0)t+i\lambda\sqrt{\varrho^2+z^2}]}}{\lambda} \, d\lambda \;,$$

(4.1.48)

and the overall potential π_z due to the rod is obtained through summation:

$$\pi_z = -I_0 \, \frac{1}{4\pi\sigma_s} \int_{z=-\infty}^{\infty} \frac{dz}{\sqrt{\varrho^2+z^2}} \, \frac{1}{\pi i} \int_{\lambda=-\infty}^{\infty} \frac{e^{-[(\lambda^2/\sigma_s\mu_0)t+i\lambda\sqrt{\varrho^2+z^2}]}}{\lambda} \, d\lambda \;.$$

(4.1.49)

After integration along z, the component π_z is obviously no more z-dependent; hence, in this case,

$$\phi = -\frac{\partial \pi_z}{\partial z} = 0 \;,$$

(4.1.50)

and the field component E_z reduces to the expression

$$E_z = -\frac{\partial V_z}{\partial t} = \sigma_s \mu_0 \frac{\partial \pi_z}{\partial t} \;.$$

(4.1.51)

Thus,

$$E_z = I_0 \, \frac{1}{4\pi\sigma_s} \int_{z=-\infty}^{\infty} \frac{dz}{\sqrt{\varrho^2+z^2}} \, \frac{1}{\pi i} \int_{\lambda=-\infty}^{\infty} \lambda e^{-[(\lambda^2/\sigma_s\mu_0)t+i\lambda\sqrt{\varrho^2+z^2}]} d\lambda \;.$$

(4.1.52)

Now [Ref. 4.3, No. 3.462−6, p. 338]

$$\int_{\lambda=-\infty}^{\infty} \lambda e^{-[(\lambda^2/\sigma_s\mu_0)t+i\lambda\sqrt{\varrho^2+z^2}]} d\lambda = \frac{i}{2} \, \frac{\sqrt{\varrho^2+z^2}}{t/\sigma_s\mu_0} \sqrt{\frac{\pi}{t}} \sqrt{\sigma_s\mu_0} \, e^{-\frac{\varrho^2+z^2}{4t}\sigma_s\mu_0} \;,$$

(4.1.53)

so that

$$\frac{1}{\pi i} \int_{\lambda=-\infty}^{\infty} \lambda e^{-[(\lambda^2/\sigma_s\mu_0)t+i\lambda\sqrt{\varrho^2+z^2}]} d\lambda$$

$$= \frac{1}{2\sqrt{\pi}} \left(\frac{\sigma_s\mu_0}{t}\right)^{3/2} e^{-\frac{\varrho^2+z^2}{4t}\sigma_s\mu_0} \sqrt{\varrho^2+z^2} \;,$$

(4.1.54)

i.e.,

$$E_z = I_0 \, \frac{1}{4\pi\sigma_s} \, \frac{1}{2\sqrt{\pi}} \left(\frac{\sigma_s\mu_0}{t}\right)^{3/2} e^{-(\sigma_s\mu_0/4t)\varrho^2} \int_{z=-\infty}^{\infty} e^{-(\sigma_s\mu_0/4t)z^2} dz \;. \quad (4.1.55)$$

Hence, finally

$$E_z = -I_0 \frac{1}{\pi \sigma_s} \frac{\sigma_s \mu_0}{4t} e^{-(\sigma_s \mu_0/4t)\varrho^2}, \quad t \geq 0 , \qquad (4.1.56)$$

which provides the answer to our question (b) – see above.

Introducing now again the dimensionless time parameter τ through the defining identity

$$\frac{\sigma_s \mu_0}{4t} \varrho^2 \equiv \frac{1}{\tau} , \qquad (4.1.57)$$

we may rewrite the absolute value $|E_z|$ of E_z as

$$|E_z| = |I_0| \frac{1}{\pi \sigma_s \varrho^2} \frac{1}{\tau} e^{-(1/\tau)} . \qquad (4.1.58)$$

The dimensionless field function

$$\eta \equiv \frac{|E_z|}{|I_0|/(\pi \sigma_s \varrho^2)} = \frac{e^{-(1/\tau)}}{\tau} \qquad (4.1.59)$$

is reproduced in Fig. 4.4. It attains its maximum value $\exp(-1)$ at $\tau = 1$.

Let us remark in passing that with $t \equiv t_0$, $\varrho \equiv \varrho_m$ for $\tau = 1$, (4.1.57) becomes

$$\varrho_m^2 = \frac{4t_0}{\sigma_s \mu_0} , \qquad (4.1.60)$$

i.e.,

$$2\varrho_m d\varrho_m = \frac{4dt_0}{\sigma_s \mu_0} , \qquad (4.1.61)$$

so that the ratio

$$\frac{d\varrho_m}{dt_0} = \frac{2}{\sigma_s \mu_0 \varrho_m} = \frac{1}{\sqrt{\sigma_s \mu_0 t_0}} \equiv v_{fr} \qquad (4.1.62)$$

may be taken as the front spread propagation velocity v_{fr} of the cylindrical surge field. Strictly speaking, v_{fr} has here a formal meaning only and is void of any physical content: taking once more $\sigma_s = 10^{-2} (\Omega m)^{-1}$ and $\varrho_m = 0.05$ m, we have $v_{fr} \sim 30 \times 10^8$ m/s, which is about ten times the phase velocity c of a plane light wave in vacuum – a contingency ruled out in the dynamic sense: it should be borne in mind that we have taken $\varepsilon_r \varepsilon_0 = 0$ and assumed arbitrary values for σ_s and ϱ_m. Therefore, as the form of the differential equation

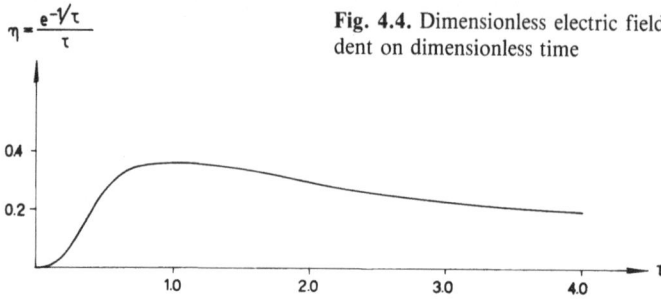

Fig. 4.4. Dimensionless electric field component as dependent on dimensionless time

(4.1.35) is that of the heat equation, the field build-up is not of dynamic nature: the artificial result obtained is due to the form $\nabla \times H = \sigma_s E$ of Maxwell's law replacing the more general form $\nabla \times H = \sigma_s E + \partial(\varepsilon_r \varepsilon_0 E)/\partial t$. In conclusion, as v_{fr} is not really essential to our understanding of the problem, we shall adhere to our approach which, as already remarked, greatly simplifies the computational labor involved; the excessive value of v_{fr} implies once more, as in the previous case, that the peak rod voltage is instantaneously attained on the time scale resolution considered.

The value $\varrho_m = 0.05$ m for the current "wave" front at $t = t_0$ was chosen with the assumed statistical width of a "stepped leader". Returning now to (4.1.58) with $(1/\tau)\exp(-1/\tau) = 1$, and defining the (absolute) vertical (subscript v) rod voltage $|V_v|$ through the relation

$$|V_v| \equiv |E_z|h \; ,\tag{4.1.63}$$

we find for its maximum absolute value $|V_v|_{\max}$ the expression

$$|V_v|_{\max} = \frac{|I_0|}{\pi \sigma_s \varrho_m^2} \, \mathrm{e}^{-1} \cdot h \; .\tag{4.1.64}$$

Identifying therefore $\mathrm{e}^{-1} h/4\pi\sigma_s\varrho_m^2$ with the vertical (subscript v) surge (subscript s) impedance $R_{s,v}$, we have

$$R_{s,v} = \frac{\mathrm{e}^{-1}\cdot 1}{\pi \times 10^{-2} \times 0.05^2} \simeq 4{,}700 \; \Omega \; .\tag{4.1.65}$$

4.1.3 Steady-State Resistance Levels of Grounding Rods

Having obtained an estimate for the surge impedance, let us turn for comparison, to the opposite end of the scale, i.e. to the appropriate steady-state resistances of the rods.

Our estimates will again refer to *Ollendorff's* book; thus, assigning to the vertical rod the distributed resistance of $R = 0.05 \; \Omega/\mathrm{m}$ (as opposed to the above approximation with $R = 0$), we obtain its specific conductivity σ_r from

$$R = \frac{1}{\sigma_r \pi \varrho_0^2} \; ,\tag{4.1.66}$$

i.e., with $\varrho_0 = 0.05$ m, we have $\sigma_r = 2.55 \times 10^3\,(\Omega \text{m})^{-1}$. The propagation resistance R_0 of the injection rod is estimated by [Ref. 4.1, p. 177, Eq. (5.6.6)],

$$R_0 = \frac{1}{2.26\,\varrho_0 \sqrt{\sigma_r \sigma_s}} \,, \tag{4.1.67}$$

yielding the value $R_0 \simeq 1.75\ \Omega$, much smaller than $R_{s,v} \simeq 4.700\ \Omega$.

As regards the horizontal rod, it should be borne in mind that, in practice, the soil is inhomogeneous due, for instance, to the presence of ground water. Applying the equation [Ref. 4.1, p. 139, Eq. (4.9.9)]

$$G = \frac{2\pi\sigma_s}{\ln\left[\left(\dfrac{4}{\pi}\dfrac{H}{\varrho_0}\right) \cdot \cot\left(\dfrac{\pi}{2}\dfrac{h}{H}\right)\right]} \,, \qquad \varrho_0 < h < H \tag{4.1.68}$$

to the schematic representation of Fig. 4.5, we obtain an estimate for the propagation conductivity G per unit length of the horizontal grounding rod. [Remark: *Ollendorff's* original equation has been adapted to our notation]. With $\varrho_0 = 0.05$ m, $h = 1$ m, $H = 2$ m, we obtain $G = 1.60 \times 10^{-2}\,(\Omega \text{m})^{-1}$. Taking for the horizontal rod the same value of R as above, there results an estimate [Ref. 4.1, p. 162, Eq. (5.3.14)] for the propagation resistance R_p of the ground conductor; thus

$$R_p = \frac{1}{2}\sqrt{\frac{R}{G}} \,, \tag{4.1.69}$$

or, in our case ($R_p = 0.88\ \Omega$) again much lower than $R_{s,h} \simeq 285\ \Omega$.

4.1.4 Discussion

Failure of equipment and bushing flashover under transient conditions due, for instance, to lightning surge currents, is at least in part to be ascribed to the

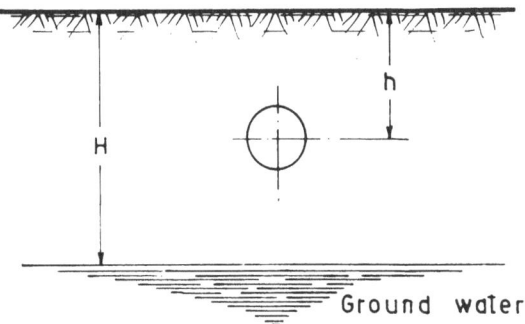

Fig. 4.5. Schematic representation of soil and ground water

high surge impedance levels exhibited in practice by grounding systems. The present section provides an outline to an engineering estimation of such impedances; although oversimplifications have here been permitted, e.g., neglect of soil displacement currents; extension of image theory; some extension of transmission line theory; assumption of validity of Ohm's law in the soil even in the case of high current lightning strokes; homogeneity of the soil, etc., it nevertheless seems − in view of the results which indeed show a marked increase of the surge impedances compared to their static counterparts − that the overall approach is quite plausible [Ref. 4.1, pp. 152−160 and 319−356]. Let us not forget to mention that various variables, e.g., return strokes, surge current peak values, waveforms and grounding electrode forms have not been considered: they all contribute, in fact, to the actual value of the surge impedance. Finally, no attention has been paid to the junction, which gives rise to secondary end effects not directly contributing insight to the posed problem [Ref. 4.1, p. 306].

4.1.5 Appendix

Combining (4.1.6 and 7), we obtain the one-dimensional scalar diffusion equation

$$\frac{\partial^2 I}{\partial x^2} = GL \frac{\partial I}{\partial t} \ . \tag{4.1.70}$$

With the complex quantity $s = u + iv$ as an independent variable, we introduce

$$I(x,t) = \frac{1}{2\pi i} \int_{-i\infty}^{i\infty} e^{st} \bar{I}(x,s)\,ds\,, \quad t > 0 \ , \tag{4.1.71}$$

and substituting (4.1.71) into (4.1.70), we have (dispensing with the partial derivative notation in the absence of ds in the expression)

$$\frac{d^2 \bar{I}}{dx^2} = GLs\bar{I} \ . \tag{4.1.72}$$

The solution of (4.1.72) for $x > 0$ is given (with the aid of an as yet undetermined complex amplitude $\bar{A}$) by

$$\bar{I}(x,s) = \bar{A}\,e^{-\sqrt{GLs}\,x} \ , \tag{4.1.73}$$

subject to

$$\mathrm{Re}\{\sqrt{s}\} > 0 \ . \tag{4.1.74}$$

Now, for $x = 0$, the imposed current is represented in the complex s-plane by

$$\bar{I}(0,s) = \frac{I_0}{s} \ . \tag{4.1.75}$$

Combining (4.1.73 and 75), we have

$$I(x, t) = I_0 \frac{1}{2\pi i} \int_{-i\infty}^{i\infty} \frac{e^{st}}{s} e^{-\sqrt{GLs}x} ds, \quad t > 0 . \tag{4.1.76}$$

In order to evaluate the integral, we first consider

$$V = -\frac{1}{G} \frac{\partial I}{\partial x}, \tag{4.1.77}$$

namely

$$V(x, t) = I_0 \frac{\sqrt{GL}}{G} \frac{1}{2\pi i} \int_{-i\infty}^{i\infty} \frac{e^{st}}{\sqrt{s}} e^{-\sqrt{GLs}x} ds, \quad t > 0 . \tag{4.1.78}$$

The sheet of the double-valued Riemann surface $\sqrt{s}$ is bounded by a branch cut $s = s_0$ which, in turn, is specified by means of the *real* variable q through

$$\sqrt{s_0} = iq; \quad s_0 = -q^2 . \tag{4.1.79}$$

The branch cut runs along the negative u-axis as shown in Fig. 4.6. In its close vicinity, namely

$$s = s_0 + i\varepsilon; \quad |\varepsilon| \ll |s_0| \tag{4.1.80}$$

we have

$$\sqrt{s} = \sqrt{s_0} \sqrt{1 + \frac{i\varepsilon}{s_0}} = iq + \frac{\varepsilon}{2q} + \dots , \tag{4.1.81}$$

and therefore

$$\varepsilon \lessgtr 0 \quad \text{for} \quad q \gtrless 0 . \tag{4.1.82}$$

For $t < 0$, the initially-open path of the integral may be closed by the half-circle $|s| \to \infty$, $u > 0$ without affecting the overall picture. The loop thus formed does not contain any singularities, hence

$$V(x, t) = 0, \quad t < 0 . \tag{4.1.83}$$

For $t > 0$, the two quarter-circles $u < 0$; $v \gtrless 0$, contribute nothing to the integral in the limit $|s| \to \infty$; accordingly, by virtue of Cauchy's theorem, the integral may be evaluated at the edges of the branch cut, so that

$$V(x, t) = I_0 \frac{\sqrt{GL}}{G} \frac{1}{\pi} \int_{q=-\infty}^{\infty} e^{-[q^2 t + iqx\sqrt{GL}]} dq, \quad t > 0 . \tag{4.1.84}$$

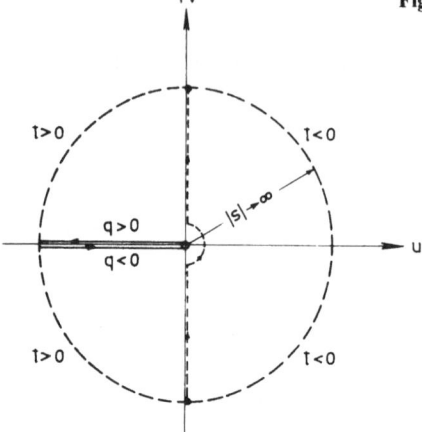

Fig. 4.6. Path of integration in complex s-plane

Resorting to the identity

$$q^2 t + iqx\sqrt{GL} \equiv \left(q\sqrt{t} + \frac{1}{2}ix\sqrt{\frac{GL}{t}}\right)^2 + \frac{x^2 GL}{4t}, \qquad (4.1.85)$$

and using the substitution

$$p = q\sqrt{t} + \frac{1}{2}ix\sqrt{\frac{GL}{t}}$$

we obtain

$$V(x,t) = I_0 \frac{\sqrt{GL}}{G} \frac{1}{\sqrt{t}} e^{(x^2 GL/4t)} \frac{1}{\pi} \int_{p=-\infty}^{\infty} e^{-p^2} dp$$

$$= \frac{I_0}{G} \sqrt{\frac{GL}{t}} e^{-(x^2 GL/4t)} \frac{1}{\sqrt{\pi}}; \qquad t>0 . \qquad (4.1.86)$$

Hence, finally,

$$V(x,t) = I_0 \sqrt{\frac{L}{G\pi t}} e^{-(x^2 GL/4t)}, \qquad t>0 . \qquad (4.1.87)$$

Now, using (4.1.7) with I' and x' as variables of integration, we get

$$\int_{I'=I_0}^{I} dI' = -GI_0 \sqrt{\frac{L}{G\pi t}} \int_{x'=0}^{x} e^{-(x'^2 GL/4t)} dx', \qquad \text{or} \qquad (4.1.88)$$

$$I = I_0 \left[1 - \Phi\left(\sqrt{\frac{GL}{4t}}x\right) \right]. \qquad (4.1.89)$$

Let us point out that (4.1.87, 89) were obtained with the implicit assumption $x > 0$ with I_0 injected at $x = 0$. The rod, however, extends along the x-axis from $-\infty$ to $+\infty$; hence, in view of symmetry, the initial surge current I_0 distributes itself uniformly in both directions $x \gtrless 0$, i.e.

$$\lim_{\Delta \to 0} I(\pm \Delta) = \pm \frac{1}{2} I_0 \ , \tag{4.1.90}$$

where Δ is an arbitrarily small extension in the close vicinity of the surge plane $x = 0$.

Bearing in mind that (4.1.12) represents an odd function, namely $\Phi(-\alpha) = -\Phi(\alpha)$, (4.1.89) is now finally converted into the more comprehensive expression

$$I = \frac{1}{2} I_0 \left[\pm 1 - \Phi \left(\sqrt{\frac{GL}{4t}} \, x \right) \right], \quad x \gtrless 0, \quad t > 0 \ , \tag{4.1.91}$$

whereas (4.1.87) is valid for both $(+x)^2$ and $(-x)^2$:

$$V(x, t) = I_0 \sqrt{\frac{L}{G4\pi t}} \, e^{-(GLx^2/4t)}, \quad x \gtrless 0, \quad t > 0 \ . \tag{4.1.92}$$

4.2 Field Transients in Saturable Cores

Current surges frequently tend to saturate ferromagnetic components of power equipment; the phenomenon, accompanied by a variety of undesirable effects as a result of drastically diminished impedance levels of saturated equipment, has been tackled in the past [4.4] and also more recently [4.5, 6]; however, with ever-increasing use of highly saturable components subject to a diversity of switching processes, it would be useful to state some of the basic premises and results. Accordingly we address ourselves in this section to the following questions:

- How does the *field* build up in the core?
- How does the *saturation effect* proper spread within the iron?

4.2.1 Formulation of Problem and Basic Assumptions

A slab of ferromagnetic material of finite width L extends to infinity along the $(\pm x)$ and $(\pm y)$ axes of a right-hand Cartesian frame of reference (Fig. 4.7) in which the time parameter t is defined. The electromagnetic field acting inside the slab is, for simplicity, taken to be confined to the region $0 < z < L$, by means of two "ideally conducting" semi-infinite short-circuiting planes

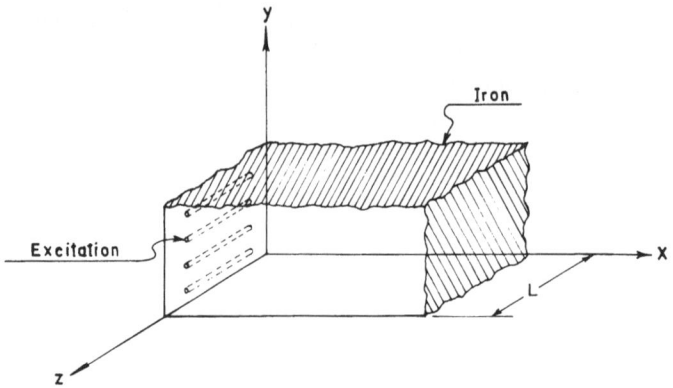

Fig. 4.7. Ferromagnetic slab

located at $z = 0$, $z = L$, respectively. Excitation, provided at the $x = 0$ plane, is modelled by an array of "internal" synchronized surface generators imposing a z-directed surface current density of amplitude A_0. The time dependence of this current is modelled by a step function $1(t)$. The magnetic properties of the iron are described by the schematic saturation curve shown in Fig. 4.8. The "break" in the magnetic characteristic is determined by the critical saturation field H_s and by the corresponding magnetic induction B_s. The unsaturated region $H < H_s$ is characterized by a constant relative permeability $\mu_r = \mu_1$ (vacuum permeability μ_0) while the saturated region is likewise characterized by constant relative permeability $\mu_r = \mu_2$, with $\mu_2 < \mu_1$. We further ascribe to this iron mass a phenomenological scalar conductivity σ [of, say, $1 \times 10^7 \, (\Omega m)^{-1}$] thus enabling a conduction current of density j to pervade it.

The meaning of the relative dielectric constant ε_r (vacuum permittivity ε_0) is not entirely clear when dealing with metals; within the Faraday-Maxwell phenomenological approach (excluding electrostatics), ε_r is taken to equal unity. For occurrences in which the magnetic energy density highly outweighs its electric counterpart, we have consistently assumed validity of the mathematical fiction $\varepsilon_r = 0$. Let us here somewhat deepen our insight into ε_r for metals, making a mental note of the following formal approach [4.7]:

– At low frequencies ω the function $\varepsilon_r = \varepsilon_r(\omega)$ is defined for metals in such a way that Maxwell's first curl relation

$$\nabla \times H = \frac{\partial D}{\partial t} , \tag{4.2.1}$$

interlinking the magnetic field H with the electric induction D, reduces to the relatively simple equation

$$\nabla \times H = \sigma E , \tag{4.2.2}$$

thus involving the electric field E instead of D.

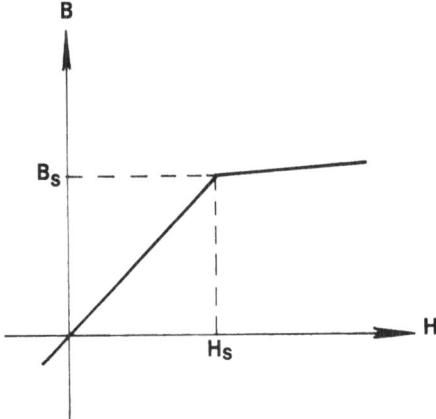

B

B$_s$

H$_s$

H

Fig. 4.8. Idealized saturation curve

– At optical frequencies, there is actually no difference between metals and dielectrics.
– At the far ultra-violet end (light elements) and in the x-ray region (heavier elements) we have

$$[1 - \varepsilon_r(\omega)] \propto 1/\omega^2 \;. \tag{4.2.3}$$

In general, however, for such high frequencies E and D acquire meanings differing from those of their quasistatic counterparts; this is quite clear from simple physical considerations, since under rapid field changes, electric polarisation processes which underlie the difference between E and D, cannot occur.

We briefly mention these formal results, as any time-varying field (in our case, the field due to the sudden excitation) may be resolved by a Fourier expansion into monochromatic oscillations for which – in phasor notation –

$$\underline{D} = \varepsilon_r(\omega)\,\varepsilon_0\underline{E} \;, \tag{4.2.4}$$

where, in turn, $\varepsilon_r(\omega)$ is obtained with the aid of an inherent time function $g(t)$ defined by

$$\varepsilon_r(\omega) \equiv 1 + \int_{t=0}^{\infty} g(t)\,e^{i\omega t}\,dt \;. \tag{4.2.5}$$

With a view to a workable phenomenological description of $\varepsilon \equiv \varepsilon_r\varepsilon_0$ for metals under transient conditions, we assume: (i) ε is in the order of about 10×10^{-12} farad/m (with $\varepsilon_0 = 8.854 \times 10^{-12}$ farad/m); (ii) the magnetic energy density w_m is much higher than its electric counterpart w_e which is equivalent to taking the magnetic diffusion time τ_d as much longer than the electric relaxation time τ_r. The typical penetration depth is, in the worst case, set here in the order of some atomic dimension a. Thus, taking $\tau_d = \mu_1\mu_0\sigma a^2$, $\tau_r = \varepsilon/\sigma$ and imposing the condition

$$\mu_1 \mu_0 \sigma a^2 \gg \frac{\varepsilon_r \varepsilon_0}{\sigma} \; , \tag{4.2.6}$$

we arrive at the requirement

$$(\sigma a) \gg \sqrt{\frac{\varepsilon_r \varepsilon_0}{\mu_r \mu_0}} \; , \tag{4.2.7}$$

or, e.g., $(10^{+7} \times 1 \times 10^{-10}) \gg 8 \times 10^{-5}$, obtained with $a \simeq 1 \times 10^{-10}$ m, $\varepsilon_r \simeq 1.0$ and $\mu_r \simeq 1000$. Experimental results having shown that $\tau_r \ll \tau_d$ for conductors, we sharpen the above requirement by also imposing

a) $\lim_{\substack{\sigma \to \infty \\ a \to 0}} (\sigma a) = \text{finite}$ \hfill (4.2.8)

and, further, for *finite current values* (implying $w_m \gg w_e$) we assume within the "iron" the mathematical fiction (Sect. 4.1.1)

b) $\lim_{\mu_r \to 1} \sqrt{\frac{\varepsilon_r}{\mu_r}} = 0$. \hfill (4.2.9)

Hence, Maxwell's two curl equations reduce (even under transient conditions) to

$$\nabla \times H = \sigma E \quad \text{and} \tag{4.2.10}$$

$$\nabla \times E = -\frac{\partial B}{\partial t} \; , \tag{4.2.11}$$

where validity of Ohm's law $j = \sigma E$ and of the constitutive equation $B = \mu_r \mu_0 H$ for the magnetic induction has been assumed. In particular, assuming in the sequel a magnetic field confined to the y-direction $H_y \equiv H$, and a similarly directed magnetic induction $B_y \equiv B$, we may write (see once more Fig. 4.8)

$$B = \mu_1 \mu_0 H, \quad 0 < H < H_s \; , \tag{4.2.12}$$

$$B = B_s + \mu_2 \mu_0 H, \quad H > H_s > 0 \; . \tag{4.2.13}$$

With the preliminary assumptions thus clarified, we now wish to determine how the saturation front advances inside the iron upon application of the above mentioned step function of excitation current density.

4.2.2 Field Solution

Under the prescribed geometry and mode of excitation, the post-switching electromagnetic field inside the iron is exclusively described by a *y-directed* component of magnetic field intensity H, and a *z-directed* electric field E. At the excitation plane $x = +0$ the magnetic field is modelled by a step function whose amplitude H_0 is prescribed by the imposed surface current density A_0, i.e. $A_0 = H_0$ thus

$$H = H_0 \cdot 1(t), \quad x = +0 , \tag{4.2.14}$$

wherein $H_0 > H_s$. (For $x < 0$, the field is taken as zero throughout.)

Saturation, setting in when $H > H_s$, prevails to an as yet undetermined distance $x = x_s$ beyond which the iron remains unsaturated, i.e., $H < H_s$. Maxwell's curl equations in conjunction with the constitutive equations presented in the preceding section now yield

$$\frac{\partial H}{\partial x} = \sigma E, \quad 0 < x , \tag{4.2.15}$$

along with

$$\frac{\partial E}{\partial x} = \mu_0 \mu_2 \frac{\partial H}{\partial t}, \quad 0 < x < x_s , \quad \text{and} \tag{4.2.16}$$

$$\frac{\partial E}{\partial x} = \mu_0 \mu_1 \frac{\partial H}{\partial t}, \quad x > x_s . \tag{4.2.17}$$

Upon elimination of E, we have

$$\frac{\partial^2 H}{\partial x^2} = \sigma \mu_0 \mu_2 \frac{\partial H}{\partial t}, \quad 0 < x < x_s , \tag{4.2.18}$$

$$\frac{\partial^2 H}{\partial x^2} = \sigma \mu_0 \mu_1 \frac{\partial H}{\partial t}, \quad x > x_s . \tag{4.2.19}$$

The particular solution of (4.2.18), reducing at the excitation plane $x = +0$ to H_0 for all times $t > 0$, is well known [4.8]; it is formulated (see also Sect. 4.1.2) with the aid of the error integral Φ of argument α, i.e.

$$\Phi(\alpha) \equiv \frac{2}{\sqrt{\pi}} \int_0^\alpha e^{-\xi^2} d\xi . \tag{4.2.20}$$

Introducing the as yet unspecified constant of integration h_2 (magnetic region μ_2!) we find

$$H = H_0 + h_2 \Phi \left(\frac{x}{2} \sqrt{\frac{\mu_0 \mu_2 \sigma}{t}} \right), \quad 0 \leqslant x < x_s , \tag{4.2.21}$$

whereas, in the μ_1 region (again resorting to an as yet unspecified constant h_1) the field reads

$$H = h_1 \left[1 - \Phi \left(\frac{x}{2} \sqrt{\frac{\mu_0 \mu_1 \sigma}{t}} \right) \right], \quad x > x_s .$$

(4.2.22)

At the plane of saturation $x = x_s$ the value of H must equal (by definition) that of H_s. Hence

$$H_0 + h_2 \Phi \left(\frac{x_s}{2} \sqrt{\frac{\mu_0 \mu_2 \sigma}{t}} \right) = H_s$$

(4.2.23)

and likewise

$$h_1 \left[1 - \Phi \left(\frac{x_s}{2} \sqrt{\frac{\mu_0 \mu_1 \sigma}{t}} \right) \right] = H_s .$$

(4.2.24)

In order to satisfy the last two requirements, the factor $(x_s/2)\sqrt{\mu_0 \sigma/t}$ appearing in the error integral must obviously have a constant value. Introducing the notation

$$\frac{x_s}{2} \sqrt{\frac{\mu_0 \sigma}{t}} \equiv k ,$$

(4.2.25)

we therefore find

$$h_2 = \frac{H_s - H_0}{\Phi(k\sqrt{\mu_2})} ; \quad h_1 = \frac{H_s}{1 - \Phi(k\sqrt{\mu_1})} ,$$

(4.2.26)

so that

$$H = H_0 + \frac{H_s - H_0}{\Phi(k\sqrt{\mu_2})} \Phi \left(\frac{x}{2} \sqrt{\frac{\mu_0 \mu_2 \sigma}{t}} \right), \quad 0 \leqslant x < x_s ,$$

(4.2.27)

$$H = \frac{H_s}{1 - \Phi(k\sqrt{\mu_1})} \left[1 - \Phi \left(\frac{x}{2} \sqrt{\frac{\mu_0 \mu_1 \sigma}{t}} \right) \right], \quad x > x_s .$$

(4.2.28)

We now turn our attention to the electric field intensity; in view of the obtained solutions for H, the values

$$E = \frac{1}{\sigma} \frac{H_s - H_0}{\Phi(k\sqrt{\mu_2})} \frac{2}{\sqrt{\pi}} \left(\frac{1}{2} \sqrt{\frac{\mu_0 \mu_2 \sigma}{t}} \right) e^{-(x^2/4)(\mu_0 \mu_2 \sigma)/t}, \quad 0 \leqslant x < x_s ,$$

(4.2.29)

$$E = -\frac{1}{\sigma}\frac{H_s}{1-\Phi(k\sqrt{\mu_1})}\frac{2}{\sqrt{\pi}}\left(\frac{1}{2}\sqrt{\frac{\mu_0\mu_1\sigma}{t}}\right)e^{-(x^2/4)(\mu_0\mu_1\sigma)/t}, \quad x > x_s$$

(4.2.30)

result in a straightforward way.

The continuity requirement for E at $x = x_s$ yields the equation

$$\frac{H_s-H_0}{\Phi(k\sqrt{\mu_2})}\sqrt{\mu_2}e^{-k^2\mu_2} = -\frac{H_s}{1-\Phi(k\sqrt{\mu_1})}\sqrt{\mu_1}e^{-k^2\mu_1},$$

(4.2.31)

i.e.,

$$\frac{H_0-H_s}{H_s} = \sqrt{\frac{\mu_1}{\mu_2}}e^{-k^2(\mu_1-\mu_2)}\frac{\Phi(k\sqrt{\mu_2})}{1-\Phi(k\sqrt{\mu_1})}.$$

(4.2.32)

Hence, k is in fact determined, and our problem is in principle solved. In order, however, to gain deeper insight into the obtained results, let us consider the practical case of $\mu_2 \ll \mu_1$. Passing to the limit $\mu_2 \to 0$, we have

$$\lim_{\mu_2\to 0}\frac{\Phi(k\sqrt{\mu_2})}{\sqrt{\mu_2}} = \lim_{\mu_2\to 0}\frac{(2/\sqrt{\pi})e^{-k^2\mu_2}k/\sqrt{4\mu_2}}{1/\sqrt{4\mu_2}} = \frac{2}{\sqrt{\pi}}k,$$

(4.2.33)

so that (4.2.32) yields the relatively simple relationship

$$\frac{H_0}{H_s} - 1 = \frac{2}{\sqrt{\pi}}e^{-k^2\mu_1}\frac{k\sqrt{\mu_1}}{1-\Phi(k\sqrt{\mu_1})}$$

(4.2.34)

which is reproduced in Fig. 4.9.

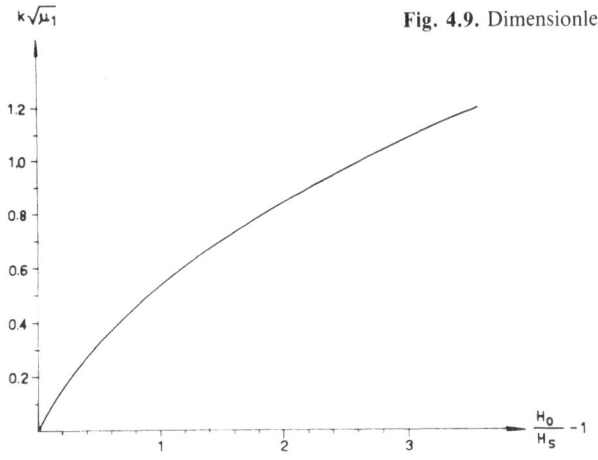

Fig. 4.9. Dimensionless saturation-ratio curve

Further, for $\mu_2 \to 0$, the magnetic field components, see (4.2.27, 28), reduce to the expressions

$$H = H_0 - (H_0 - H_s)\frac{x}{x_s}, \quad 0 \leqslant x < x_s \tag{4.2.35}$$

$$H = H_s \frac{1 - \Phi\left(\frac{x}{2}\sqrt{\frac{\mu_0\mu_1\sigma}{t}}\right)}{1 - \Phi(k\sqrt{\mu_1})}, \quad x > x_s, \tag{4.2.36}$$

so that the electric field component equation $E = (1/\sigma)(\partial H/\partial x)$, in turn, yields

$$E = -\frac{1}{\sigma}\frac{H_0 - H_s}{x_s}, \quad 0 \leqslant x < x_s, \tag{4.2.37}$$

and, see (4.2.30), once more, for completeness,

$$E = -\frac{1}{\sigma}\frac{H_s}{1 - \Phi(k\sqrt{\mu_1})}\frac{2}{\sqrt{\pi}}\left(\frac{1}{2}\sqrt{\frac{\mu_0\mu_1\sigma}{t}}\right)e^{-(x^2/4)(\mu_0\mu_1\sigma)/t},$$

$$x > x_s. \tag{4.2.38}$$

We now turn our attention to the inherent power losses p per unit volume due to the finite conductivity of the iron; in the saturated region $x > x_s$, we shall ascribe to p the subscript s, whereas in the unsaturated region – the subscript u. Thus,

$$p_s = \sigma E^2 = \sigma\frac{1}{\sigma^2}\frac{(H_0 - H_s)^2}{x_s^2}, \quad 0 \leqslant x < x_s. \tag{4.2.39}$$

Resorting to (4.2.25), the explicit time dependence emerges:

$$p_s = \mu_0\frac{(H_0 - H_s)^2}{4k^2 t}, \quad 0 \leqslant x < x_s. \tag{4.2.40}$$

Multiplying the numerator and the denominator by μ_1 and, further introducing the notation $B_s \equiv \mu_1\mu_0 H_s$, we readily find

$$p_s = \frac{1}{4}\frac{B_s H_s}{t}\frac{1}{(k\sqrt{\mu_1})^2}(H_0/H_s - 1)^2, \quad 0 \leqslant x < x_s. \tag{4.2.41}$$

With the aid of (4.2.34), we obtain an expression comprising well known functions, namely

$$p_s = \frac{1}{4}\frac{B_s H_s}{t}\frac{((2/\sqrt{\pi})e^{-(k\sqrt{\mu_1})^2})^2}{[1 - \Phi(k\sqrt{\mu_1})]^2}, \quad 0 \leqslant x < x_s. \tag{4.2.42}$$

Now $(2/\sqrt{\pi})\exp(-\alpha^2)$ is conventionally denoted by $\Phi_1(\alpha)$, representing the first derivative of the error integral, whereas the difference $[1 - \Phi(\alpha)]$ is usually tabulated under the standard notation $\mathrm{erfc}(\alpha)$; hence, the dimensionless loss curve (Fig. 4.10) is given by

$$\frac{p_s}{\dfrac{B_s H_s}{4t}} = \left(\frac{\Phi_1(k\sqrt{\mu_1})}{\mathrm{erfc}(k\sqrt{\mu_1})}\right)^2, \qquad x < x_s \tag{4.2.43}$$

which monotonically increases with $(k\sqrt{\mu_1})$: the larger the field ratio H_0/H_s, the higher the specific loss rate for the same instant of time t. Resorting once more to (4.2.25), we obtain for the unsaturated region

$$p_u = \frac{1}{\pi} \frac{B_s H_s}{t} \frac{1}{[1 - \Phi(k\sqrt{\mu_1})]^2} e^{-2(x^2/x_s^2)(k\sqrt{\mu_1})^2}, \qquad x > x_s \tag{4.2.44}$$

or, in a form similar to (4.2.43)

$$\frac{p_u}{\dfrac{B_s H_s}{4t}} = \frac{4}{\pi} \frac{1}{[\mathrm{erfc}(k\sqrt{\mu_1})]^2} e^{-2(x^2/4)(\mu_0\mu_1\sigma)/t}, \qquad x > x_s . \tag{4.2.45}$$

It is quite obvious that, for an imposed field ratio H_0/H_s, the volume loss rate p_u at a specified point in space-time is directly obtainable through Fig. 4.9 and (4.2.45).

It should be re-emphasized that the relatively simple field solutions obtained in terms of tabulated functions were arrived at by means of certain engineering approximations, the most pertinent of which are:
1) a highly simplified saturation curve;
2) "infiniteness" of the slab;
3) total neglect of any electric energy density, even though we are dealing with transients.

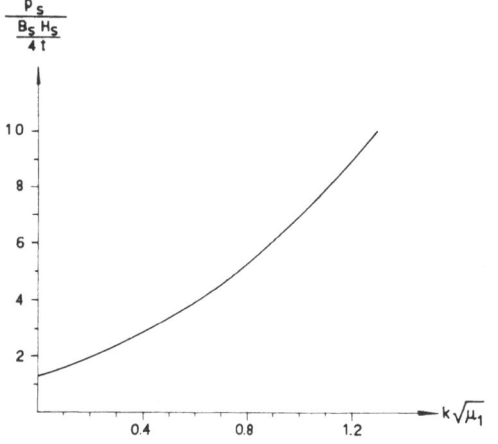

Fig. 4.10. Dimensionless saturation-loss curve

These simplifications necessarily have analytic consequences, of which it is pertinent to enumerate the following:

    – The phase velocity v of propagation of "light" inside the iron is infinite; this is due to the fact that the field is here governed by the heat equation rather than by the wave equation so that its build-up is not of dynamic nature, see Sect. 4.1.2 following (4.1.62). As, however, "light" cannot in reality propagate inside the opaque material (finite σ), it would perhaps be more accurate to say that Sommerfeld's forerunner – if defined here at all – is assumed to propagate across it with an infinite velocity; moreover, the said slab of extension $l \gg L$ is not really infinite along the x-axis. These two assumptions should, moreover, be reconciled with the fact that choice of a unit step function ascribes to the current generator a rise time $T_r = 0$. For internal coherence, all three approximations should be unified in the following two requirements:

$$\lim_{\substack{l \to \infty \\ v \to \infty}} \frac{l}{v} = \text{finite} \equiv T_1 \quad \text{and} \tag{4.2.46}$$

$$T_r \ll T_1 . \tag{4.2.47}$$

    – Although we are dealing with a transient field, the reduction of its degrees of freedom through the choice $\varepsilon_r \to 0$ leads to a stationary field inside the fictitious saturated region $\mu_2 \to 0$, i.e.

$$\nabla \times H = j , \quad 0 < x < x_s , \tag{4.2.48}$$

$$\nabla \times E = 0 , \quad 0 < x < x_s . \tag{4.2.49}$$

While there exists a Poynting vector S having only an x-component S, the field comprises no momentum density; furthermore, the energy propagation velocity $v_E = S/(w_e + w_m)$ is here also infinite, by definition. On the other hand, the divergence of S directly yields the required specific losses p_s

$$\frac{\partial}{\partial x} \left\{ -\left(\frac{1}{\sigma} \frac{H_0 - H_s}{x_s} \right) \left[H_0 - (H_0 - H_s) \frac{x}{x_s} \right] \right\} = -p_s , \quad 0 < x < x_s \tag{4.2.50}$$

or, see (4.2.39),

$$p_s = \frac{1}{\sigma} \frac{(H_0 - H_s)^2}{x_s^2} , \quad 0 < x < x_s . \tag{4.2.51}$$

    – Iron losses formally increase beyond all bounds; thus, the energy density w_s required to heat the saturated region reads, see (4.2.43),

$$w_s = \frac{B_s H_s}{4} \left(\frac{\Phi_1 (k\sqrt{\mu_1})}{\mathrm{erfc}(k\sqrt{\mu_1})} \right)^2 \int_{T_r}^{t} \frac{dt'}{t'}, \qquad 0 < x < x_s , \tag{4.2.52}$$

(as remarked, T_r stands for the generator rise time), whereas its counterpart w_u in the unsaturated region reads[1], see (4.2.45),

$$w_u = \frac{B_s H_s}{4} \frac{4/\pi}{[\mathrm{erfc}(k\sqrt{\mu_1})]^2} \int_{T_r}^{t} \frac{e^{-2(x^2/4)(\mu_0\mu_1\sigma)/t'}}{t'} dt', \qquad x > x_s . \tag{4.2.53}$$

Explicitly,

$$\frac{w_s}{\frac{B_s H_s}{4}} = \lim_{t\to\infty} \left(\frac{\Phi_1 (k\sqrt{\mu_1})}{\mathrm{erfc}(k\sqrt{\mu_1})} \right)^2 \ln \frac{t}{T_r} , \tag{4.2.54}$$

and (Sect. 4.2.5)

$$\frac{w_u}{B_s H_s/4} = \lim_{t\to\infty} \frac{4/\pi}{[\mathrm{erfc}(k\sqrt{\mu_1})]^2} \left[-E_i \left(-\frac{x^2 \mu_0 \mu_1 \sigma}{2t} \right) \right] , \tag{4.2.55}$$

where

$$-E_i(-\alpha) \equiv \int_{\alpha}^{\infty} \frac{e^{-\xi}}{\xi} d\xi . \tag{4.2.56}$$

For $\alpha \to 0$, we find [4.9] $-E_i(-\alpha) \to +\infty$.

The limiting requirements may be further sharpened: for (4.2.54), by replacing the operative fraction t/T_r by its equivalent $(t/T_1)/(T_r/T_1)$, the limit as $t\to\infty$ being obtained as

$$\lim_{t\to\infty} \frac{t}{t_r} = \lim_{\substack{t/T_1\to\infty \\ T_r/T_1\to0}} \frac{t/T_1}{T_r/T_1} = \infty , \tag{4.2.57}$$

for (4.2.55), by multiplying the numerator and denominator of the operative fraction by a^2, namely

$$\frac{x^2}{a^2} \frac{\mu_0\mu_1\sigma a^2}{t} = \frac{x^2}{a^2} \frac{\tau_d}{t} \tag{4.2.58}$$

---

[1] The lower limit ("light velocity" $v = \infty$) is to a certain extent arbitrary and was adopted here in analogy with (4.2.52); interest centers on the upper limit which indicates that the integral (4.2.53) diverges.

the limit at $x \to l \to \infty$ being obtained as

$$\lim_{t \to \infty} \frac{x^2 \mu_0 \mu_1 \sigma}{t} = \lim_{\substack{x \to \infty \\ t \to \infty}} \frac{x^2}{a^2} \frac{\tau_d}{t} = 0 \ . \tag{4.2.59}$$

Having pinpointed some of the analytical constraints and emphasized the deficiencies of the solution, we may regard (4.2.35 – 38), and their corollaries (4.2.43, 45), as feasible engineering approximations.

4.2.3 Velocity of Saturation Front

Because of the sharp break in the fictitious saturation curve, we may ascribe to the plane $x = x_s$ a definite "front[2] velocity" v_{fr} which is represented by the derivative [compare again Sect. 4.1.2, following (4.1.62)],

$$v_{fr} \equiv \frac{dx_s}{dt}, \quad t > 0 \ , \tag{4.2.60}$$

and thus equals

$$v_{fr} = \frac{k \sqrt{\mu_1}}{\sqrt{\mu_0 \mu_1 \sigma t}}, \quad t > 0 \ . \tag{4.2.61}$$

Taking an imposed field amplitude $H_0 \simeq 1.2\, H_s$, we find from Fig. 4.9 that $k \sqrt{\mu_1} \simeq 0.15$; hence, for the assumed values of $\mu_1 = 1000$ and $\sigma = 1 \times 10^{+7} (\Omega m)^{-1}$ we obtain, e.g., at $t = 0.01$ s, for v_{fr} the value of 13 mm/s; at $\mu_0 \sigma t = 1.0$, v_{fr} equals[3] k, i.e., the front is still creeping on at an instantaneous speed of about 5 mm/s $(= 0.15/\sqrt{\mu_1})$.

We thus gain some idea of the duration of the field build-up; in our case, it takes about 0.08 s $(= 1/\mu_0 \sigma)$ for the rate of penetration to decrease by a factor of about 2.6 $(= 13/5)$.

For a better grasp of the magnitude of v_{fr}, let us consider the so-called[4] phase velocity v_{ph} of a plane electromagnetic wave field propagating in the iron characterized by the above-mentioned values of μ_1 and σ. Thus, introducing the skin depth δ at the circular supply frequency ω, we have

$$v_{ph} = \omega \delta \ , \tag{4.2.62}$$

where

---

[2] Our use of the term "front" here is admittedly unconventional: normally confined to the context of motion vs. rest [4.10], it is used here to describe the boundary between saturation and non-saturation.

[3] Thus, incidentally, providing an interpretation for the parameter k: it stands for the numerical value of v_{fr} at the numerical instant $t = (\mu_0 \sigma)^{-1}$.

[4] The phrasing "so-called" is in order, as v_{ph} is, strictly speaking, defined only for undamped, traveling monochromatic waves.

$$\delta = \sqrt{\frac{2}{\omega \mu_0 \mu_1 \sigma}} , \qquad (4.2.63)$$

or, with the period $T = 2\pi/\omega$:

$$v_{\mathrm{ph}} = \frac{2\sqrt{\pi}}{\sqrt{\mu_0 \mu_1 \sigma T}} . \qquad (4.2.64)$$

For the low-frequency case relevant to power systems, let us take, e.g., $\omega = 314 \ \mathrm{s}^{-1}$, so that $v_{\mathrm{ph}} \simeq 0.22 \ \mathrm{m/s}$; this value may be compared with the formerly quoted one of v_{fr}.

In more general terms, comparing (4.2.61 and 64), we have

$$\frac{v_{\mathrm{fr}}}{v_{\mathrm{ph}}} = \frac{k\sqrt{\mu_1}}{2\sqrt{\pi}} \sqrt{\frac{T}{t}} , \qquad (4.2.65)$$

and arbitrarily taking $T/t = 1.0$ for the purpose comparison, yields the simple relation

$$v_{\mathrm{fr}} = \frac{k\sqrt{\mu_1}}{2\sqrt{\pi}} v_{\mathrm{ph}} . \qquad (4.2.66)$$

Even for the excessively high – in fact fictitious – quotient $(H_0/H_s) \simeq 3.7$, the parameter $k\sqrt{\mu_1}$ attains the value of about 1.0 only, so that $v_{\mathrm{fr}} \simeq 0.28 \ v_{\mathrm{ph}}$, which is quite low.

4.2.4 Discussion

Field build-up in the presence of saturable iron was investigated. Excitation was modelled by a current step, this choice being governed by practical considerations: whereas the coil-winding distribution may be realized (subject to technological limitations) almost at will, prescribed electric field conditions are very hard to realize.

The field inside the core is divided by the saturation plane into two distinct regions; the propagation velocity of the field proper is unspecified due to neglect of the electric energy density inside the ferromagnetic core; however, the front velocity (for times $t > 0$) is well defined. Calculations show that this velocity is quite small compared, e.g., to the phase velocity of a steady ac field pervading the investigated conductor.

Network transients often entail impedance level reductions in power circuits comprising *saturable* equipment; this reduction may be rather heavy and lead to harmonic or subharmonic oscillations, or to instabilities and/or electromechanical chain transients; nevertheless, the saturation process proper is relatively slow when reckoning in terms of supply periods: a clue to this circumstance is provided by the relative slowness of propagation of the saturation front discussed above.

4.2.5 Appendix

Performance of Integration – Eq. (4.2.53)

We wish to evaluate the integral

$$I \equiv \int_{T_r}^{t} \frac{e^{-(\zeta/t')}}{t'} \, dt' \tag{4.2.67}$$

wherein

$$\zeta = 2(x^2/4)(\mu_0 \mu_1 \sigma) \ . \tag{4.2.68}$$

Introducing the new variable τ through the defining relation

$$\tau \equiv \frac{\zeta}{t'} \ , \tag{4.2.69}$$

we have

$$I = \int_{\tau = \zeta/T_r}^{\zeta/t} \frac{e^{-\tau}}{\zeta/\tau} \left(-\frac{\zeta}{\tau^2} \, d\tau \right) \ , \tag{4.2.70}$$

i.e.,

$$I = - \int_{\tau = \zeta/T_r}^{\zeta/t} \frac{e^{-\tau}}{\tau} \, d\tau \ . \tag{4.2.71}$$

Resorting now to the notation

$$\theta \equiv -\tau \tag{4.2.72}$$

one finds that

$$I = - \int_{\theta = -(\zeta/T_r)}^{-(\zeta/t)} \frac{e^{+\theta}}{-\theta} (-d\theta) = + \int_{\theta = -(\zeta/T_r)}^{-(\zeta/t)} \frac{e^{+\theta}}{-\theta} \, d\theta \ , \tag{4.2.73}$$

or, with $T_r \to 0$:

$$I = \int_{\theta = -\infty}^{-(\zeta/t)} \frac{e^{+\theta}}{-\theta} \, d\theta = -E_i\left(-\frac{\zeta}{t} \right) \ . \tag{4.2.74}$$

4.3 Field Switching in the Presence of Superconducting Material

With the advent of the superconductor technology, devices based on this property are finding ever-growing applications, such as light-weight rotating machines, dc transmission of power, zero-field chambers, rf cavities with very high Q, electromagnetic standards, cryogenic signal transmission lines, etc.

Superconductors are best known to electrical engineers by virtue of their ability to carry current at extremely low loss and to exclude magnetic flux; because of this they are frequently described as "perfect diamagnets", having an "induced magnetic moment" equal to and opposite to the imposed magnetic field.

While thorough understanding of superconductivity is only possible through quantum mechanics, the London equations, in conjunction with Maxwell's field equations provide a convenient means for insight into the mechanism of superconducting devices. Although the London two-fluid "local" model of superconductivity is the least exact of several proposed models, it has the advantage of relative mathematical simplicity – in contrast to, say, Pippard's "non-local" approach [4.11]. As the discrepancy between the two theories is reasonably small for many cases of practical interest, we shall adopt the London approach for investigating transient-field penetration in superconductors. To this end, we consider switch-on of the external excitation and the attendant field build-up.

4.3.1 Formulation of Problem and Approach to Solution

We assume that a slab of superconducting material occupies the entire region $(0 < x < \infty, \ -\infty < y < \infty, \ 0 < z < L)$. The excitation is modeled by an array of generators flush with the $x = 0$ plane (see Fig. 4.7, with the iron mentally replaced by a superconductor). In practice, the slab under consideration has a *finite residual resistance* (see below); by contrast, we assume the existence of a fictitious, ideally superconducting material which we shall characterize as "ideally conducting"; it is supposed that the field inside the slab is confined at $z = 0$ and $z = L$ by two such "ideally conducting" planes. How does the electromagnetic field evolve inside the superconducting slab when the array of generators is instantaneously put to work at a given moment of time?

Disregarding the microscopic behavior of electrons and electron pairs inside the material, we base our analysis on a set of conventional phenomenological equations, similar to the well-known Maxwell equations for a nonmagnetic substance. It is assumed, therefore, that the electric field E, the magnetic field H and the conduction current density j are interrelated inside the material at each moment of time t by Maxwell's equations

$$\nabla \times H = \varepsilon_0 \frac{\partial E}{\partial t} + j \ , \tag{4.3.1}$$

$$\nabla \times E = -\mu_0 \frac{\partial H}{\partial t} \ . \tag{4.3.2}$$

We assume here from the outset the relative permittivity $\varepsilon_r = 1$ (as customary in Maxwell's theory for all metallic conductors) although it is not yet entirely clear whether the concept of permittivity is at all meaningful for superconductors; further, we take $\mu_r = 1$ which, in turn, is a direct result of the London theory. Also, the conduction current density j is assumed [4.12] to consist of (a) the "normal", "ohmic" component $j_n = \sigma E$, where σ is the electrical conductivity of the material, and (b) the "superconducting" component j_{sc}.

The first London equation relates the time variation of j_{sc} to the local electric field intensity by means of a so-called "penetration depth" λ

$$\frac{\partial j_{sc}}{\partial t} = \frac{1}{\mu_0 \lambda^2} E \ , \tag{4.3.3}$$

which depends only on the material and on its absolute temperature.

Applying the curl operation to (4.3.3), and introducing on the right-hand side the induction law (4.3.2), we obtain

$$\nabla \times \frac{\partial j_{sc}}{\partial t} = -\frac{1}{\lambda^2} \frac{\partial H}{\partial t} \ , \tag{4.3.4}$$

or, after integration and introduction of an (as yet undetermined) time independent field H_0:

$$\nabla \times j_{sc} = -\frac{1}{\lambda^2}(H - H_0) \ . \tag{4.3.5}$$

For a superconductor, the Meissner effect implies $H_0 = 0$. Hence the second London equation reads

$$\nabla \times j_{sc} = -\frac{1}{\lambda^2} H \ . \tag{4.3.6}$$

The electromagnetic field is now taken to derive from a magnetic vector potential V. A suitable gauge (Sect. 4.3.4A) is introduced, so that we may take

$$E = -\frac{\partial V}{\partial t} \tag{4.3.7}$$

$$\mu_0 H = \nabla \times V \ . \tag{4.3.8}$$

Assuming that no current existed inside the slab prior to the onset of the field, we find that

$$j_{sc} = -\frac{1}{\mu_0 \lambda^2} V \ . \tag{4.3.9}$$

We now rewrite (4.3.1) in the form

$$\nabla \times \left(\frac{\nabla \times V}{\mu_0} \right) = \varepsilon_0 \frac{\partial}{\partial t} \left(-\frac{\partial V}{\partial t} \right) + \sigma \left(-\frac{\partial V}{\partial t} \right) + \left(-\frac{1}{\mu_0 \lambda^2} V \right) \ . \tag{4.3.10}$$

Now, although the superconducting current density j_{sc} is carried by electrons condensed into *pairs* with opposite spin, charge accumulation is excluded in a macroscopic treatment, and neutrality is assumed throughout. Hence, we take $\nabla \cdot j_{sc} = 0$ and accordingly (4.3.9) yields $\nabla \cdot V = 0$. Resorting to the identity $\nabla \times (\nabla \times V) = \nabla(\nabla \cdot V) - \nabla^2 V$, (4.3.10) finally reads (Sect. 4.3.4A)

$$\nabla^2 V = \mu_0 \varepsilon_0 \frac{\partial^2 V}{\partial t^2} + \mu_0 \sigma \frac{\partial V}{\partial t} + \frac{1}{\lambda^2} V . \qquad (4.3.11)$$

For simplicity, the geometry of the device and its excitation are chosen so that the electric field comprises a z-component only, i.e. $E = E_z 1_z$, whereas the magnetic field is given by $H = H_y 1_y$. Accordingly, we assume also for the vector potential V a z-component only, i.e., we take $V = V 1_z$, so that the former vector equation reduces to the scalar equation

$$\nabla^2 V = \mu_0 \varepsilon_0 \frac{\partial^2 V}{\partial t^2} + \mu_0 \sigma \frac{\partial V}{\partial t} + \frac{1}{\lambda^2} V . \qquad (4.3.12)$$

For $\lambda^2 \to \infty$, (4.3.11) [or (4.3.12)] reduces to the classical equation of field penetration in a conductor[5]. We may therefore assert that the London relations enter into our considerations through the last term of the above equations.

In the plane $x = 0$, the excitation is chosen as a step function of time, of amplitude V_a

$$V = V_a \cdot 1(t), \quad x = 0 . \qquad (4.3.13)$$

We wish to determine, under the given assumptions, the electromagnetic field inside the superconducting slab.

4.3.2 Field Analysis

As shown in Sect. 4.3.4B, the electromagnetic field inside the superconducting (subscript sc, i.e. $E_{z,sc}$; $H_{y,sc}$) slab is represented by

$$E_{z,sc} = - V_a \sqrt{\left(\frac{\sigma}{2\varepsilon_0}\right)^2 - \left(\frac{c}{\lambda}\right)^2} \exp\left[-(\sigma/2\varepsilon_0) t\right]$$

$$\times \frac{x/c}{\sqrt{t^2 - x^2/c^2}} \left[-i J_1\left(i \sqrt{\left(\frac{\sigma}{2\varepsilon_0}\right)^2 - \left(\frac{c}{\lambda}\right)^2} \sqrt{t^2 - \frac{x^2}{c^2}}\right)\right],$$

$$t > \frac{x}{c} . \qquad (4.3.14)$$

---

[5] Hitherto, in our approximations we solved the diffusion equation for conductors, disregarding the product $\mu_0 \varepsilon_0$. Here, however, in view of the basic difference in the quality of the conductor, we preferred to consider the formal solution with the product included.

$$H_{y,\,sc} = V_a \sqrt{\left(\frac{\sigma}{2\,\varepsilon_0}\right)^2 - \left(\frac{c}{\lambda}\right)^2} \sqrt{\frac{\varepsilon_0}{\mu_0}}$$

$$\times \left\{ \frac{\sigma/2\,\varepsilon_0}{\sqrt{(\sigma/2\,\varepsilon_0)^2 - (c/\lambda)^2}} \exp\left[-(\sigma/2\,\varepsilon_0)\,t\right] \right.$$

$$\times J_0 \left(i \sqrt{\left(\frac{\sigma}{2\,\varepsilon_0}\right)^2 - \left(\frac{c}{\lambda}\right)^2} \sqrt{t^2 - \frac{x^2}{c^2}} \right)$$

$$+ \exp\left[-(\sigma/2\,\varepsilon_0)\,t\right] \frac{t}{\sqrt{t^2 - \dfrac{x^2}{c^2}}}$$

$$\times \left[-i J_1 \left(i \sqrt{\left(\frac{\sigma}{2\,\varepsilon_0}\right)^2 - \left(\frac{c}{\lambda}\right)^2} \sqrt{t^2 - \frac{x^2}{c^2}} \right) \right]$$

$$+ \frac{(c/\lambda)^2}{\sqrt{(\sigma/2\,\varepsilon_0)^2 - (c/\lambda)^2}} \int_{t'=(x/c)}^{t} \exp\left[-(\sigma/2\,\varepsilon_0)\,t'\right]$$

$$\left. \times J_0 \left(i \sqrt{\left(\frac{\sigma}{2\,\varepsilon_0}\right)^2 - \left(\frac{c}{\lambda}\right)^2} \sqrt{t'^2 - \frac{x^2}{c^2}} \right) dt' \right\} \,, \quad t > \frac{x}{c} \,. \quad (4.3.15)$$

Introducing now the notation

$$\frac{\sigma}{\varepsilon_0} \equiv \frac{1}{T} \,; \quad \frac{\sigma t}{2\,\varepsilon_0} \equiv \tau \,; \quad \frac{\sigma x}{2\,\varepsilon_0} \frac{1}{c} \equiv \xi \,; \quad \frac{c}{\lambda} \equiv \frac{1}{T_s} \,,$$

$$\frac{E_{z,\,sc}}{V_a \sqrt{(\sigma/2\,\varepsilon_0)^2 - (c/\lambda)^2}} \equiv e_{sc} \,; \quad \frac{H_{y,\,sc}}{V_a \sqrt{(\sigma/2\,\varepsilon_0)^2 - (c/\lambda)^2} \sqrt{\varepsilon_0/\mu_0}} \equiv h_{sc} \,,$$

(4.3.16)

Eqs. (4.3.14, 15) may be rewritten in dimensionless form as

$$e_{sc} = -\frac{\xi}{\sqrt{\tau^2 - \xi^2}} e^{-\tau} [-i J_1 (i \sqrt{1 - 4\,T^2/T_s^2} \sqrt{\tau^2 - \xi^2})] \,, \quad \tau > \xi \,;$$

(4.3.17)

$$h_{sc} = \frac{1}{\sqrt{1 - 4\,T^2/T_s^2}} e^{-\tau} J_0 (i \sqrt{1 - 4\,T^2/T_s^2} \sqrt{\tau^2 - \xi^2})$$

$$+ e^{-\tau} \frac{\tau}{\sqrt{(\tau^2 - \xi^2)}} [-i J_1 (i \sqrt{1 - 4\,T^2/T_s^2} \sqrt{\tau^2 - \xi^2})]$$

$$+ \frac{4\,T^2/T_s^2}{\sqrt{1 - 4\,T^2/T_s^2}} \int_{\tau'=\xi}^{\tau} e^{-\tau'} J_0 (i \sqrt{1 - 4\,T^2/T_s^2} \sqrt{\tau'^2 - \xi^2}) d\tau' \,,$$

$$\tau > \xi \,.$$

(4.3.18)

Even though $(T^2/T_s^2) \ll 1$, we retain the term for formal reasons, as it leads, see (4.3.114, 115), to convergence of the integral in (4.3.18).

From (4.3.3 and 14), we find for the superconducting z-directed current density $j_{sc, z}$ that

$$j_{sc, z} = -\frac{1}{\mu_0 \lambda^2} V_a \sqrt{\left(\frac{\sigma}{2\varepsilon_0}\right)^2 - \left(\frac{c}{\lambda}\right)^2} \int_{t'=(x/c)}^{t} \frac{\frac{x}{c}\exp[-(\sigma/2\varepsilon_0)t]}{\sqrt{t'^2 - x^2/c^2}}$$

$$\times \left[-iJ_1\left(i\sqrt{\left(\frac{\sigma}{2\varepsilon_0}\right)^2 - \left(\frac{c}{\lambda}\right)^2}\sqrt{t'^2 - x^2/c^2}\right)\right] dt', \quad t' > \frac{x}{c} .$$

$$(4.3.19)$$

Defining now

$$\frac{j_{sc, z}}{\dfrac{V_a}{\mu_0 \lambda^2} \dfrac{\sqrt{(\sigma/2\varepsilon_0)^2 - (c/\lambda)^2}}{\sigma/2\varepsilon_0}} \equiv \imath ,$$

$$(4.3.20)$$

we have in dimensionless units

$$\imath = -\int_{\tau'=\xi}^{\tau} \frac{\xi}{\sqrt{\tau'^2 - \xi^2}} e^{-\tau'}[-iJ_1(i\sqrt{1 - 4T^2/T_s^2}\sqrt{\tau'^2 - \xi^2})]\,d\tau', \quad \tau' > \xi .$$

$$(4.3.21)$$

Once more, the factor $\sqrt{1 - 4(T/T_s)^2}$ has been retained for the purpose of considering limiting values of the integral, see (4.3.117), etc.

With increasing values of τ, the electric field as well as the conduction current density $\sigma E_{sc, z}$ vanishes; this is also the case with the first two terms of the magnetic field. Its third term, and the superconducting current density $j_{sc, z}$, however, never vanish. Thus, evaluating, see (4.3.18),

$$4\frac{T^2}{T_s^2}\frac{1}{\sqrt{1 - 4T^2/T_s^2}}\int_{\tau'=\xi}^{\infty} e^{-\tau'}J_0(i\sqrt{1 - 4T^2/T_s^2}\sqrt{\tau'^2 - \xi^2})\,d\tau' \equiv h_\infty$$

$$(4.3.22)$$

and

$$-\int_{\tau'=\xi}^{\infty} \frac{\xi}{\sqrt{\tau'^2 - \xi^2}} e^{-\tau'}[-iJ_1(i\sqrt{1 - 4T^2/T_s^2}\sqrt{\tau'^2 - \xi^2})]\,d\tau' \equiv \imath_\infty , \quad (4.3.23)$$

we find (Sect. 4.3.4C)

$$h_\infty = \frac{2T/T_s}{\sqrt{1 - 4T^2/T_s^2}}\exp[-\xi 2(T/T_s)] ,$$

$$(4.3.24)$$

or, with the definition $\lim_{t\to\infty} H_{y, sc} \equiv H_{y\infty}$ −

$$H_{y\infty} = \frac{V_a}{\mu_0 \lambda}\exp(-x/\lambda) .$$

$$(4.3.25)$$

Similarly (Sect. 4.3.4C)

$$l_\infty = \frac{1}{\sqrt{1-4T^2/T_s^2}}\{\exp(-\xi)-\exp[-2\xi(T/T_s)]\} ,$$ (4.3.26)

so that with the additional definition $\lim_{t\to\infty} j_{sc,z} \equiv j_{sc\infty}$ we obtain

$$j_{sc\infty} = \frac{V_a}{\mu_0\lambda^2}\{\exp[-(\sigma x/2\varepsilon_0 c)]-\exp[-(x/\lambda)]\} .$$ (4.3.27)

It is of interest to note that the steady-state solutions (4.3.25, 27) defined for the limiting value $\tau'\to\infty$ satisfy neither Maxwell's curl equation (4.3.1), nor London's equation (4.3.6) and this owing to the term $\exp[-(\sigma x/2\varepsilon_0 c)]$: while Maxwell's equations account for field energy interaction through time derivatives appearing in both curl equations, the London equations (4.3.3,6) do so through a single time rate, thereby acquiring in fact a quasistatic character.

Furthermore, for a superconductor carrying "indefinitely" an intermittent current without any external excitation, the ordinary steady-state definition through passage to the limit $t\to\infty$ – which is routinely undertaken within the framework of Maxwell's equations – is rather questionable.

However, taking for a practical superconductor [4.13] the values $\sigma \sim 1 \times 10^9 (\Omega m)^{-1}$ and $\lambda \sim 5 \times 10^{-8}$ m, we obtain $T/T_s \sim 5 \times 10^{-5}$ and therefore in (4.3.27) $\exp(-\xi) \ll \exp[-2\xi(T/T_s)]$ for $\xi > 0$. With this approximation we now find that

$$j_{sc\infty} = -\frac{V_a}{\mu_0\lambda^2}\exp(-x/\lambda) ;$$ (4.3.28)

hence, (4.3.25, 28) are now both compatible with Maxwell's equation (4.3.1) and London's equation (4.3.6).

Field and current variations, (4.3.17, 18 and 21), are plotted in Figs. 4.11 to 13. Obviously, the time unit T as well as the times $t = \tau T$ are exceedingly small; our intention was, however, to portray phenomenological field build-up under a step function and this was obtained.

Let us now compare the transient electromagnetic field inside the superconductor to that inside an "ordinary" conductor with the same (fictitious) conductivity (i.e., assumed parameters σ, $\varepsilon_0\mu_0$).

We obtain the field of such an "ordinary" conductor by the limit $\lambda\to\infty$; (4.3.11) thus reads

$$\nabla^2 V = \mu_0\varepsilon_0\frac{\partial^2 V}{\partial t^2}+\mu_0\sigma\frac{\partial V}{\partial t} ,$$ (4.3.29)

and similarly, in (4.3.16) we find that $1/T_s\to 0$.

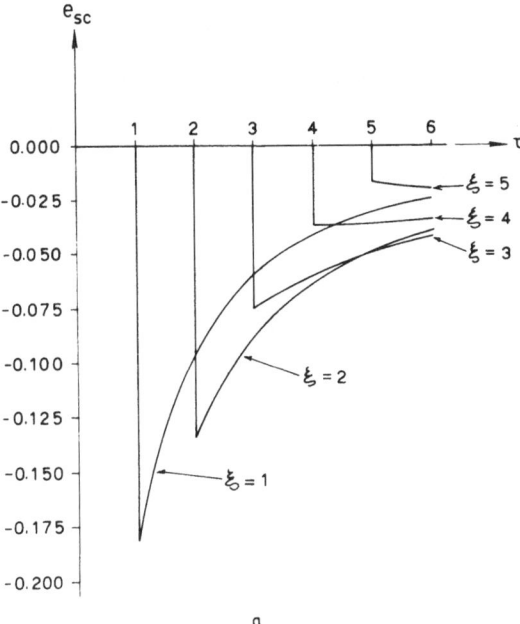

a

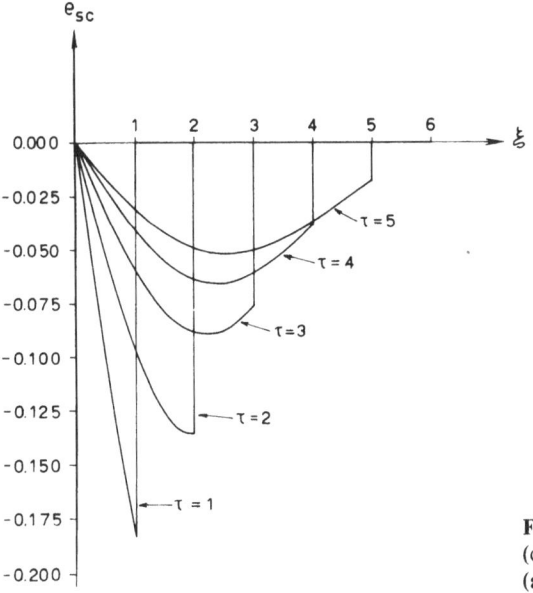

b

Fig. 4.11a, b. Electric field variation (dimensionless units). (a) at different locations; (b) at different times

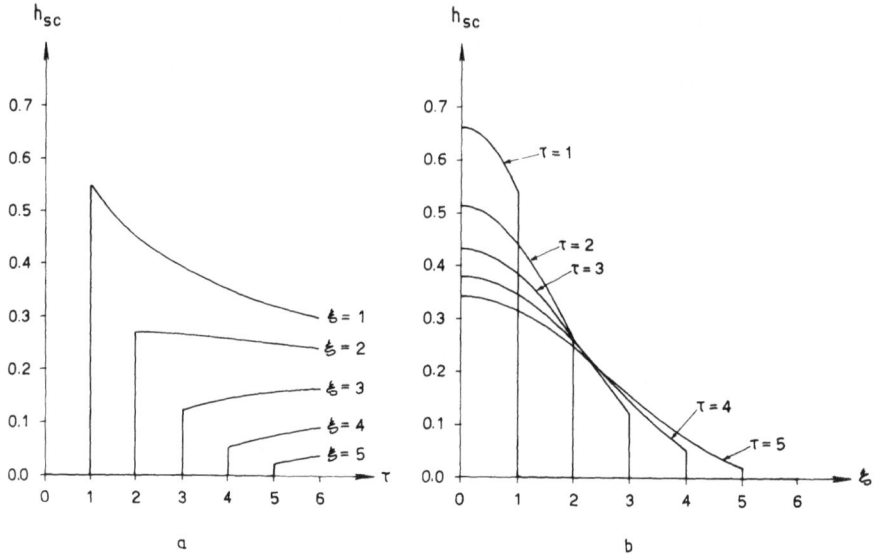

Fig. 4.12a,b. Magnetic field variation (dimensionless units): **(a)** at different locations; **(b)** at different times

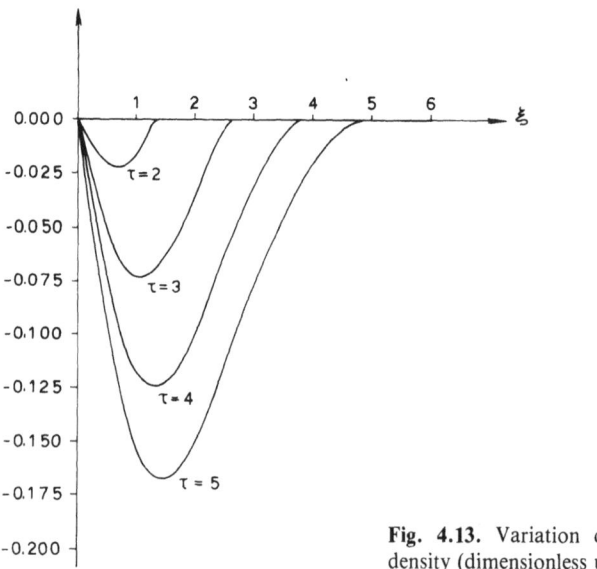

Fig. 4.13. Variation of superconducting current density (dimensionless units)

Hence, in the case of an "ordinary" conductor (field components E_z, H_y, and subscript co) the dimensionless variables $e_{co} \equiv E_z / [V_a(\sigma/2\,\varepsilon_0)]$, $h_{co} \equiv H_y / [V_a(\sigma/2\,\varepsilon_0)\sqrt{\varepsilon_0/\mu_0}]$ are given by

$$e_{co} = -\frac{\xi}{\sqrt{\tau^2 - \xi^2}}\, e^{-\tau}[-iJ_1(i\sqrt{\tau^2 - \xi^2})], \quad \tau > \xi ; \tag{4.3.30}$$

$$h_{co} = e^{-\tau}J_0(i\sqrt{\tau^2 - \xi^2}) + e^{-\tau}\frac{\tau}{\sqrt{\tau^2 - \xi^2}}[-iJ_1(i\sqrt{\tau^2 - \xi^2})], \quad \tau > \xi . \tag{4.3.31}$$

Due to the fact that the value of $4(T/T_s)^2 \simeq 10^{-8}$ is exceedingly small compared with unity, (4.3.17 and 30) are identical; the same holds true for the two first terms of (4.3.18 and 31). The difference between the superconducting and "ordinary material" with the same values of σ is therefore due only to the third term in (4.3.18), which we denote by Δh. Thus

$$\Delta h \equiv h_{sc} - h_{co} = \frac{4T^2/T_s^2}{\sqrt{1 - 4T^2/T_s^2}} \int_{\tau'=\xi}^{\tau} e^{-\tau'}J_0(i\sqrt{1 - 4(T^2/T_s^2)}\sqrt{\tau'^2 - \xi^2})d\tau',$$

$$\tau' > \xi . \tag{4.3.32}$$

i.e., with (4.3.22),

$$\lim_{\tau \to \infty} \Delta h = h_\infty . \tag{4.3.33}$$

The difference Δj_z between the total current density $(j_{sc,z} + \sigma E_{z,sc})$ inside the superconductor and that inside the "ordinary conductor" (of admittedly extremely high conductivity) is obviously given by

$$\Delta j_z \equiv (j_{sc,z} + \sigma E_{z,sc}) - \sigma E_z = j_{sc,z} . \tag{4.3.34}$$

Now, resorting to (4.3.14, 19), we find

$$\frac{j_{sc,z}}{\sigma E_{z,sc}} = 2\frac{1}{(\mu_0/\varepsilon_0)(\sigma\lambda)^2}\frac{l}{e_{sc}} . \tag{4.3.35}$$

In addition, from (4.3.16) we have

$$\frac{1}{\sqrt{\mu_0/\varepsilon_0}(\sigma\lambda)} = \frac{T}{T_s} , \tag{4.3.36}$$

and therefore

$$\frac{\Delta j_z}{\sigma E_{z,sc}} = \frac{j_{sc,z}}{\sigma E_{z,sc}} = \frac{1}{2}4\left(\frac{T}{T_s}\right)^2\frac{l}{e_{sc}} . \tag{4.3.37}$$

The ratio $\Delta j_z/(\sigma E_{z,sc})$, which may be taken as measure of the comparative behavior in the two cases, is reproduced in Fig. 4.14.

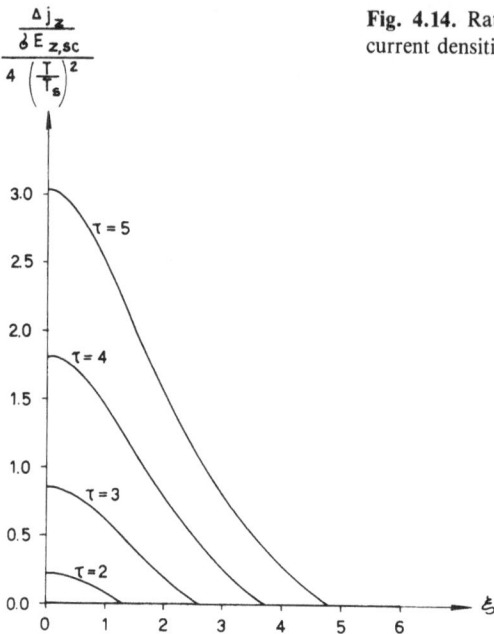

Fig. 4.14. Ratio between superconducting and ohmic current densities

4.3.3 Concluding Remarks

In the present section, transient field build-up inside a superconductor is investigated on the basis of formal application of macroscopic electrodynamics; although understanding of superconducting phenomena may only be attained through quantum mechanics, the outlined phenomenological approach leads at least to a qualitative description of field penetration of a superconductor.

Let us remark in this context that the quotient $\sqrt{\mu_0/\varepsilon_0} : (1/\sigma\lambda)$ – which is very large compared to unity – indicates (Sect. 1.5) that a much shorter magneto-quasistatic solution might in fact have been attempted; nevertheless the more detailed approach was preferred, with a view to better insight into the compatibility of the Maxwell and London equations. It goes without saying that once deeper insight into the permittivity of superconductors has been gained – their transient behavior will also be better understood.

4.3.4 Appendix

A) Some Remarks on Gauge Transformations in the Presence of Superconducting Material

Under the macroscopic phenomenological approach we assume neutrality throughout inside the material. Hence, we postulate that behavior of the electromagnetic field inside the superconductor is governed by the equations

$$\nabla \times H = \varepsilon_0 \frac{\partial E}{\partial t} + \sigma E + j_{sc} \ , \tag{4.3.38}$$

$$\nabla \times E = -\mu_0 \frac{\partial H}{\partial t} \ , \tag{4.3.39}$$

$$\nabla \cdot \varepsilon_0 E = 0 \ , \tag{4.3.40}$$

$$\nabla \cdot \mu_0 H = 0 \ , \tag{4.3.41}$$

along with the London equations

$$\frac{\partial j_{sc}}{\partial t} = \frac{1}{\mu_0 \lambda^2} E \ , \tag{4.3.42}$$

$$\nabla \times j_{sc} = -\frac{1}{\lambda^2} H \ . \tag{4.3.43}$$

Introducing the vector potential V according to the relationship

$$\mu_0 H = \nabla \times V \tag{4.3.44}$$

and substituting the last expression in (4.3.39) we find, in conjunction with an electric scalar potential ϕ, that

$$E = -\frac{\partial V}{\partial t} - \nabla \phi \ . \tag{4.3.45}$$

Now (4.3.38) yields

$$\nabla \times (\nabla \times V) = \mu_0 \varepsilon_0 \frac{\partial}{\partial t} \left(-\frac{\partial V}{\partial t} - \nabla \phi \right) + \mu_0 \sigma \left(-\frac{\partial V}{\partial t} - \nabla \phi \right) + \mu_0 j_{sc} \ , \tag{4.3.46}$$

or, resorting (in Cartesian coordinates) to the identity $\nabla \times (\nabla \times V) = \nabla(\nabla \cdot V) - \nabla^2 V$, we have

$$\nabla(\nabla \cdot V) - \nabla^2 V = -\mu_0 \varepsilon_0 \frac{\partial^2 V}{\partial t^2} - \mu_0 \sigma \frac{\partial V}{\partial t} - \nabla \left(\varepsilon_0 \mu_0 \frac{\partial \phi}{\partial t} + \mu_0 \sigma \phi \right) + \mu_0 j_{sc} \ . \tag{4.3.47}$$

In view of (4.3.43, 44) there results

$$\nabla \times j_{sc} = -\frac{1}{\lambda^2} \frac{1}{\mu_0} \nabla \times V \ , \tag{4.3.48}$$

and hence, introducing a scalar stream function Λ, we find

$$j_{sc} = -\frac{1}{\mu_0 \lambda^2} V + \nabla \Lambda \ . \tag{4.3.49}$$

Reverting to (4.3.47), we obtain

$$\nabla(\nabla \cdot V) - \nabla^2 V = -\mu_0 \varepsilon_0 \frac{\partial^2 V}{\partial t^2} - \mu_0 \sigma \frac{\partial V}{\partial t} - \frac{1}{\lambda^2} V$$

$$-\nabla\left(\varepsilon_0 \mu_0 \frac{\partial \phi}{\partial t} + \mu_0 \sigma \phi - \mu_0 \Lambda\right) . \qquad (4.3.50)$$

Up to now, only the curl of V was specified. Imposing the additional condition

$$\nabla \cdot V = -\left(\varepsilon_0 \mu_0 \frac{\partial \phi}{\partial t} + \mu_0 \sigma \phi - \mu_0 \Lambda\right) \qquad (4.3.51)$$

(4.3.50) reduces to

$$\nabla^2 V = \mu_0 \varepsilon_0 \frac{\partial^2 V}{\partial t^2} + \mu_0 \sigma \frac{\partial V}{\partial t} + \frac{1}{\lambda^2} V . \qquad (4.3.52)$$

Reverting to (4.3.40) and introducing the field value from (4.3.45), the charge neutrality condition yields

$$\nabla^2 \phi + \frac{\partial}{\partial t} \nabla \cdot V = 0 , \qquad (4.3.53)$$

i.e., from (4.3.51):

$$\nabla^2 \phi + \frac{\partial}{\partial t}\left(-\varepsilon_0 \mu_0 \frac{\partial \phi}{\partial t} - \sigma \mu_0 \phi + \mu_0 \Lambda\right) = 0 . \qquad (4.3.54)$$

We still must investigate the relationship (4.3.42). Thus

$$\frac{\partial}{\partial t}\left(-\frac{1}{\mu_0 \lambda^2} V + \nabla \Lambda\right) = \frac{1}{\mu_0 \lambda^2}\left(-\frac{\partial V}{\partial t} - \nabla \phi\right) , \qquad \text{or} \qquad (4.3.55)$$

$$-\frac{\partial}{\partial t} \nabla \Lambda = \frac{1}{\mu_0 \lambda^2} \nabla \phi . \qquad (4.3.56)$$

This expression is identically satisfied for $\partial \Lambda / \partial t = -(\mu_0 \lambda^2)^{-1} \phi$; hence, the differential equation for the electric scalar potential ϕ now reads

$$\nabla^2 \phi = \mu_0 \varepsilon_0 \frac{\partial^2 \phi}{\partial t^2} + \mu_0 \sigma \frac{\partial \phi}{\partial t} + \frac{1}{\lambda^2} \phi . \qquad (4.3.57)$$

As is well known, the electromagnetic field remains unchanged if two different potential functions V_0 and ϕ_0 are introduced in conjunction with a scalar function ψ, so that

$$V_0 = V + \nabla \psi \;, \tag{4.3.58}$$

$$\phi_0 = \phi - \frac{\partial \psi}{\partial t} \;. \tag{4.3.59}$$

Thus,

$$E_0 = -\frac{\partial V_0}{\partial t} - \nabla \phi_0 = -\frac{\partial}{\partial t}(V + \nabla \psi) - \nabla \left(\phi - \frac{\partial \psi}{\partial t} \right) = E \;, \tag{4.3.60}$$

and

$$\mu_0 H_0 = \nabla \times V_0 = \nabla \times (V + \nabla \psi) = \mu_0 H \;, \tag{4.3.61}$$

i.e., the new electromagnetic field E_0, H_0 is identical to the previous one, E, H.

Reverting to (4.3.38), we find that

$$\nabla \times (\nabla \times V_0) = \mu_0 \varepsilon_0 \frac{\partial}{\partial t} \left(-\frac{\partial V_0}{\partial t} - \nabla \phi_0 \right) + \mu_0 \sigma \left(-\frac{\partial V_0}{\partial t} - \nabla \phi_0 \right) + \mu_0 j_{sc} \;. \tag{4.3.62}$$

The expression (4.3.43) now reads

$$\nabla \times j_{sc} = -\frac{1}{\lambda^2} \frac{1}{\mu_0} \nabla \times (V_0 - \nabla \psi) = -\frac{1}{\lambda^2} \frac{1}{\mu_0} \nabla \times V_0 \;. \tag{4.3.63}$$

Resorting to a (still unspecified) scalar function Σ, we therefore find that j_{sc} may now be expressed through

$$j_{sc} = -\frac{1}{\mu_0 \lambda^2} V_0 + \nabla \Sigma \;. \tag{4.3.64}$$

Hence

$$\nabla(\nabla \cdot V_0) - \nabla^2 V_0 = -\mu_0 \varepsilon_0 \frac{\partial^2 V_0}{\partial t^2} - \mu_0 \sigma \frac{\partial V_0}{\partial t} - \frac{1}{\lambda^2} V_0$$
$$- \nabla \left(\varepsilon_0 \mu_0 \frac{\partial \phi_0}{\partial t} + \mu_0 \sigma \phi_0 - \mu_0 \Sigma \right) \;. \tag{4.3.65}$$

Imposing the requirement

$$\nabla \cdot V_0 = -\mu_0 \varepsilon_0 \frac{\partial \phi_0}{\partial t} - \mu_0 \sigma \phi_0 + \mu_0 \Sigma \tag{4.3.66}$$

we obtain the differential equation for V_0

$$\nabla^2 V_0 = \mu_0 \varepsilon_0 \frac{\partial^2 V_0}{\partial t^2} + \mu_0 \sigma \frac{\partial V_0}{\partial t} + \frac{1}{\lambda^2} V_0 \ . \tag{4.3.67}$$

Charge neutrality, see again (4.3.40), yields, through

$$\nabla^2 \phi_0 + \nabla \cdot \frac{\partial V_0}{\partial t} = 0 \tag{4.3.68}$$

the partial differential equation

$$\nabla^2 \phi_0 = \mu_0 \varepsilon_0 \frac{\partial^2 \phi_0}{\partial t^2} + \mu_0 \sigma \frac{\partial \phi_0}{\partial t} - \mu_0 \frac{\partial \Sigma}{\partial t} \ . \tag{4.3.69}$$

Now the relationship (4.3.42) leads to

$$\frac{\partial}{\partial t} \left(-\frac{1}{\mu_0 \lambda^2} V_0 + \nabla \Sigma \right) = \frac{1}{\mu_0 \lambda^2} \left(-\frac{\partial V_0}{\partial t} - \nabla \phi_0 \right) \ , \tag{4.3.70}$$

i.e.,

$$\frac{\partial}{\partial t} \nabla \Sigma = -\frac{1}{\mu_0 \lambda^2} \nabla \phi_0 \ . \tag{4.3.71}$$

This expression is identically satisfied (see above) if we choose $\partial \Sigma / \partial t = -(\mu_0 \lambda^2)^{-1} \phi_0$, and so (4.3.69) reduces to

$$\nabla^2 \phi_0 = \mu_0 \varepsilon_0 \frac{\partial^2 \phi_0}{\partial t^2} + \mu_0 \sigma \frac{\partial \phi_0}{\partial t} + \frac{1}{\lambda^2} \phi_0 \ . \tag{4.3.72}$$

To obtain the differential equation for ψ, we proceed as follows: Combining the last equation with (4.3.59), we obtain

$$\nabla^2 \left(\phi - \frac{\partial \psi}{\partial t} \right) = \mu_0 \varepsilon_0 \frac{\partial^2}{\partial t^2} \left(\phi - \frac{\partial \psi}{\partial t} \right)$$
$$+ \mu_0 \sigma \frac{\partial}{\partial t} \left(\phi - \frac{\partial \psi}{\partial t} \right) + \frac{1}{\lambda^2} \left(\phi - \frac{\partial \psi}{\partial t} \right) \ . \tag{4.3.73}$$

Substituting the differential equation (4.3.57), there results

$$\nabla^2 \left(\frac{\partial \psi}{\partial t} \right) = \mu_0 \varepsilon_0 \frac{\partial^2}{\partial t^2} \left(\frac{\partial \psi}{\partial t} \right) + \mu_0 \sigma \frac{\partial}{\partial t} \left(\frac{\partial \psi}{\partial t} \right) + \frac{1}{\lambda^2} \left(\frac{\partial \psi}{\partial t} \right) \ , \tag{4.3.74}$$

which is identically satisfied if ψ is subject to

$$\nabla^2 \psi = \mu_0 \varepsilon_0 \frac{\partial^2 \psi}{\partial t^2} + \mu_0 \sigma \frac{\partial \psi}{\partial t} + \frac{1}{\lambda^2} \psi \; . \tag{4.3.75}$$

Hence, in principle, solving (4.3.52, 57 and 75) subject to the additional requirement $\phi = \partial \psi / \partial t$, we find

$$V_0 = V + \nabla \psi \; , \tag{4.3.76}$$

$$\phi_0 = \phi - \frac{\partial \psi}{\partial t} = 0 \; , \quad \text{i.e.} \tag{4.3.77}$$

$$E = -\frac{\partial V_0}{\partial t} \; , \tag{4.3.78}$$

$$\mu_0 H = \nabla \times V_0 \; , \tag{4.3.79}$$

$$j_{sc} = -\frac{1}{\mu_0 \lambda^2} V_0 \; . \tag{4.3.80}$$

Dropping the zero subscript, we find that the differential equation (4.3.52) governs the behaviour of the whole electromagnetic field inside the superconducting slab.

B) Field Evaluation

In order to solve the partial differential equation

$$\frac{\partial^2 V}{\partial x^2} = \frac{1}{c^2} \frac{\partial^2 V}{\partial t^2} + \sigma \mu_0 \frac{\partial V}{\partial t} + \frac{1}{\lambda^2} V \tag{4.3.81}$$

subject to the initial condition

$$V(t) = V_a \cdot 1(t), \quad x = 0 \; ; \tag{4.3.82}$$

we resort to the complex plane $s = u + iv$, and introduce the complex spectral density $\bar{V}(x, s)$ of the magnetic vector potential component by means of

$$V(x, t) = \frac{1}{2\pi i} \int_{-i\infty}^{i\infty} \bar{V}(x, s) e^{st} ds, \quad t > 0 \; . \tag{4.3.83}$$

Reverting to (4.3.81) we now obtain

$$\frac{d^2 \bar{V}}{dx^2} = \frac{1}{c^2} \left(s^2 + \frac{\sigma}{\varepsilon_0} s + \frac{c^2}{\lambda^2} \right) \bar{V} \; . \tag{4.3.84}$$

The solution of this equation will be chosen so as to represent an outgoing wave, propagating in the $+x$-direction; thus

$$\bar{V}(x,s) = \bar{V}(0,s)\exp[-(x/c)\sqrt{s^2+(\sigma/\varepsilon_0)s+(c^2/\lambda^2)}] \qquad (4.3.85)$$

for $\lim_{x\to\infty} \mathrm{Re}[-x\sqrt{s^2+(\sigma/\varepsilon_0)s+(c^2/\lambda^2)}] > 0$.

In view of (4.3.82) the exciting field may be written in the form

$$V(0,t) = \frac{1}{2\pi i}\int_{-i\infty}^{i\infty}\frac{V_a}{s}e^{st}ds, \quad t>0 . \qquad (4.3.86)$$

Hence,

$$\bar{V}(x,s) = V_a\frac{\exp[-(x/c)\sqrt{s^2+(\sigma/\varepsilon_0)s+(c^2/\lambda^2)}]}{s} . \qquad (4.3.87)$$

The space-time dependence of the magnetic vector potential $V(x,t)$ is therefore finally obtained from the integral

$$V(x,t) = V_a\frac{1}{2\pi i}\int_{-i\infty}^{i\infty}\frac{\exp[st-(x/c)\sqrt{s^2+(\sigma/\varepsilon_0)s+(c^2/\lambda^2)}]}{s}ds . \qquad (4.3.88)$$

Now $E_z = -(\partial V/\partial t)$ and therefore

$$E_z = -V_a\frac{1}{2\pi i}\int_{-i\infty}^{i\infty}\exp[st-(x/c)\sqrt{s^2+(\sigma/\varepsilon_0)s+(c^2/\lambda^2)}]ds$$

$$= V_a\frac{\partial}{\partial x}\frac{1}{2\pi i}\int_{-i\infty}^{i\infty}\frac{\exp[st-(x/c)\sqrt{s^2+(\sigma/\varepsilon_0)s+(c^2/\lambda^2)}]}{\sqrt{s^2+(\sigma/\varepsilon_0)s+(c^2/\lambda^2)}}ds . \qquad (4.3.89)$$

Similarly, as $H_y = -(1/\mu_0)(\partial V/\partial x)$, we find

$$H_y = \frac{V_a}{\mu_0}\frac{1}{c}\frac{1}{2\pi i}\int_{-i\infty}^{i\infty}\frac{\exp[st-(x/c)\sqrt{s^2+(\sigma/\varepsilon_0)s+(c^2/\lambda^2)}]}{s}$$

$$\times\frac{s^2+(\sigma/\varepsilon_0)s+(c^2/\lambda^2)}{\sqrt{s^2+(\sigma/\varepsilon_0)s+(c^2/\lambda^2)}}ds$$

$$= \frac{V_a}{\mu_0}\frac{1}{c}\left(\frac{\partial}{\partial t}+\frac{\sigma}{\varepsilon_0}+\frac{c^2}{\lambda^2}\int dt\right)$$

$$\times\frac{1}{2\pi i}\int_{-i\infty}^{i\infty}\frac{\exp[st-(x/c)\sqrt{s^2+(\sigma/\varepsilon_0)s+(c^2/\lambda^2)}]}{\sqrt{s^2+(\sigma/\varepsilon_0)s+(c^2/\lambda^2)}}ds . \qquad (4.3.90)$$

The problem is now reduced to the evaluation of the integral

$$\frac{1}{2\pi i} \int_{-i\infty}^{i\infty} \frac{\exp[st-(x/c)\sqrt{s^2+(\sigma/\varepsilon_0)s+(c^2/\lambda^2)}]}{\sqrt{s^2+(\sigma/\varepsilon_0)s+(c^2/\lambda^2)}} \, ds \equiv f(x,t) \ . \qquad (4.3.91)$$

We introduce the notation

$$s' \equiv s + \frac{\sigma}{2\varepsilon_0} \ , \qquad -\Omega^2 \equiv -\left(\frac{\sigma}{2\varepsilon_0}\right)^2 + \frac{c^2}{\lambda^2} \ . \qquad (4.3.92)$$

Hence

$$f(x,t) = \frac{\exp[-(\sigma/2\varepsilon_0)t]}{2\pi i} \int_{-i\infty}^{i\infty} \frac{\exp[s't-(x/c)\sqrt{s'^2-\Omega^2}]}{\sqrt{s'^2-\Omega^2}} \, ds' \ . \qquad (4.3.93)$$

Further, we resort to the transformation

$$s' = -\Omega\cos\gamma, \qquad \gamma = \alpha + i\beta \ , \qquad (4.3.94)$$

which introduces, through the auxiliary notation $s' = u' + iv'$, the elliptic coordinates α, β, namely

$$\left.\begin{array}{l} u' = -\Omega\cos\alpha\cosh\beta, \\ v' = \Omega\sin\alpha\sinh\beta \end{array}\right\} \ . \qquad (4.3.95)$$

It is a posteriori evident that these elliptic coordinates may be restricted to the values

$$-\pi < \alpha < \pi, \qquad \beta > 0 \ . \qquad (4.3.96)$$

With the requirement

$$\lim_{s'\to\infty} \frac{\sqrt{s'^2-\Omega^2}}{s'} = +1 \qquad (4.3.97)$$

in mind, we obtain

$$\sqrt{s'^2-\Omega^2} = +i\Omega\sin\gamma \ . \qquad (4.3.98)$$

Now what is the path of integration of (4.3.93) in the plane α, β?

In order to answer this question, we resort to Fig. 4.15 and to (4.3.94, 95); thus, if $\alpha = -(\pi/2)$ and β decreases from $\beta\to\infty$ to $\beta\to\varepsilon$ (where, in turn, $\varepsilon\to 0$), the path $-\infty < v' < 0$ is traced in the u', v'-plane.

Keeping now ε fixed, and varying α between the limits $-\pi < \alpha < -\pi/2$, we obtain the path $0 < u' < \Omega$. The lower half (AB) of the integration contour

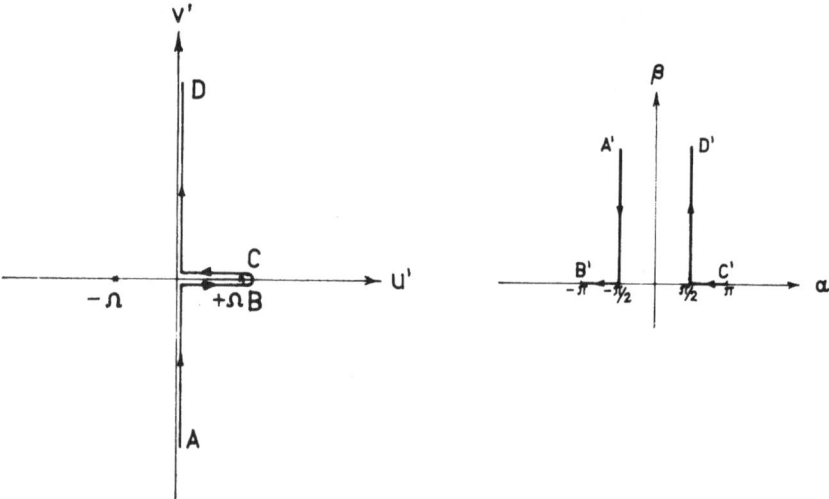

Fig. 4.15. Integration path in $u'+iv'$ plane

Fig. 4.16. Transformation of path of integration in $\alpha+i\beta$ plane

in the u', v'-plane is therefore represented by the path $A'B'$ (Fig. 4.16) in the α, β-plane. Similarly, the contour CD transforms into $C'D'$.

In order to effect the integration along the new contour, we must distinguish between two cases: (a) $t < (x/c)$; (b) $t > (x/c)$.

a) $t < (x/c)$. Defining the argument θ through

$$\tanh \theta \equiv \frac{ct}{x} < 1 \tag{4.3.99}$$

we obtain

$$\mathrm{Re}\{s't - x/c\sqrt{s'^2 - \Omega^2}\} = +\Omega\sqrt{x^2/c^2 - t^2}\sinh(\beta - \theta)\cos\alpha . \tag{4.3.100}$$

Closure of the path of integration at $\beta \to \infty$ is possible only for the values $-\pi < \alpha < -\pi/2$; $\pi/2 < \alpha < \pi$, as one obtains $\mathrm{Re}\{s't - (x/c)\sqrt{s'^2 - \Omega^2}\} < 0$ only for these values of α. The closed contour of Fig. 4.17 comprises no singularities; hence, according to Cauchy's theorem, we obtain

$$f(x, t) = 0; \quad t < (x/c) . \tag{4.3.101}$$

b) $t > (x/c)$. Instead of θ, we define in this case

$$\tanh \chi = \frac{x}{ct} \tag{4.3.102}$$

and obtain

$$\mathrm{Re}\{s't - (x/c)\sqrt{s'^2 - \Omega^2}\} = -\Omega\sqrt{t^2 - x^2/c^2}\cosh(\beta - \chi)\cos\alpha . \tag{4.3.103}$$

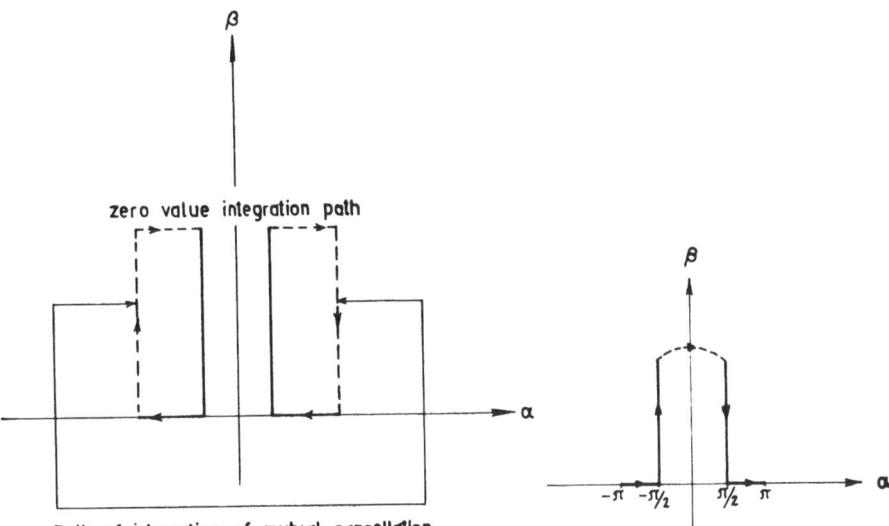

Fig. 4.17. Integration contour for $t < x/c$

Fig. 4.18. Integration contour for $t > x/c$

Convergence for $\beta \to \infty$ is now ensured for $\cos \alpha > 0$, i.e. for the interval $-\pi/2 < \alpha < \pi/2$ (Fig. 4.18). The direction of the original path of integration is traced in Fig. 4.15; reversing this direction, we obtain the path traced in Fig. 4.18, and the result $-f(x, t)$ instead of $+f(x, t)$.

Hence,

$$-f(x, t) = \exp[-(\sigma/2\varepsilon_0)t] \frac{1}{2\pi i} \int_{+i\infty}^{-i\infty} \frac{\exp[s't - (x/c)\sqrt{s'^2 - \Omega^2}]}{\sqrt{s'^2 - \Omega^2}}\, ds'$$

$$= -\exp[-(\sigma/2\varepsilon_0)t] \frac{1}{2\pi} \int_{-\pi}^{\pi} \exp[-\Omega t(\cos\gamma + i(x/ct)\sin\gamma)]\, d\gamma \ .$$

$$(4.3.104)$$

Now

$$t[\cos\gamma + i(x/ct)\sin\gamma] = \sqrt{t^2 - x^2/c^2}\,(\cosh\chi \cos\gamma + i \sinh\chi \sin\gamma)$$

$$= \sqrt{t^2 - x^2/c^2}\,\cos(\gamma - i\chi), \quad t > (x/c) \ , \quad (4.3.105)$$

and therefore

$$f(x, t) = \exp[-(\sigma/2\varepsilon_0)t]\frac{1}{2\pi}\int_{-\pi}^{\pi}\exp[-\Omega\sqrt{t^2 - x^2/c^2}\cos(\gamma - i\chi)]\, d\gamma,$$

$$t > (x/c) \ . \quad (4.3.106)$$

Introducing the new variable $\phi = \gamma - i\chi$, we have (in view of the periodicity of the integrand) that

$$f(x,t) = \exp[-(\sigma/2\,\varepsilon_0)\,t]\,\frac{1}{\pi}\int_0^\pi \exp(-\Omega\sqrt{t^2-x^2/c^2}\cos\phi)\,d\phi\,, \qquad t > x/c\,.$$

$$(4.3.107)$$

or, finally, resorting to Sommerfeld's integral representation [4.14]

$$f(x,t) = \exp[-(\sigma/2\,\varepsilon_0)\,t]\,J_0(\mathrm{i}\,\Omega\sqrt{t^2-x^2/c^2})$$

$$= \exp[-(\sigma/2\,\varepsilon_0)\,t]\,J_0(\mathrm{i}\sqrt{(\sigma/2\,\varepsilon_0)^2-(c/\lambda)^2}\cdot\sqrt{t^2-x^2/c^2})\,, \quad t > x/c\,.$$

$$(4.3.108)$$

Hence, see (4.3.89, 90),

$$E_z = -V_a\sqrt{(\sigma/2\,\varepsilon_0)^2-(c/\lambda)^2}\,\exp[-(\sigma/2\,\varepsilon_0)\,t]\,\frac{x/c}{\sqrt{t^2-x^2/c^2}}$$

$$\times[-\mathrm{i}\,J_1\,(\mathrm{i}\sqrt{(\sigma/2\,\varepsilon_0)^2-(c/\lambda)^2}\sqrt{t^2-x^2/c^2})]\,, \qquad t > x/c\,, \qquad (4.3.109)$$

$$H_y = V_a\sqrt{\varepsilon_0/\mu_0}\sqrt{(\sigma/2\,\varepsilon_0)^2-(c/\lambda)^2}\left\{\frac{\sigma/2\,\varepsilon_0}{\sqrt{(\sigma/2\,\varepsilon_0)^2-(c/\lambda)^2}}\,\exp[-(\sigma/2\,\varepsilon_0)\,t]\right.$$

$$\times J_0(\mathrm{i}\sqrt{(\sigma/2\,\varepsilon_0)^2-(c/\lambda)^2}\sqrt{t^2-x^2/c^2})$$

$$+\exp[-(\sigma/2\,\varepsilon_0)\,t]\,\frac{t}{\sqrt{t^2-x^2/c^2}}$$

$$\times[-\mathrm{i}\,J_1\,(\mathrm{i}\sqrt{(\sigma/2\,\varepsilon_0)^2-(c/\lambda)^2}\sqrt{t^2-x^2/c^2})]$$

$$+\frac{(c/\lambda)^2}{\sqrt{(\sigma/2\,\varepsilon_0)^2-(c/\lambda)^2}}\int_{t'=x/c}^{t}\exp[-(\sigma/2\,\varepsilon_0)\,t']$$

$$\left.\times J_0(\mathrm{i}\sqrt{(\sigma/2\,\varepsilon_0)^2-(c/\lambda)^2}\sqrt{t'^2-x^2/c^2})\,dt'\right\}\,, \qquad t > x/c\,. \quad (4.3.110)$$

C) Evaluation of Integrals

a) We wish to evaluate, see (4.3.22), the integral

$$h_\infty = 4\,\frac{T^2}{T_s^2}\,\frac{1}{\sqrt{1-4T^2/T_s^2}}\int_{\tau'=\xi}^{\infty}\mathrm{e}^{-\tau'}J_0(\mathrm{i}\sqrt{1-4T^2/T_s^2}\sqrt{\tau'^2-\xi^2})\,d\tau'\,.$$

$$(4.3.111)$$

Introducing the abbreviation

$$\eta \equiv \sqrt{1-4T^2/T_s^2}\,, \tag{4.3.112}$$

we must find the value of

$$\xi \int\limits_{(\tau'/\xi)=1}^{\infty} \exp[-\xi(\tau'/\xi)] J_0(i\eta\xi\sqrt{(\tau'/\xi)^2-1}) d(\tau'/\xi) \ . \tag{4.3.113}$$

Now, in general, [Ref. 4.3, No. 6.616−2, p. 710]

$$\int\limits_{1}^{\infty} e^{-\alpha x} J_0(\beta\sqrt{x^2-1}) dx = \frac{1}{\sqrt{\alpha^2+\beta^2}} \exp(-\sqrt{\alpha^2+\beta^2}) \ . \tag{4.3.114}$$

Hence,

$$\xi \int\limits_{1}^{\infty} \exp[-\xi(\tau'/\xi)] J_0(i\eta\xi\sqrt{(\tau'/\xi)^2-1}) d(\tau'/\xi)$$

$$= \frac{1}{\sqrt{1-\eta^2}} \exp(-\xi\sqrt{1-\eta^2}) \tag{4.3.115}$$

and therefore

$$h_\infty = \left[4\frac{T^2}{T_s^2} \frac{1}{\sqrt{1-4T^2/T_s^2}} \right] \frac{1}{2T/T_s} \exp[-\xi 2T/T_s]$$

$$= \frac{2T/T_s}{\sqrt{1-4T^2/T_s^2}} \exp(-x/\lambda) \ . \tag{4.3.116}$$

b) Let us now evaluate, see (4.3.23), the integral

$$l_\infty = -\int\limits_{\tau'=\xi}^{\infty} \frac{\xi}{\sqrt{\tau'^2-\xi^2}} e^{-\tau'}\{-iJ_1(i\sqrt{1-4T^2/T_s^2}\sqrt{\tau'^2-\xi^2})\}d\tau' \ . \tag{4.3.117}$$

We resort to the result [Ref. 4.3, No. 6.645−1, p. 721]

$$\int\limits_{1}^{\infty} (x^2-1)^{-\frac{1}{2}} e^{-\alpha x} J_\nu(\beta\sqrt{x^2-1}) dx$$

$$= I_{1/2\nu}\left[\frac{1}{2}(\sqrt{\alpha^2+\beta^2}-\alpha)\right] K_{1/2\nu}\left[\frac{1}{2}(\sqrt{\alpha^2+\beta^2}+\alpha)\right] \ . \tag{4.3.118}$$

Introducing once more the notation (4.3.112) we find

$$l_\infty = -\xi \int\limits_{1}^{\infty} \frac{1}{\sqrt{(\tau'/\xi)^2-1}} \exp[-\xi(\tau'/\xi)]$$

$$\times [-iJ_1(i\eta\xi\sqrt{(\tau'/\xi)^2-1})] d(\tau'/\xi) \ , \tag{4.3.119}$$

i.e.

$$
\begin{aligned}
l_\infty &= i\,\zeta I_{1/2}\left(\frac{1}{2}(\sqrt{\xi^2-\eta^2\xi^2}-\xi)\right)K_{1/2}\left(\frac{1}{2}(\sqrt{\xi^2-\eta^2\xi^2}+\xi)\right) \\
&= i\,\zeta I_{1/2}\left(\frac{\xi}{2}[2(T/T_\mathrm{s})-1]\right)K_{1/2}\left(\frac{\xi}{2}[2(T/T_\mathrm{s})+1]\right).
\end{aligned}
\tag{4.3.120}
$$

Now, the modified spherical Bessel function $I_{1/2}(z)$ of the first kind, order 1/2 and argument z may be expressed [Ref. 4.14, No. 10.2.13, p. 443] by means of the relationship

$$
\sqrt{\frac{\pi}{2z}}\,I_{1/2}(z) = \frac{\sinh z}{z},
\tag{4.3.121}
$$

whereas for the modified spherical Bessel function $K_{1/2}(z)$ of the third kind, order 1/2 and argument z we have [Ref. 4.14, No. 10.2.17, p. 444]

$$
\sqrt{\frac{\pi}{2z}}\,K_{1/2}(z) = \frac{\pi}{2z}\,e^{-z}.
\tag{4.3.122}
$$

Hence

$$
\begin{aligned}
l_\infty = i\,\zeta &\sqrt{\frac{2\dfrac{\xi}{2}[2(T/T_\mathrm{s})-1]}{\pi}}\;\frac{\sinh\dfrac{\xi}{2}[2(T/T_\mathrm{s})-1]}{\dfrac{\xi}{2}[2(T/T_\mathrm{s})-1]} \\[2ex]
&\times\sqrt{\frac{2\dfrac{\xi}{2}[2(T/T_\mathrm{s})+1]}{\pi}}\;\frac{\pi}{2\dfrac{\xi}{2}[2(T/T_\mathrm{s})+1]}\;\exp\left\{-\frac{\xi}{2}[2(T/T_\mathrm{s})+1]\right\} \\[2ex]
&= \frac{1}{\sqrt{1-4T^2/T_\mathrm{s}^2}}\{\exp(-\xi)-\exp[2\xi(T/T_\mathrm{s})]\} \\[2ex]
&= \frac{1}{\sqrt{1-4T^2/T_\mathrm{s}^2}}\left(e^{-\frac{\sigma}{2\varepsilon_0 c}x}-e^{-\frac{x}{\lambda}}\right).
\end{aligned}
\tag{4.3.123}
$$

4.4 Electromechanical Transients in Liquid Metal upon Field Disruption

Liquid metals are widely used in modern engineering and technology: alkali metals serving as a cooling medium in atomic reactors permit higher working temperatures than are possible with water cooling; certain liquid alloys are used in machinery designed to replace steam engines and turbines; very high purity is obtained in various metal production processes involving levitation melting; finally, the production processes of certain construction materials involve work with molten silicates which exhibit electric conductivity at high temperatures and, in this respect, resemble liquid metals. All these applications entail remote action on the metal by electromagnetic means; flow devices based on this principle have been specially designed for liquid zinc, sodium, potassium, indium-gallium alloys, magnesium and aluminum, and agitators and pumps of the same type have found increasing use in recent years.

In these contexts, the main practical aspect is the interaction between a conducting liquid and an applied magnetic field; such a problem is treated in the present section, namely, that of a force field acting on a mercury layer upon disruption of an imposed magnetic field.

4.4.1 Formulation of the Problem

The system considered here is a very thin, infinitely extending sheet of liquid metal residing in the narrow air gap (height $2b$) of a circular cylindrical electromagnet (diameter $2R$), which is identical with that of Fig. 3.1, except that the solid sheet (thickness Δ) is replaced by a layer of mercury with the origin now on its undersurface.

The following (by now quite standard) simplifications are adopted:
– The iron is replaced by a fictitious material of infinite relative permeability, vanishingly small electric specific conductivity and zero dielectric constant.

– The exciting magnetic induction is taken to be restricted to the air gap, and for the very narrow region confined to $r < R$ it is taken to comprise – in the central plane of the mercury dish – an axial component only. It should be noted that disregard of any magnetic induction for the region beyond R is mathematically equivalent to assumption of a zero relative magnetic permeability for this so-called "air"; furthermore, we assume vanishingly small electric specific conductivity and zero dielectric constant for $r > R$, throughout.

– The air gap proper $(-b < z < b,\ r < R)$ is replaced by a fictitious material of relative permeability $\mu_r = 1.0$, yet again with zero dielectric constant, i.e. according to Faraday's law, the electric field induced in the surroundings is, in turn, not linked to any electric flux density.

– Within the mercury layer, the time rate of the *electrical* flux density is at all times completely negligible with respect to its local electric conduction current counterpart.

The question we address now is: What is the transient motion of the mercury caused by sudden disruption of the magnetic field?

4.4.2 Approach to the Solution

In order to clarify the conditions imposed upon the system when the field is switched on, we further assume as follows:

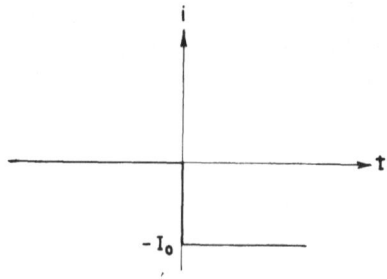

Fig. 4.19. Assumed model for current excitation

– At steady state, prior to disruption, the coils of the electromagnet are under continuous excitation by a constant, direct current $i_1 = I_0$ linked within the air gap to a steady axial field component H_0.

– Switch-off (disruption) is modelled (Fig. 4.19) by an instantaneous series connection (at the instant $t = 0$) of the exciting coils to an additional, identical, constant current source yielding a current $i_2 = -I_0 \cdot 1(t)$ with polarity opposite to that of i_1. Hence, an axial component $-H_0 \cdot 1(t)$ is set up in the air gap, in addition to H_0 which had been established in the "remote" past, i.e. at $t \to -\infty$.

The resulting annihilation of the magnetic free energy within the air gap is accompanied by generation of an electric field which forces an electric current through the mercury; this current, in turn, interacts with the attendant magnetic flux density, and the resulting force actuates locally the liquid metal layer, causing it to undergo decaying oscillatory motion.

Quantitative analysis of these oscillations will now be undertaken in two distinct steps: (a) solution of the electrical problem concerning the electric transients; (b) solution of the mechanical problem concerning the motion of the mercury. Calculations will accordingly be performed on a so-called "ballistic" basis, first solving part (a) with the mercury assumed stationary, and only then turning to (b) with the assumption that the imposed magnetic field has by now vanished. This approach is warranted by the large discrepancy between the electrical and mechanical characteristic time constants.

4.4.3 Electrical Problem

Resorting to an analysis resembling that outlined in Sect. 3.1, two distinct field zones are visualized: the inner zone $0 \leqslant r \leqslant R$ in which the magnetic flux density linked to the current within the mercury threads the poles of the electromagnet, and the outer zone $R \leqslant r < \infty$ in which the magnetic flux density (although not the field) is assumed to vanish.

Having disregarded the time rate of the electric induction – which is inherently nonexistent due to the assumption of a zero dielectric constant – the magnetic field is obtainable in the fictitious "air" from a magnetic scalar potential which satisfies the Laplace equation. As already pointed out (Sect. 3.1.5), an exact solution of this equation is unobtainable in the present case; hence we again resort to an approximation by solving first for the inner zone and then coupling the result to the outer region.

A) Inner Zone

Maxwell's two curl equations and Ohm's law applied to a metal with conductivity σ yield the field build-up according to the differential equation

$$\nabla \times (\nabla \times H) = -\sigma \mu_0 \frac{\partial H}{\partial t} \ . \tag{4.4.1}$$

This relation is valid in principle for conductors whose dimensions are relatively large, e.g., when compared to a relevant skin depth, at an appropriate working frequency.

In order to adapt the equation to the thin layer of mercury, we specifically consider the axial field component H_z, i.e.,

$$\nabla \left(\frac{\partial H_z}{\partial z} \right) - \nabla^2 H_z = -\sigma \mu_0 \frac{\partial H_z}{\partial t} \ . \tag{4.4.2}$$

With axial variation neglected in view of the narrowness of the layer, (4.4.2) reduces to

$$\nabla^2 H_z = \sigma \mu_0 \frac{\partial H_z}{\partial t} \ , \tag{4.4.3}$$

the Laplacian operator being now defined for two dimensions only. Integrating (4.4.3) with respect to the z-axis within the air gap and along a path along the z-axis (threading the poles) we obtain (for the case of circular symmetry) the partial differential equation

$$\frac{\partial^2 H_z}{\partial r^2} + \frac{1}{r} \frac{\partial H_z}{\partial r} - \frac{\sigma \Delta \mu_0}{2b} \frac{\partial H_z}{\partial t} = 0 \ . \tag{4.4.4}$$

Here we introduce with the aid of the complex variable s, the one-sided Laplace transform $\bar{H}_z(r, s)$ of $H_z(r, t)$ through the formula

$$\bar{H}_z(r, s) = \int_0^\infty H_z(r, t) e^{-st} dt \ , \tag{4.4.5}$$

along with its inverse

$$H_z(r, t) = \frac{1}{2\pi i} \int_{-i\infty}^{i\infty} \bar{H}_z(r, s) e^{st} ds \ . \tag{4.4.6}$$

Substitution of the last expression in (4.4.4) yields the ordinary differential equation

$$\frac{d^2\bar{H}_z}{dr^2} + \frac{1}{r}\frac{d\bar{H}_z}{dr} - \frac{\sigma\Delta\mu_0 s}{2b}\bar{H}_z = 0 \ , \tag{4.4.7}$$

whose solution (subject to convergence conditions) is given by a Bessel function of the first kind, order zero; hence, denoting the complex field component $\bar{H}_z(r)$ at the inner circumference $r = R(-0)$ by $\bar{H}_z(R)$, we obtain

$$\bar{H}_z(r) = \bar{H}_z(R)\frac{J_0\left(i\sqrt{\dfrac{\sigma\Delta\mu_0 s}{2b}}\,r\right)}{J_0\left(i\sqrt{\dfrac{\sigma\Delta\mu_0 s}{2b}}\,R\right)} \ . \tag{4.4.8}$$

As the imposed magnetic time dependence is modelled by the step function $-H_0 \cdot 1(t)$ whose Laplace transform is $-H_0/s$, we replace in (4.4.8) the complex amplitude $\bar{H}_z(R)$ by the latter expression, our approximation now disregarding the electric current in the region $r > R$ (Sect. 3.1).

Denoting the roots of the Bessel function $J_0(u)$ by u_k and introducing the notation

$$T \equiv \frac{\sigma\Delta\mu_0 R^2}{2b} \ , \tag{4.4.9}$$

we finally obtain (Sect. 4.4.6 A) that the time-varying, resultant (superscript r) axial flux density pervading the mercury is given by

$$\mu_0 H_z^{(r)}(r, t) = \mu_0 H_0 2 \sum_{k=1}^{\infty} \frac{J_0[(r/R)u_k]}{u_k J_1(u_k)} \exp[-u_k^2(t/T)] \cdot 1(t) \ . \tag{4.4.10}$$

This functional dependence is reproduced in Fig. 4.20.

Once $H_z(r, t)$ is known, the circular current density j_α permeating the mercury is found to read [6]

$$j_\alpha = H_0\frac{4b}{\Delta}\frac{1}{R}\sum_{k=1}^{\infty}\frac{J_1[(r/R)u_k]}{J_1(u_k)}\exp[-u_k^2(t/T)] \cdot 1(t) \ , \tag{4.4.11}$$

so that the radial component f_r of the volume force density is given by

$$f_r = j_\alpha\mu_0 H_z \ , \tag{4.4.12}$$

i.e.,

$$f_r = \mu_0 H_0^2\frac{4b}{\Delta}\frac{2}{R}\sum_{k=1}^{\infty}\sum_{m=1}^{\infty}\frac{J_0[(r/R)u_k]}{u_k J_1(u_k)}\frac{J_1[(r/R)u_m]}{J_1(u_m)}$$
$$\times\exp[-(u_k^2+u_m^2)(t/T)] \cdot 1(t) \ . \tag{4.4.13}$$

---

[6] As j_α is confined to a very narrow layer Δ within the air gap $2b$, (4.4.11) is obtained by integration of the circumferential component of Maxwell's first curl equation along the z-axis, for $-b \leqslant z \leqslant b$.

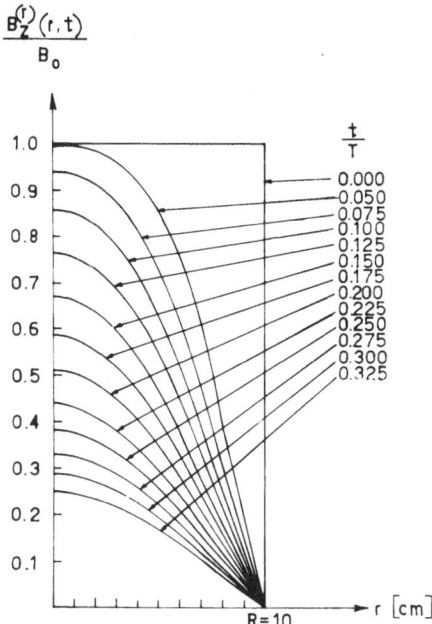

$$\frac{B_z^{(r)}(r,t)}{B_0}$$

1.0

0.9

0.8

0.7

0.6

0.5

0.4

0.3

0.2

0.1

$\frac{t}{T}$

0.000
0.050
0.075
0.100
0.125
0.150
0.175
0.200
0.225
0.250
0.275
0.300
0.325

R = 10

r [cm]

B) Outer Zone

Although the influence of the sheet current in the outer zone $R \leqslant r < \infty$ is disregarded here, we shall nevertheless provide for completeness an estimate for the field in this region (for a different approximation, see Sect. 3.1.8); thus by Faraday's law the rate of change of the magnetic flux ϕ determines the circular electric field component E_α at the pole circumference via

$$2\pi R E_\alpha = -\frac{d\phi}{dt}, \quad r = R \ . \tag{4.4.14}$$

However, in view of Ohm's law $j_\alpha = \sigma E_\alpha$, along with (4.4.11), we have

$$E_\alpha = H_0 \frac{4b}{R} \frac{1}{\sigma \Delta} \sum_{k=1}^{\infty} \exp[-u_k^2(t/T)] \cdot 1(t), \quad r = R \ , \tag{4.4.15}$$

so that

$$-\frac{d\phi}{dt} = H_0 8\pi b \frac{1}{\sigma \Delta} \sum_{k=1}^{\infty} \exp[-u_k^2(t/T)] \cdot 1(t), \quad r = R \ . \tag{4.4.16}$$

Hence, as $E_\alpha = -(d\phi/dt)/2\pi r$ at any radius $r > R$, we finally obtain

$$E_\alpha = H_0 \frac{4b}{r} \frac{1}{\sigma \Delta} \sum_{k=1}^{\infty} \exp[-u_k^2(t/T)] \cdot 1(t), \quad r > R \ . \tag{4.4.17}$$

4.4.4 Mechanical Problem

Having provided a solution to the electrical problem, we turn our attention to the electromechanical interaction between the field and the liquid metal. The latter (of specific density ϱ) is assumed incompressible and inviscid, with all surface-tension phenomena totally disregarded[7]. Each incremental volume element of the fluid, moving in the instantaneous velocity u, is characterized by the *material* acceleration Du/Dt, so that the relevant specific inertial resistance is, in turn, determined by the product $(\varrho Du/Dt)$; hence, under the action of an external pressure p and an applied volume force density f, Euler's equation of motion for a perfect fluid reads

$$\varrho \left[\frac{\partial u}{\partial t} + (u \cdot \nabla)u \right] + \nabla p = f \ . \tag{4.4.18}$$

Now f comprises two terms: one, f_g due to gravity, i.e.,

$$f_g = -g1_z \ , \tag{4.4.19}$$

(with 1_z denoting the unit vector) and the other f_m due to the magnetic interaction, i.e.,

$$f_m = j \times (\mu_0 H) \ . \tag{4.4.20}$$

Henceforth we omit second-order terms in the velocity variation; introducing the condition of incompressibility

$$\nabla \cdot u = 0 \ , \tag{4.4.21}$$

and taking into account that, due to symmetry, u lacks a circular component u_α, Euler's equations of motion finally read

$$\varrho \frac{\partial u}{\partial t} + \nabla p = -\varrho g 1_z + f \ , \tag{4.4.22}$$

$$\frac{1}{r} \frac{\partial}{\partial r}(ru_r) + \frac{\partial u_z}{\partial z} = 0 \ . \tag{4.4.23}$$

Within the very shallow layer, we assume the axial variation of u_r to be negligible; hence

$$u_z = -z \frac{1}{r} \frac{\partial}{\partial r}(ru_r) \ . \tag{4.4.24}$$

---

[7] In plane waves, the combined effect of gravity and capillarity produces a minimum point v_m (about 19 cm/s for mercury) in the wave velocity [4.15]. This minimum is calculated from the surface tension T_s, the gravity acceleration g, and the density ϱ, through the equation $v_m^2 = 2\sqrt{T_s g/\varrho}$. For a known ratio v/v_m, the influence of capillarity on the wave velocity may be estimated [4.16]. In our case, the thickness of the mercury layer is $\Delta \approx 5.5 \times 10^{-3}$ m, and accordingly (see quoted references) capillarity is disregarded: in view of the overall simplifications, this approximation seems quite legitimate.

Denoting now the axial displacement of the mercury from static equilibrium by η, and taking $\eta \ll \Delta$, we have

$$z = \Delta + \eta \simeq \Delta ,$$ (4.4.25)

so that

$$u_z \simeq -\Delta \frac{1}{r} \frac{\partial}{\partial r} (r u_r) .$$ (4.4.26)

Along with η, we introduce the radial displacement ξ from static equilibrium; writing now

$$u_z = \frac{\partial \eta}{\partial t} \quad \text{and}$$ (4.4.27)

$$u_r = \frac{\partial \xi}{\partial t} ,$$ (4.4.28)

we find, after substitution in (4.4.26), now regarded as an exact equality, and differentiation with respect to t, the partial differential equation

$$\frac{\partial^2 \eta}{\partial t^2} = -\Delta \frac{1}{r} \frac{\partial}{\partial r} \left(r \frac{\partial^2 \xi}{\partial t^2} \right) .$$ (4.4.29)

Let us now return to (4.4.22); once $\partial p / \partial r$ is known, the radial acceleration is prescribed by the radial magnetic force density f_r:

$$\frac{\partial u_r}{\partial t} = -\frac{1}{\varrho} \frac{\partial p}{\partial r} + \frac{f_r}{\varrho} .$$ (4.4.30)

In order to determine the radial pressure variation $\partial p / \partial r$, let us formally note that at any elevation z_p within the shallow layer, the difference between p and the ambient pressure p_0 is given by

$$p - p_0 = \varrho g (\Delta + \eta - z_p) .$$ (4.4.31)

Hence,

$$\frac{\partial p}{\partial r} = \varrho g \frac{\partial \eta}{\partial r} ,$$ (4.4.32)

and therefore (4.4.30) becomes

$$\frac{\partial^2 \xi}{\partial t^2} = -g \frac{\partial \eta}{\partial r} + \frac{f_r}{\varrho} .$$ (4.4.33)

Introducing the characteristic speed

$$v \equiv \sqrt{g\Delta} \,, \tag{4.4.34}$$

the partial differential equation (4.4.29) is finally rewritten in the form

$$\frac{1}{r}\frac{\partial}{\partial r}\left(r\frac{\partial \eta}{\partial r}\right) - \frac{1}{v^2}\frac{\partial^2 \eta}{\partial t^2} = \frac{1}{\varrho g}\frac{1}{r}\frac{\partial}{\partial r}(rf_r) \,, \tag{4.4.35}$$

where f_r is given by (4.4.13). As already pointed out, solution of this equation is obtained "ballistically", i.e. it is assumed that the magnetic force density f_r had practically vanished before any change occurred in the shape of the free surface.

Introducing the dimensionless magnetic force number $S = (\tfrac{1}{2}\mu_0 H_0^2)/(\tfrac{1}{2}\varrho v^2)$ along with the magnetic Reynolds number $R_m = \sigma \Delta \mu_0 v$, the solution of (4.4.35) is given (Sect. 4.4.6B) by

$$\eta(r,t) = -\frac{4}{\pi} R S R_m$$

$$\times \int_{\gamma=0}^{\pi} \sum_{k=1}^{\infty} \sum_{m=1}^{\infty} \frac{(u_m - u_k \cos\gamma)\{[\sin(\Lambda vt/R)J_0(\Lambda r/R)]/\Lambda\}}{(u_m^2 + u_k^2)(u_k^2/u_m)J_1(u_k)J_1(u_m)}d\gamma \tag{4.4.36}$$

where

$$\Lambda^2 = u_k^2 + u_m^2 - 2u_k u_m \cos\gamma \,. \tag{4.4.37}$$

The shape function of the free surface $\eta(r)$ was plotted (Fig. 4.21) for three different values of t, with k- and m-terms taken to an accuracy of 5%.

Qualitative agreement was found on comparing the graphs with experimental results (Sect. 4.4.6C) for a mercury layer of thickness $\Delta \approx 5.5 \times 10^{-3}$ m in the air gap of an electromagnet with pole radius $R = 1 \times 10^{-1}$ m (magnetic flux density 0.1 Vs/m^2).

Plots of $\eta(t)$ at $r = $ const were similarly in qualitative agreement with experiment, and moreover showed that the general time-dependent behavior of the solution is essentially determined by the first term in (4.4.36). Accordingly, this term (an almost periodic, low-frequency, slowly decaying oscillation) was plotted separately (Fig. 4.22), and again proved to be in qualitative agreement with experimental results[8].

4.4.5 Discussion

The "ballistic" oscillation of a layer of mercury upon sudden disruption of an externally imposed magnetic field has been described. The relevant eddy-

---

[8] In fact, the degree of agreement observed is remarkable, considering the extensive simplifications introduced throughout the analysis and the obvious diversity of experimental conditions.

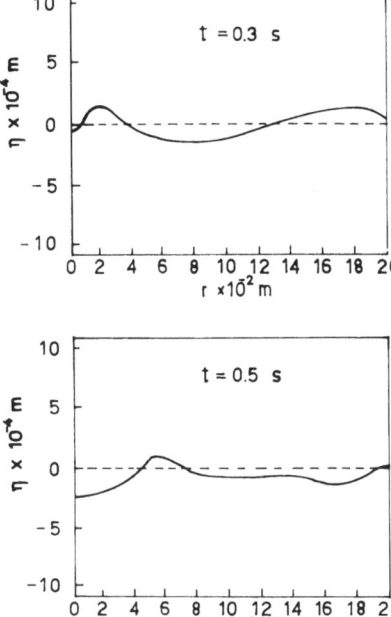

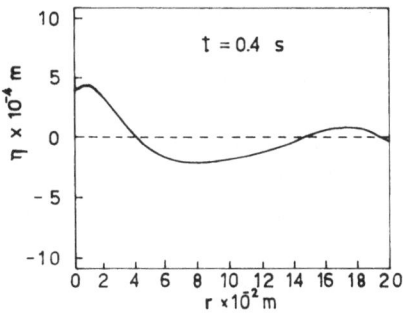

Fig. 4.21. Shape of mercury surface for different instants of time after switch-off

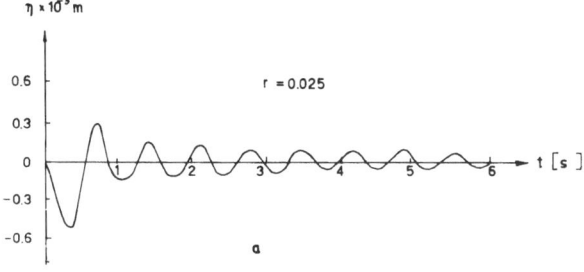

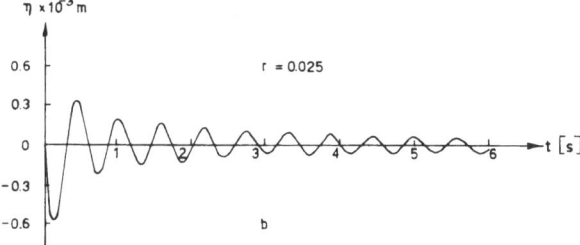

Fig. 4.22a, b. Elevation of free surface as function of time at r = const: (a) Experimental; (b) First term of (4.4.36)

current problem was formulated and solved, and the expression for the force densities acting on the liquid metal was obtained. The solution is a (double) infinite series, but numerical calculation and comparison with experimental results indicate that the *first* term suffices for practical evaluation of the time dependence of the induced oscillations. Motion of the mercury surface was recorded on a specially designed instrument, and both theory and experiment indicate almost-periodic, low-frequency, slowly decaying oscillation.

4.4.6 Appendix

A) Transient Fields

With (4.4.8), the complex amplitude $\mu_0 \bar{H}_z(r)$ of the (Laplace-transformed) time-dependent axial magnetic flux density is given by

$$\mu_0 \bar{H}_z(r) = -\mu_0 H_0 \frac{1}{s} J_0 \left(ir \sqrt{\frac{\sigma \Delta \mu_0 s}{2b}} \right) \Big/ J_0 \left(iR \sqrt{\frac{\sigma \Delta \mu_0 s}{2b}} \right). \quad (4.4.38)$$

Hence, in integral representation,

$$\mu_0 H_z(r, t) = -\mu_0 H_0 \frac{1}{2\pi i} \int_{-i\infty}^{i\infty} \left[J_0 \left(ir \sqrt{\frac{\sigma \Delta \mu_0 s}{2b}} \right) \Big/ \right.$$

$$\left. J_0 \left(iR \sqrt{\frac{\sigma \Delta \mu_0 s}{2b}} \right) \right] \frac{e^{st}}{s} ds . \quad (4.4.39)$$

Applying now the residue theorem at the poles $s_0 = 0$; $s_k = -u_k^2 2b/\sigma \Delta \mu_0 R^2$, u_k being the kth root of the equation $J_0(u) = 0$, we obtain, with

$$T \equiv \frac{\sigma \Delta \mu_0 R^2}{2b} , \quad (4.4.40)$$

the flux density

$$\mu_0 H_z(r, t) = -\mu_0 H_0 \left\{ 1 - 2 \sum_{k=1}^{\infty} \frac{J_0((r/R)u_k)}{u_k J_1(u_k)} \exp[-u_k^2(t/T)] \right\}, \quad t > 0 . \quad (4.4.41)$$

Hence, the resultant axial magnetic flux density $\mu_0 H^{(r)}$ in the air gap is obtained as the sum of the predisruption steady-state density $\mu_0 H_0$ and $\mu_0 H_z(r, t)$, i.e.,

$$\mu_0 H_z^{(r)}(r, t) = \mu_0 H_0 + \mu_0 H_z(r, t)$$

$$= \begin{cases} \mu_0 H_0, & t < 0 , \\[2mm] \mu_0 H_0 2 \sum_{k=1}^{\infty} \dfrac{J_0((r/R)u_k)}{u_k J_1(u_k)} \exp[-u_k^2(t/T)] , & t > 0 . \end{cases}$$

(4.4.42)

B) Mechanical Vibration

We seek the solution of (4.4.35).

Recalling that f_r is of extremely short duration compared with the (experimentally recorded) period of oscillation of the mercury surface, we integrate (4.4.35) between $t = 0$ and $t = t_1$: at the latter instant, f_r is regarded as having practically vanished. As t_1 denotes, per definition, a very short time interval, we assume that no sensible change has occurred in the value of η; furthermore, because for $t > t_1$ the right member of (4.4.35) is zero, we may in fact set — in view of our "ballistic" assumption — the interval from $t = 0$ to $t = \infty$, obtaining

$$-\frac{1}{v^2} \frac{\partial \eta}{\partial t}(r, 0) = \mu_0 H_0^2 \frac{4b}{\Delta} \frac{2}{R} \frac{1}{\varrho g} \frac{1}{r} \frac{d}{dr} \left\{ r \sum_{k=1}^{\infty} \sum_{m=1}^{\infty} T \frac{1}{u_k^2 + u_m^2} \right.$$

$$\left. \times \frac{J_0((r/R)u_k)}{u_k J_1(u_k)} \frac{J_1((r/R)u_m)}{J_1(u_m)} \right\} .$$

(4.4.43)

Now the homogeneous equation

$$\frac{1}{r} \frac{\partial}{\partial r} \left(r \frac{\partial \eta}{\partial r} \right) - \frac{1}{v^2} \frac{\partial^2 \eta}{\partial t^2} = 0$$

(4.4.44)

describes the motion of the liquid metal for times $t > t_1$, at which the magnetic force densities are non-existent. The solution of this partial differential equation adapted to our infinitely extending sheet is conveniently formulated in integral representation. For this purpose we introduce a dummy variable of integration λ and a still unspecified spectral density $\alpha(\lambda)$, writing

$$\eta(r, t) = \int_{\lambda=0}^{\infty} \alpha(\lambda) \sin(\lambda v t) J_0(\lambda r) d\lambda , \qquad t > t_1 , \qquad t_1 \to 0$$

(4.4.45)

so that

$$\lim_{t \to 0} \frac{\partial \eta}{\partial t}(r, t) = \int_{\lambda=0}^{\infty} \lambda v \alpha(\lambda) J_0(\lambda r) d\lambda .$$

(4.4.46)

The fact that (4.4.43 and 46) refer to the same phenomenon provides the condition for determination of the spectral density $\alpha(\lambda)$. Accordingly, we first rewrite the time-derivative $\partial \eta / \partial t$ from (4.4.43) in the form

$$\frac{\partial \eta}{\partial t}(r,0) = -4vRSR_m \frac{1}{r} \frac{d}{dr} \left\{ r \sum_{k=1}^{\infty} \sum_{m=1}^{\infty} \frac{1}{u_k^2 + u_m^2} \right.$$

$$\left. \times \frac{J_0((r/R)u_k)}{u_k J_1(u_k)} \frac{J_1((r/R)u_m)}{J_1(u_m)} \right\} \equiv g(r) \tag{4.4.47}$$

(recalling that $S = \mu_0 H_0^2/\varrho v^2$; $R_m = \mu_0 \sigma \Delta v$; $T = \sigma \Delta \mu_0 R^2/2b$) and then resort to Fourier-Bessel representation of $g(r) \equiv \partial \eta/\partial t$, namely

$$g(r) = \int_{\lambda=0}^{\infty} \lambda\, d\lambda\, J_0(\lambda r) \int_{\varrho=0}^{\infty} g(\xi) J_0(\lambda \xi)\xi\, d\xi . \tag{4.4.48}$$

The integral

$$\int_{\xi=0}^{\infty} \xi\, d\xi \frac{1}{\xi} \frac{d}{d\xi} \left[\xi J_0\left(\frac{\xi}{R} u_k\right) J_1\left(\frac{\xi}{R} u_m\right) \right] J_0(\lambda \xi) \equiv I \tag{4.4.49}$$

must be evaluated for each term of the double series in (4.4.47). Integration by parts yields

$$I = \int_{\xi=0}^{\infty} \frac{d}{d\xi} \left[\xi J_0\left(\frac{\xi}{R} u_k\right) J_1\left(\frac{\xi}{R} u_m\right) \right] J_0(\lambda \xi)\, d\xi$$

$$= \left[\xi J_0\left(\frac{\xi}{R} u_k\right) J_1\left(\frac{\xi}{R} u_m\right) \right] J_0(\lambda \xi) \Bigg|_0^{\infty}$$

$$+ \lambda \int_{\xi=0}^{\infty} \xi J_0\left(\frac{\xi}{R} u_k\right) J_1\left(\frac{\xi}{R} u_m\right) J_1(\lambda \xi)\, d\xi$$

$$= \lambda \int_{\xi=0}^{\infty} \xi J_0\left(\frac{\xi}{R} u_k\right) J_1\left(\frac{\xi}{R} u_m\right) J_1(\lambda \xi)\, d\xi , \tag{4.4.50}$$

or [Ref. 4.3, No. 6.578 – 8, p. 695]

$$I = \frac{1}{\sqrt{2\pi}(u_k/R)} (\sin \theta)^{-1/2} P_{1/2}^{1/2}(\cos \theta) \tag{4.4.51}$$

where the angle θ is defined by

$$2\frac{u_m}{R} \lambda \cos \theta = \left(\frac{u_m}{R}\right)^2 + \lambda^2 - \left(\frac{u_k}{R}\right)^2 , \tag{4.4.52}$$

subject to

$$\frac{|u_k - u_m|}{R} < \lambda < \frac{u_k + u_m}{R} \tag{4.4.53}$$

and $P_{1/2}^{1/2}(\cos\theta)$ is the associated Legendre function of the first kind. As the ·
latter may, in turn, be expressed by the relation [Ref. 4.14, No. 8.6.12, p. 334]

$$P_{1/2}^{1/2}(\cos\theta) = (\pi/2)^{-1/2}(\sin\theta)^{-1/2}\cos\theta , \tag{4.4.54}$$

we see that $g(r) \equiv \partial\eta/\partial t$ is given[9] by

$$g(r) = -\frac{4}{\pi}vR^2SR_m \sum_{k=1}^{\infty}\sum_{m=1}^{\infty}\int_{\lambda=|u_k-u_m|/R}^{(u_k+u_m)/R}\lambda\, d\lambda\, J_0(\lambda r)$$

$$\times\frac{\cot\theta}{(u_k^2+u_m^2)u_k^2 J_1(u_k)J_1(u_m)} . \tag{4.4.55}$$

Comparing now with (4.4.46), we have

$$\alpha(\lambda) = -\frac{4}{\pi}R^2SR_m \sum_{k=1}^{\infty}\sum_{m=1}^{\infty}\frac{\cot\theta}{(u_k^2+u_m^2)u_k^2 J_1(u_k)J_1(u_m)} \tag{4.4.56}$$

and therefore

$$\eta(r,t) = -\frac{4}{\pi}R^2SR_m \sum_{k=1}^{\infty}\sum_{m=1}^{\infty}\int_{\lambda=|u_k-u_m|/R}^{(u_k+u_m)/R}\frac{\cot\theta}{(u_k^2+u_m^2)u_k^2 J_1(u_k)J_1(u_m)}$$

$$\times \sin(\lambda vt)J_0(\lambda r)\, d\lambda . \tag{4.4.57}$$

Seeking to calculate $\eta(r,t)$, we now proceed as follows: with $a \equiv u_k/R$, $b \equiv u_m/R$
we first rewrite the definition of the angle θ, namely

$$\cos\theta = \frac{b^2 + \lambda^2 - a^2}{2b\lambda} . \tag{4.4.58}$$

Resorting further to an auxiliary angle γ given by

$$\cos\gamma \equiv \frac{b^2 + a^2 - \lambda^2}{2ab} , \tag{4.4.59}$$

we obtain

$$\cos\theta = \frac{b - a\cos\gamma}{\lambda} , \tag{4.4.60}$$

---

[9] The integrand is convergent at the limits of the integration interval: accordingly, we take
$|u_k - u_m| \leqslant \lambda R \leqslant (u_k + u_m)$ instead of $|u_k - u_m| < \lambda R < (u_k + u_m)$.

along with

$$\frac{\sin \gamma}{\sin \theta} = \frac{\lambda}{a} \ .$$

(4.4.61)

After these preliminary steps, we are in a position to tackle the integral, see (4.4.57),

$$\int_{\lambda = |a-b|}^{a+b} \cot \theta \sin(\lambda v t) \, J_0(\lambda r) \, d\lambda \equiv I_1 \ .$$

(4.4.62)

Using the definition (4.4.59), I_1 may be integrated with respect to the variable γ, i.e.

$$I_1 = \int_{\gamma=0}^{\pi} \cos \theta \sin(\lambda v t) \, J_0(\lambda r) \frac{ab}{\lambda} \frac{\sin \gamma}{\sin \theta} \, d\gamma$$

$$= \int_{\gamma=0}^{\pi} \frac{b - a \cos \gamma}{\lambda} \sin(\lambda v t) \, J_0(\lambda r) \, b \, d\gamma$$

(4.4.63)

and therefore, with $\Lambda \equiv \lambda R$

$$\eta(r, t) = -\frac{4}{\pi} R S R_m$$

$$\times \int_{\gamma=0}^{\pi} \sum_{k=1}^{\infty} \sum_{m=1}^{\infty} \frac{(u_m - u_k \cos \gamma) \{ [\sin(\Lambda v t/R) \, J_0(\Lambda r/R)]/\Lambda \}}{(u_k^2 + u_m^2)(u_k^2/u_m) \, J_1(u_k) \, J_1(u_m)} \, d\gamma \ .$$

(4.4.64)

C) Experimental

a) *Magnetic Flux Density.* Mapping of the axial magnetic flux density B_z in the air gap and around the poles was effected with the aid of a magnetic probe mounted on a mechanical scanning device: one of the obtained curves – normalized with respect to the highest value of B_z, i.e. $B_{z_{mx}}$ – is reproduced in Fig. 4.23 a.

   Time-dependence of the coil current under disruption is shown in Fig. 4.23 b; it was experimentally established that the step-function pattern is adequate for flux densities up to 0.1 Vs/m², but not for higher levels, due to pronounced arcing.

b) *Recording of Mercury Oscillation.* Mercury oscillations were followed (Fig. 4.24), with the aid of a purposely designed and constructed 8-channel capacitive detector, through the changes induced by them in the capacitance between the capacitive probes a and the surface of the metal b. These changes were translated into high-frequency voltage fluctuations in the MHz range

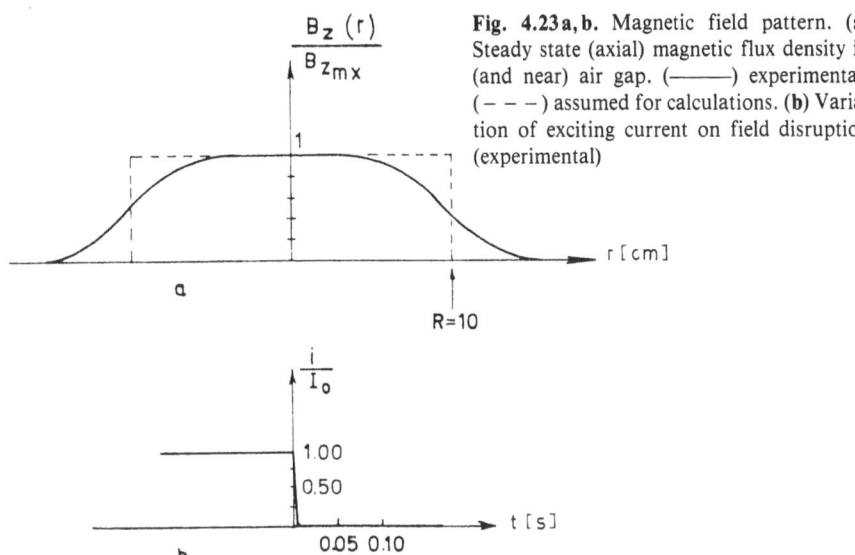

Fig. 4.23a,b. Magnetic field pattern. (a) Steady state (axial) magnetic flux density in (and near) air gap. (———) experimental; (– – –) assumed for calculations. (b) Variation of exciting current on field disruption (experimental)

(a slightly different frequency being used for each channel as a precaution against coupling), with the voltage amplified and converted into a rectified current, which was, in turn, also amplified, recorded and evaluated by means of calibration curves, using a suitable correction formula to allow for the finite diameter of the probes.

The comparison in Fig. 4.22 is based on such results.

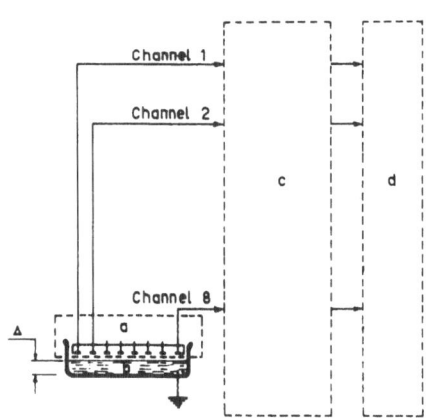

Fig. 4.24. Recording system (*a*: capacitive probes, *b*: mercury, *c*: capacitive detector, *d*: recorder)

5. Dynamic Phenomena

The first two sections are concerned with electromagnetic waves and their interaction with moving sheets. Insight into practical aspects of radiation pressure is thus gained, as well as in the attendant propulsion mechanism; it is expected that, with the advent of high-power laser and wave sources, these topics will gain in practical importance. The last two sections deal with questions of electromagnetic screening; such questions have an impact at high as well as at low frequencies and the transition to relatively low frequencies is emphasized: dynamics is thus smoothly fused into quasistatics.

5.1 Wave Interaction with Moving Layers

With the advent of modern high-power laser sources, the so-called "light pressure" has reached levels capable of inducing propulsion of matter, an achievement which opens the way for precision micromanipulation of particles [5.1 – 5], including separation, optical levitation and acceleration. The velocity that small particles may attain by this means under the continuous action of an electromagnetic field is quite large and motion, in turn, affects their response to the field; a case in point − other than laboratory micromanipulation − is continuous exposure of cosmic debris to radiation pressure from high-luminosity stars.

To gain some insight into the underlying phenomena, we replace such a particle by a plane sheet of infinite dimensions, thereby excluding edge effects. Although the resulting model is highly idealized, it brings out salient features of the interaction mechanism. Alternatively, the sheet may be considered as representing a finely dispersed mixture of gliding particles interacting with a wave field.

Two distinct cases are considered in the sequel:
− a dielectric sheet;
− a conducting sheet.

5.1.1 Dielectric Sheet

A plane layer of infinite horizontal dimensions and finite thickness $\varDelta$ moves at a constant speed v along x in a right-handed Cartesian frame of reference

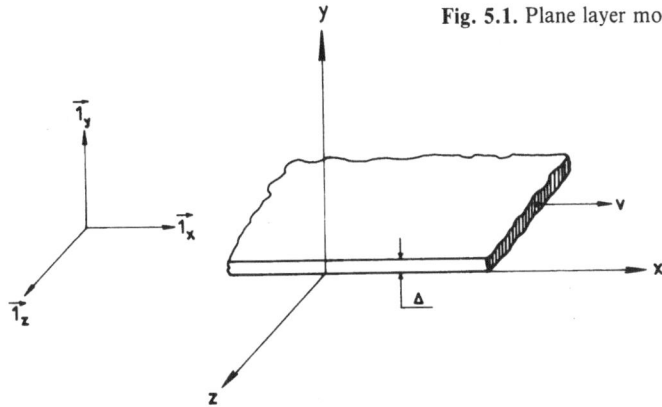

Fig. 5.1. Plane layer moving with speed v

x, y, z, t. The layer (Fig. 5.1) exhibits in its rest frame a vanishingly small electric conductivity, a finite relative dielectric coefficient ε_r and relative magnetic permeability unity; this layer is continuously irradiated by a plane electromagnetic wave described by a vector potential V. The vector potential, observed in the frame of reference x, y, z, t, comprises predominantly a z-component V_z which oscillates at the angular frequency ω; the homogeneous plane wave in question impinges on the sheet at an angle of incidence $0 < \alpha < \pi/2$. What are the reflection factor ϱ and the transmission factor τ exhibited by the moving sheet with respect to the wave?

A) Solution

Denoting by E_0 the real amplitude of the exciting electromagnetic field, we have

$$V_z = \frac{E_0}{i\omega} \exp\left[-i\omega\left(t - \frac{x\cos\alpha + y\sin\alpha}{c}\right)\right] . \tag{5.1.1}$$

Hence, the incident field components (in the absence of the gliding sheet) read:

$$E_z = -\frac{\partial V_z}{\partial t} = E_0 \exp\left[-i\omega\left(t - \frac{x\cos\alpha + y\sin\alpha}{c}\right)\right] , \tag{5.1.2}$$

$$\mu_0 H_x = \frac{\partial V_z}{\partial y} = \frac{E_0}{c}\sin\alpha\exp\left[-i\omega\left(t - \frac{x\cos\alpha + y\sin\alpha}{c}\right)\right] , \tag{5.1.3}$$

$$\mu_0 H_y = -\frac{\partial V_z}{\partial x} = -\frac{E_0}{c}\cos\alpha\exp\left[-i\omega\left(t - \frac{x\cos\alpha + y\sin\alpha}{c}\right)\right] . \tag{5.1.4}$$

In the presence of the sheet, the field pattern is altered; this fact is accounted for by introducing a reflected wave for the $y < 0$ region, and a transmitted wave for $y > \Delta$ according to the statements

$$V_z = \frac{E_0}{i\omega} \exp\left[-i\omega\left(t - \frac{x\cos\alpha}{c}\right)\right]$$

$$\times \left[\exp\left(i\omega\frac{y\sin\alpha}{c}\right) + \varrho\exp\left(-i\omega\frac{y\sin\alpha}{c}\right)\right], \quad y<0, \quad (5.1.5)$$

and

$$V_z = \tau\frac{E_0}{i\omega} \exp\left[-i\omega\left(t - \frac{x\cos\alpha}{c}\right)\right]\left[\exp\left(i\omega\frac{y\sin\alpha}{c}\right)\right], \quad y>\Delta,$$
$$(5.1.6)$$

respectively. The reflection factor ϱ and the transmission factor τ are obviously still unknown.

Next, a primed frame of reference x', y', z', t', attached to the sheet, is introduced, which is linked with its unprimed counterpart through the Lorentz transformation (2.1.23), whose inverse reads

$$x = \frac{x'+vt'}{\sqrt{1-\beta^2}}, \quad y=y', \quad z=z', \quad t = \frac{t'+(v/c^2)x'}{\sqrt{1-\beta^2}}, \quad \beta^2 = \frac{(-v)^2}{c^2}.$$
$$(5.1.7)$$

Now, phase invariance prescribes a primed angular frequency ω' and a primed angle of incidence α', which are related to their unprimed counterparts by

$$\omega\left(t - \frac{x\cos\alpha + y\sin\alpha}{c}\right) = \omega\left(\frac{t'+(v/c^2)x'}{\sqrt{1-\beta^2}}\right.$$

$$\left. - \frac{(x'+vt')\cos\alpha + y'\sqrt{1-\beta^2}\sin\alpha}{c\sqrt{1-\beta^2}}\right)$$

$$= \omega'\left(t' - \frac{x'\cos\alpha' + y'\sin\alpha'}{c}\right) \qquad (5.1.8)$$

whence

$$\omega' = \omega\frac{1-\beta\cos\alpha}{\sqrt{1-\beta^2}} \quad \text{and} \qquad\qquad\qquad (5.1.9)$$

$$\sin\alpha' = \frac{\omega}{\omega'}\sin\alpha = \frac{\sqrt{1-\beta^2}\sin\alpha}{1-\beta\cos\alpha},$$
$$(5.1.10)$$

$$\cos\alpha' = \frac{\omega}{\omega'}\frac{\cos\alpha-\beta}{\sqrt{1-\beta^2}} = \frac{\cos\alpha-\beta}{1-\beta\cos\alpha}.$$

Expressions (5.1.9, 10) yield, respectively, the Doppler effect and the wave aberration phenomenon.

Now – in analogy with (2.1.16) – the four-vector Γ is defined, comprising the Cartesian magnetic vector potential components V_x, V_y, V_z and the electric scalar potential ϕ, i.e.,

$$\Gamma = \left(V_x, V_y, V_z, \frac{i}{c} \phi \right) . \tag{5.1.11}$$

Here $V_x = V_y = 0$ and similarly $\phi = 0$; the finite component V_z, however, does not change under a Lorentz transformation; its primed counterpart V_z' therefore readily follows from (2.1.23) or even more formally from (2.1.25). Thus,

$$V_z' = \frac{E_0}{i\omega} \exp\left[-i\omega'\left(t' - \frac{x'\cos\alpha'}{c} \right) \right]$$

$$\times \left[\exp\left(i\omega' \frac{y'\sin\alpha'}{c} \right) + \varrho\exp\left(-i\omega' \frac{y'\sin\alpha'}{c} \right) \right], \quad y' < 0 , \tag{5.1.12}$$

hence the primed field components:

$$E_z' = E_0 \frac{1 - \beta\cos\alpha}{\sqrt{1 - \beta^2}} \exp\left[-i\omega'\left(t' - \frac{x'\cos\alpha'}{c} \right) \right]$$

$$\times \left[\exp\left(i\omega' \frac{y'\sin\alpha'}{c} \right) + \varrho\exp\left(-i\omega' \frac{y'\sin\alpha'}{c} \right) \right], \quad y' < 0 . \tag{5.1.13}$$

$$\mu_0 H_x' = \frac{E_0}{c} \sin\alpha \exp\left[-i\omega'\left(t' - \frac{x'\cos\alpha'}{c} \right) \right]$$

$$\times \left[\exp\left(i\omega' \frac{y'\sin\alpha'}{c} \right) - \varrho\exp\left(-i\omega' \frac{y'\sin\alpha'}{c} \right) \right], \quad y' < 0 . \tag{5.1.14}$$

$$\mu_0 H_y' = -\frac{E_0}{c} \frac{\cos\alpha - \beta}{\sqrt{1 - \beta^2}} \exp\left[-i\omega'\left(t' - \frac{x'\cos\alpha'}{c} \right) \right]$$

$$\times \left[\exp\left(i\omega' \frac{y'\sin\alpha'}{c} \right) + \varrho\exp\left(-i\omega' \frac{y'\sin\alpha'}{c} \right) \right], \quad y' < 0 . \tag{5.1.15}$$

We now turn to the region $y' > \Delta$; here we find

$$V_z' = \tau \frac{E_0}{i\omega} \exp\left[-i\omega'\left(t' - \frac{x'\cos\alpha'}{c} \right) \right] \exp\left(i\omega' \frac{y'\sin\alpha'}{c} \right), \quad y' > \Delta , \tag{5.1.16}$$

so that

$$E'_z = \tau E_0 \frac{1-\beta\cos\alpha}{\sqrt{1-\beta^2}} \exp\left[-i\omega'\left(t'-\frac{x'\cos\alpha'}{c}\right)\right]$$

$$\times \exp\left(i\omega'\frac{y'\sin\alpha'}{c}\right), \quad y' > \varDelta, \tag{5.1.17}$$

$$\mu_0 H'_x = \tau \frac{E_0}{c}\sin\alpha \exp\left[-i\omega'\left(t'-\frac{x'\cos\alpha'}{c}\right)\right]$$

$$\times \exp\left(i\omega'y'\frac{\sin\alpha'}{c}\right), \quad y' > \varDelta, \tag{5.1.18}$$

$$\mu_0 H'_y = -\tau \frac{E_0}{c}\frac{\cos\alpha-\beta}{\sqrt{1-\beta^2}} \exp\left[-i\omega'\left(t'-\frac{x'\cos\alpha'}{c}\right)\right]$$

$$\times \exp\left(i\omega'y'\frac{\sin\alpha'}{c}\right), \quad y' > \varDelta. \tag{5.1.19}$$

Because the sheet is at rest in the primed frame of reference, the component V'_z of the vector potential inside the sheet satisfies the wave equation

$$\frac{\partial^2 V'_z}{\partial x'^2} + \frac{\partial^2 V'_z}{\partial y'^2} = \frac{\varepsilon_r}{c^2}\frac{\partial^2 V'_z}{\partial t'^2}. \tag{5.1.20}$$

Introducing the index of refraction $n = \sqrt{\varepsilon_r}$, we formulate the relevant, particular solution of (5.1.20) with the aid of two as yet unspecified complex, dimensionless constants K'_1 and K'_2, namely,

$$V'_z = \frac{E_0}{i\omega} \exp\left[-i\omega'\left(t'-\frac{x'\cos\alpha'}{c}\right)\right]$$

$$\times \left[K'_1 \exp\left(i\frac{\omega'}{c}y'\sqrt{n^2-\cos^2\alpha'}\right)\right.$$

$$\left. + K'_2 \exp\left(-i\frac{\omega'}{c}y'\sqrt{n^2-\cos^2\alpha'}\right)\right]. \tag{5.1.21}$$

Hence

$$E'_z = E_0 \frac{1 - \beta \cos \alpha}{\sqrt{1 - \beta^2}} \exp \left[-i\omega' \left(t' - \frac{x' \cos \alpha'}{c} \right) \right]$$

$$\times \left[K'_1 \exp \left(i \frac{\omega'}{c} y' \sqrt{n^2 - \cos^2 \alpha'} \right) \right.$$

$$\left. + K'_2 \exp \left(-i \frac{\omega'}{c} y' \sqrt{n^2 - \cos^2 \alpha'} \right) \right] , \qquad (5.1.22)$$

$$\mu_0 H'_x = \frac{E_0}{c} \frac{\sqrt{n^2(1 - \beta \cos \alpha)^2 - (\cos \alpha - \beta)^2}}{\sqrt{1 - \beta^2}} \exp \left[-i\omega' \left(t' - \frac{x' \cos \alpha'}{c} \right) \right]$$

$$\times \left[K'_1 \exp \left(i \frac{\omega'}{c} y' \sqrt{n^2 - \cos^2 \alpha'} \right) \right.$$

$$\left. - K'_2 \exp \left(-i \frac{\omega'}{c} y' \sqrt{n^2 - \cos^2 \alpha'} \right) \right] , \qquad (5.1.23)$$

$$\mu_0 H'_y = -\frac{E_0}{c} \frac{\cos \alpha - \beta}{\sqrt{1 - \beta^2}} \exp \left[-i\omega' \left(t' - \frac{x' \cos \alpha'}{c} \right) \right]$$

$$\times \left[K'_1 \exp \left(i \frac{\omega'}{c} y' \sqrt{n^2 - \cos^2 \alpha'} \right) \right.$$

$$\left. + K'_2 \exp \left(-i \frac{\omega'}{c} y' \sqrt{n^2 - \cos^2 \alpha'} \right) \right] . \qquad (5.1.24)$$

Continuity of the tangential electric and magnetic field components at $y' = \Delta$ dictates the equations

$$\tau \exp \left(i\omega' \frac{\Delta \sin \alpha'}{c} \right) = K'_1 \exp \left(i\omega' \frac{\Delta}{c} \sqrt{n^2 - \cos^2 \alpha'} \right)$$

$$+ K'_2 \exp \left(-i\omega' \frac{\Delta}{c} \sqrt{n^2 - \cos^2 \alpha'} \right) , \qquad (5.1.25)$$

and

$$\tau \exp \left(i \frac{\omega' \Delta}{c} \sin \alpha' \right) \frac{\sqrt{1 - \beta^2} \sin \alpha}{\sqrt{n^2(1 - \beta \cos \alpha)^2 - (\cos \alpha - \beta)^2}}$$

$$= K'_1 \exp \left(i \frac{\omega' \Delta}{c} \sqrt{n^2 - \cos^2 \alpha'} \right) - K'_2 \exp \left(-i \frac{\omega' \Delta}{c} \sqrt{n^2 - \cos^2 \alpha'} \right) ,$$

$$(5.1.26)$$

which, in turn, yield

$$K_1' = \frac{\tau \exp\left(i\frac{\omega \Delta}{c}\sin\alpha\right)}{2}$$

$$\times \exp\left[-i\frac{\omega \Delta}{c}\sqrt{\frac{n^2(1-\beta\cos\alpha)^2-(\cos\alpha-\beta)^2}{1-\beta^2}}\right]$$

$$\times\left(1+\frac{\sqrt{1-\beta^2}\sin\alpha}{\sqrt{n^2(1-\beta\cos\alpha)^2-(\cos\alpha-\beta)^2}}\right), \tag{5.1.27}$$

along with

$$K_2' = \frac{\tau \exp\left(i\frac{\omega \Delta}{c}\sin\alpha\right)}{2}$$

$$\times \exp\left[i\frac{\omega \Delta}{c}\sqrt{\frac{n^2(1-\beta\cos\alpha)^2-(\cos\alpha-\beta)^2}{1-\beta^2}}\right]$$

$$\times\left(1-\frac{\sqrt{1-\beta^2}\sin\alpha}{\sqrt{n^2(1-\beta\cos\alpha)^2-(\cos\alpha-\beta)^2}}\right). \tag{5.1.28}$$

Field continuity at the interface $y' = 0$ dictates also the equalities:

$$1+\varrho = K_1'+K_2' = \tau\exp\left(i\frac{\omega \Delta}{c}\sin\alpha\right)$$

$$\times\left[\cos\left(\frac{\omega \Delta}{c}\sqrt{\frac{n^2(1-\beta\cos\alpha)^2-(\cos\alpha-\beta)^2}{1-\beta^2}}\right)\right.$$

$$-i\frac{\sqrt{1-\beta^2}\sin\alpha}{\sqrt{n^2(1-\beta\cos\alpha)^2-(\cos\alpha-\beta)^2}}$$

$$\times\left.\sin\left(\frac{\omega \Delta}{c}\sqrt{\frac{n^2(1-\beta\cos\alpha)^2-(\cos\alpha-\beta)^2}{1-\beta^2}}\right)\right], \tag{5.1.29}$$

along with

$$1 - \varrho = (K_1' - K_2') \frac{\sqrt{n^2(1 - \beta\cos\alpha)^2 - (\cos\alpha - \beta)^2}}{\sqrt{1 - \beta^2}\sin\alpha}$$

$$= \tau\exp\left(i\frac{\omega\Delta}{c}\sin\alpha\right)\left[\cos\left(\frac{\omega\Delta}{c}\sqrt{\frac{n^2(1 - \beta\cos\alpha)^2 - (\cos\alpha - \beta)^2}{1 - \beta^2}}\right)\right.$$

$$- i\frac{\sqrt{n^2(1 - \beta\cos\alpha)^2 - (\cos\alpha - \beta)^2}}{\sqrt{1 - \beta^2}\sin\alpha}$$

$$\left. \times \sin\left(\frac{\omega\Delta}{c}\sqrt{\frac{n^2(1 - \beta\cos\alpha)^2 - (\cos\alpha - \beta)^2}{1 - \beta^2}}\right)\right]. \qquad (5.1.30)$$

Introducing the notation

$$M \equiv \cos\left(\frac{\omega\Delta}{c}\sqrt{\frac{n^2(1 - \beta\cos\alpha)^2 - (\cos\alpha - \beta)^2}{1 - \beta^2}}\right)$$

$$- \frac{i}{2}\left(\frac{\sqrt{1 - \beta^2}\sin\alpha}{\sqrt{n^2(1 - \beta\cos\alpha)^2 - (\cos\alpha - \beta)^2}}\right.$$

$$+ \left.\frac{\sqrt{n^2(1 - \beta\cos\alpha)^2 - (\cos\alpha - \beta)^2}}{\sqrt{1 - \beta^2}\sin\alpha}\right)$$

$$\times \sin\left(\frac{\omega\Delta}{c}\sqrt{\frac{n^2(1 - \beta\cos\alpha)^2 - (\cos\alpha - \beta)^2}{1 - \beta^2}}\right), \qquad (5.1.31)$$

we finally obtain, for the reflection factor,

$$\varrho = -\frac{i}{2M}\left(\frac{\sqrt{1 - \beta^2}\sin\alpha}{\sqrt{n^2(1 - \beta\cos\alpha)^2 - (\cos\alpha - \beta)^2}}\right.$$

$$- \left.\frac{\sqrt{n^2(1 - \beta\cos\alpha)^2 - (\cos\alpha - \beta)^2}}{\sqrt{1 - \beta^2}\sin\alpha}\right)$$

$$\times \sin\left(\frac{\omega\Delta}{c}\sqrt{\frac{n^2(1 - \beta\cos\alpha)^2 - (\cos\alpha - \beta)^2}{1 - \beta^2}}\right), \qquad (5.1.32)$$

and, for the transmission factor,

$$\tau = \frac{\exp\left(-i\frac{\omega\Delta}{c}\sin\alpha\right)}{M}. \qquad (5.1.33)$$

The factor ϱ vanishes for

$$\frac{\omega \Delta}{c}\sqrt{\frac{n^2(1-\beta\cos\alpha)^2-(\cos\alpha-\beta)^2}{1-\beta^2}} = m\pi, \quad m=1,2,3 . \quad (5.1.34)$$

B) Graphic Results

Absolute values of reflection and transmission factors are plotted in Figs. 5.2 to 5, for $n = 2.0$. With increase of frequency (lower values of λ) − for a given index of refraction − multiple resonances occur, both for ϱ and τ. Further, the influence of motion upon ϱ and τ is quite clearly seen.

Similar resonance-peaked curves are also obtained for the arguments of ϱ and τ.

Let us point out that the results put forward are also of interest in the relatively new [5.5] technique of "force spectroscopy" − a technique which is intended to provide an additional means of studying the deposition of monolayers.

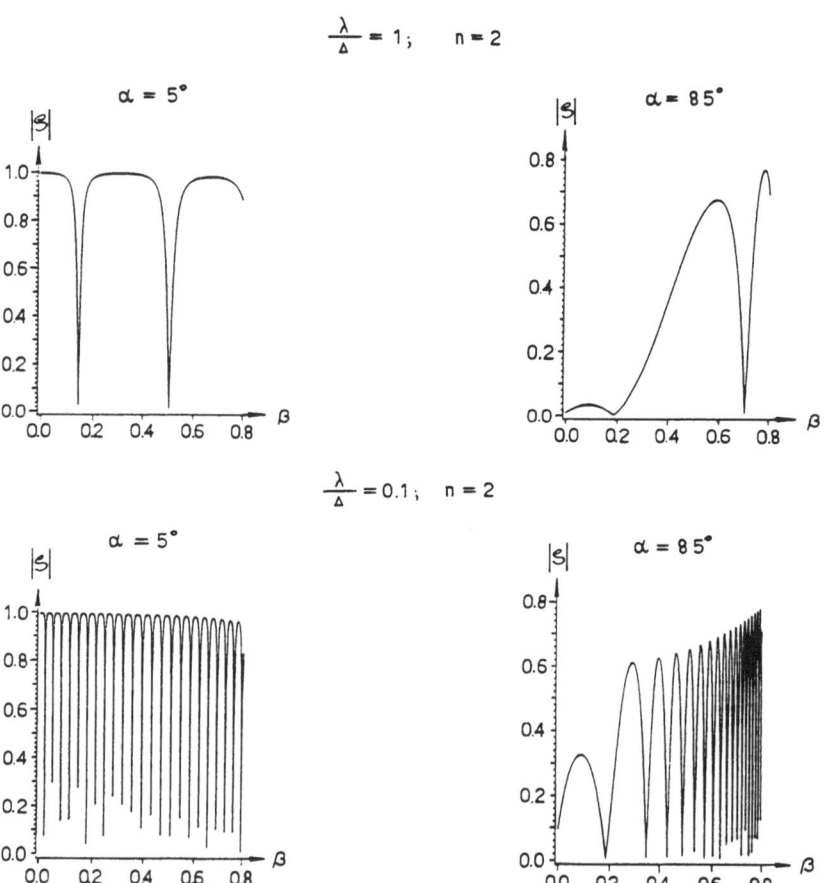

Fig. 5.2. Variation of absolute value of ϱ as function of β, for two values of α and λ/Δ

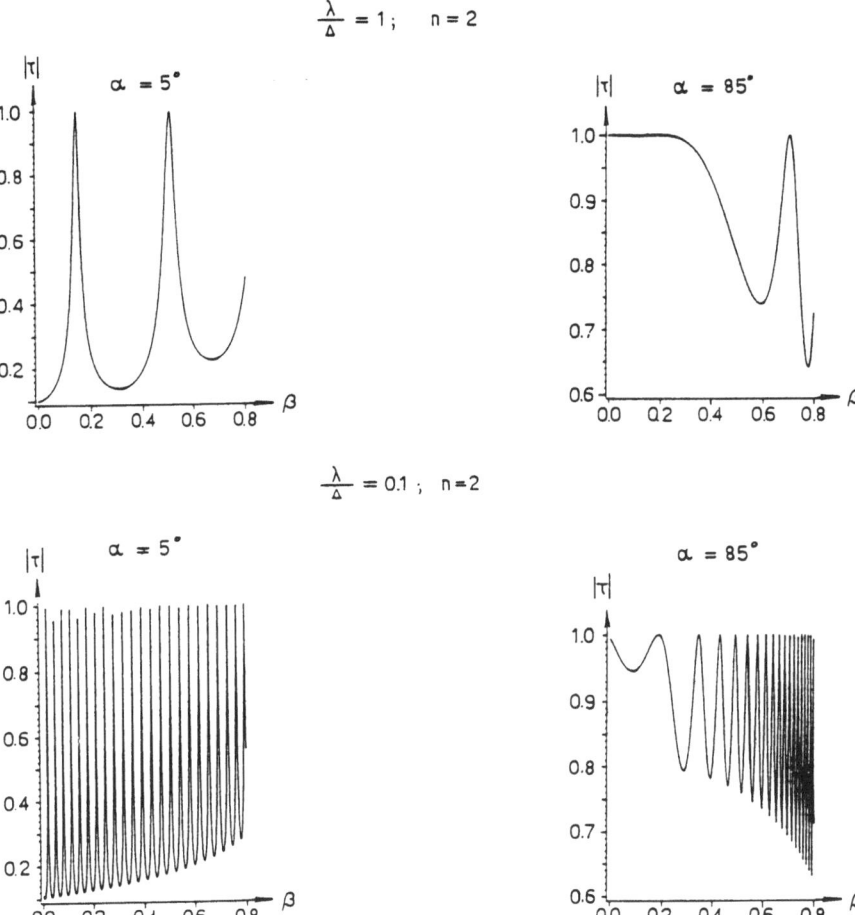

Fig. 5.3. Variation of absolute value of τ as function of β, for two values of α and λ/Δ

5.1.2 Conducting Sheet

The problem is identical with that considered in Sect. 5.1.1, with the dielectric layer now replaced by a gliding metal sheet of specific conductivity σ; its surface conductivity $\varkappa$ is given by the limiting value of the product $\sigma\Delta$, defined as usual for a highly conducting layer of very small thickness; i.e. – for $\sigma \to \infty$ as $\Delta \to 0$.

Hence, (5.1.13 – 15) for the region $y' < \Delta$ and (5.1.16 – 18) apply once more, except that ϱ and τ obviously differ from the previous results and expressions (5.1.16 – 18) apply, in fact, for $y' > 0$, as $\Delta \simeq 0$. Accordingly, continuity of the tangential electric field component of the sheet interface imposes the condition

$$1 + \varrho = \tau \; , \tag{5.1.35}$$

$$\frac{\lambda}{\Delta} = 1; \qquad n = 2$$

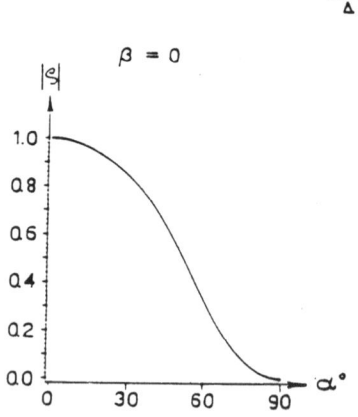

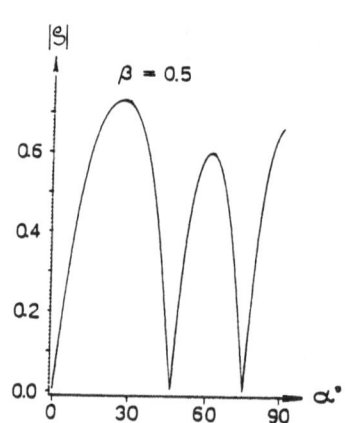

$$\frac{\lambda}{\Delta} = 0.1; \qquad n = 2$$

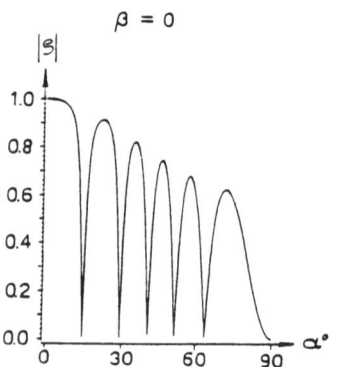

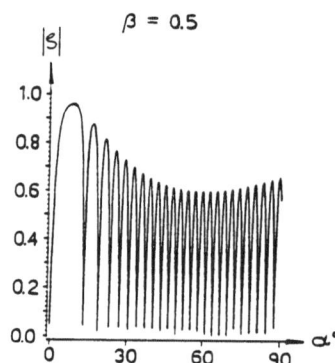

Fig. 5.4. Variation of absolute value of ϱ as function of α, for two values of β and λ/Δ

and discontinuity of the tangential magnetic field component is offset by the primed surface current density A_z', i.e.

$$A_z' = E_0 \sqrt{\frac{\varepsilon_0}{\mu_0}} \sin \alpha [1 - (\varrho + \tau)] \exp\left[-i\omega'\left(t - \frac{x'\cos\alpha'}{c}\right)\right]. \qquad (5.1.36)$$

Ohm's law

$$A_z' = \varkappa E_z' \qquad (5.1.37)$$

provides the required additional relationship, yielding

$$\sqrt{\frac{\varepsilon_0}{\mu_0}} \sin \alpha [1 - (\varrho + \tau)] = \varkappa \tau \frac{1 - \beta \cos \alpha}{\sqrt{1 - \beta^2}}. \qquad (5.1.38)$$

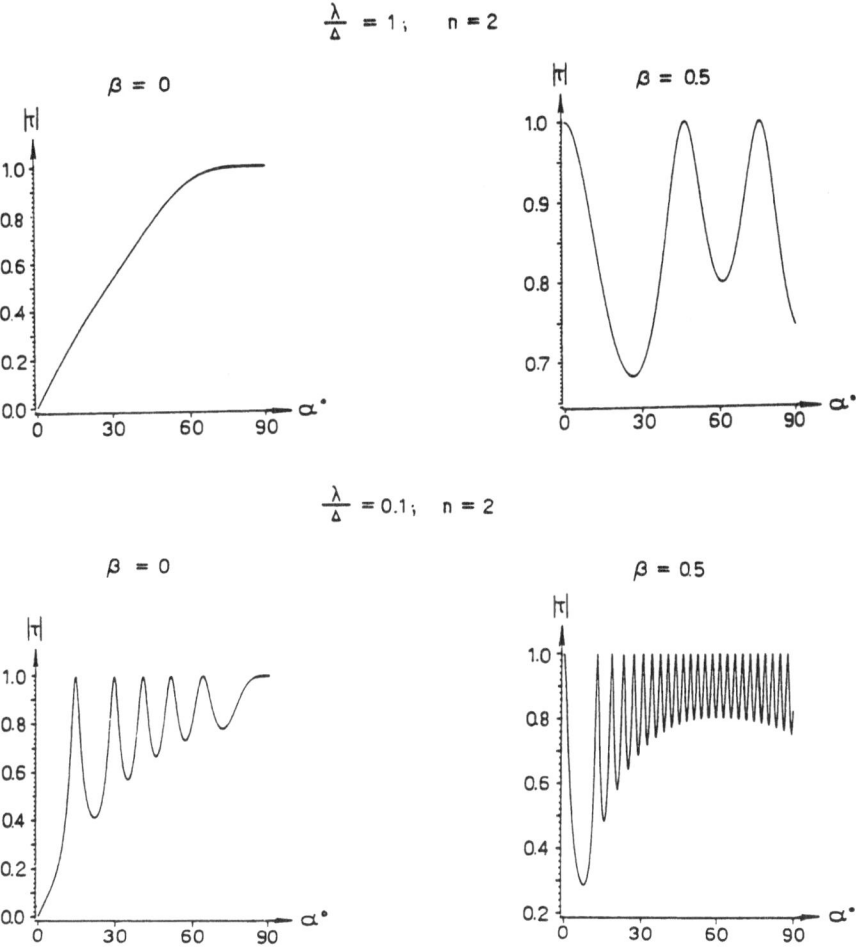

Fig. 5.5. Variation of absolute value of τ as function of α, for two values of β and λ/Δ

Hence,

$$\varrho = -\frac{1}{1+\dfrac{2}{\varkappa}\sqrt{\dfrac{\varepsilon_0}{\mu_0}}\dfrac{\sqrt{1-\beta^2}\sin\alpha}{1-\beta\cos\alpha}} \qquad \text{and} \qquad (5.1.39)$$

$$\tau = \frac{1}{1+\dfrac{\varkappa}{2}\sqrt{\dfrac{\mu_0}{\varepsilon_0}}\dfrac{1-\beta\cos\alpha}{\sqrt{1-\beta^2}\sin\alpha}} . \qquad (5.1.40)$$

These coefficients are reproduced in Fig. 5.6 for a specific value of $\varkappa$. The surface current density therefore reads

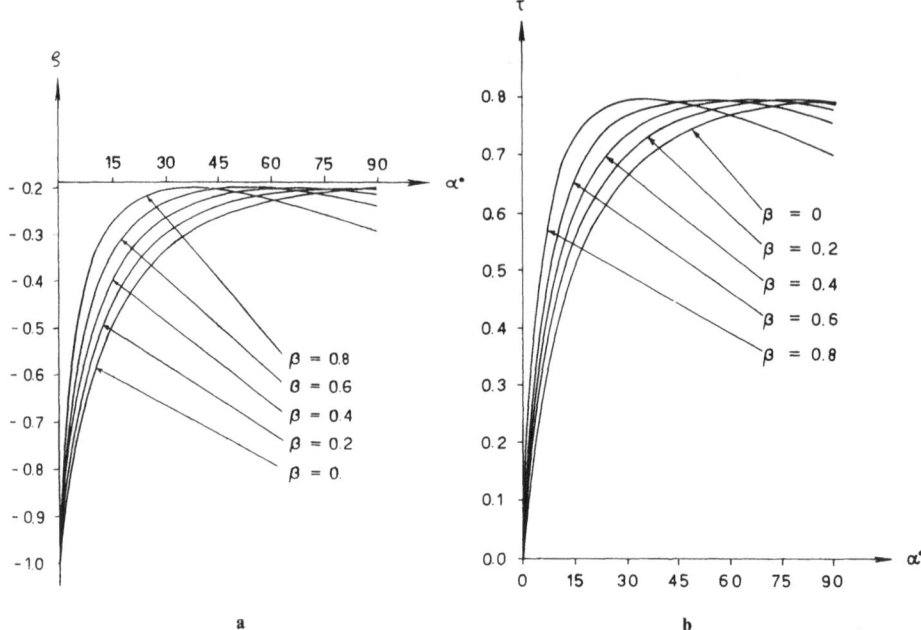

Fig. 5.6a. Reflection factor $\left.\begin{array}{l} \\ \\ \end{array}\right\}$ for $\varkappa = \dfrac{1}{2}\sqrt{\dfrac{\varepsilon_0}{\mu_0}}$
b. Transmission factor

$$
A_z' = E_0 \sqrt{\frac{\varepsilon_0}{\mu_0}}\sin\alpha \frac{2}{1+\dfrac{2}{\varkappa}\sqrt{\dfrac{\varepsilon_0}{\mu_0}}\dfrac{\sqrt{1-\beta^2}\sin\alpha}{1-\beta\cos\alpha}}
$$

$$
\times \exp\left[-\mathrm{i}\,\omega'\left(t'-\frac{x'\cos\alpha'}{c}\right)\right]. \tag{5.1.41}
$$

As A_z' does not change when referred to the unprimed frame of reference [see (2.2.79) with $A_z' = j_z' \Delta'$; $\Delta' = \Delta$], we obtain the surface current A_z

$$
A_z = E_0 \sqrt{\frac{\varepsilon_0}{\mu_0}}\sin\alpha \frac{1}{1+\dfrac{2}{\varkappa}\sqrt{\dfrac{\varepsilon_0}{\mu_0}}\dfrac{\sqrt{1-\beta^2}\sin\alpha}{1-\beta\cos\alpha}}
$$

$$
\times \exp\left[-\mathrm{i}\,\omega\left(t-\frac{x\cos\alpha}{a}\right)\right]. \tag{5.1.42}
$$

We are now able to calculate also the force densities actuating the conducting sheet; these are obviously obtained by means of the Lorentz law of force, and in order to apply it, we express the magnetic induction components in unprimed coordinates, i.e.,

$$\mu_0 H_x \big|_{y=0} = \frac{1}{2} \lim_{y \to 0} [\mu_0 H_x(x, y, t) + \mu_0 H_x(x, -y, t)]$$

$$= \frac{E_0}{c} \sin \alpha \exp \left[-i\omega \left(t - \frac{x \cos \alpha}{c} \right) \right] , \qquad (5.1.43)$$

$$\mu_0 H_y \big|_{y=0} = -\frac{E_0}{c} \cos \alpha \frac{\dfrac{2}{\varkappa} \sqrt{\dfrac{\varepsilon_0}{\mu_0}} \dfrac{\sqrt{1 - \beta^2} \sin \alpha}{1 - \beta \cos \alpha}}{1 + \dfrac{2}{\varkappa} \sqrt{\dfrac{\varepsilon_0}{\mu_0}} \dfrac{\sqrt{1 - \beta^2} \sin \alpha}{1 - \beta \cos \alpha}}$$

$$\times \exp \left[-i\omega \left(t - \frac{x \cos \alpha}{c} \right) \right] . \qquad (5.1.44)$$

Hence, the sheet is propelled by the horizontal time-average thrust (per unit surface)

$$\langle f_x \rangle = \frac{\frac{1}{2} \varepsilon_0 E_0^2 \sin 2\alpha}{\left(\sqrt{\dfrac{2}{\varkappa}} \sqrt{\dfrac{\varepsilon_0}{\mu_0}} \dfrac{\sqrt{1 - \beta^2} \sin \alpha}{1 - \beta \cos \alpha} + \sqrt{\dfrac{\varkappa}{2}} \sqrt{\dfrac{\mu_0}{\varepsilon_0}} \dfrac{1 - \beta \cos \alpha}{\sqrt{1 - \beta^2} \sin \alpha} \right)^2} ,$$

$$(5.1.45)$$

and actuated along the y-axis by the time-averaged surface force density

$$\langle f_y \rangle = \frac{\frac{1}{2} \varepsilon_0 E_0^2 \sin^2 \alpha}{1 + \dfrac{2}{\varkappa} \sqrt{\dfrac{\varepsilon_0}{\mu_0}} \dfrac{\sqrt{1 - \beta^2} \sin \alpha}{1 - \beta \cos \alpha}} . \qquad (5.1.46)$$

The last two expressions are plotted in Fig. 5.7.

5.1.3 Discussion

Insight was provided into the field-matter interaction for gliding sheets exposed to continuous, steady state, monochromatic irradiation. Calculations were performed for two distinct gliding sheets, namely a dielectric one and a

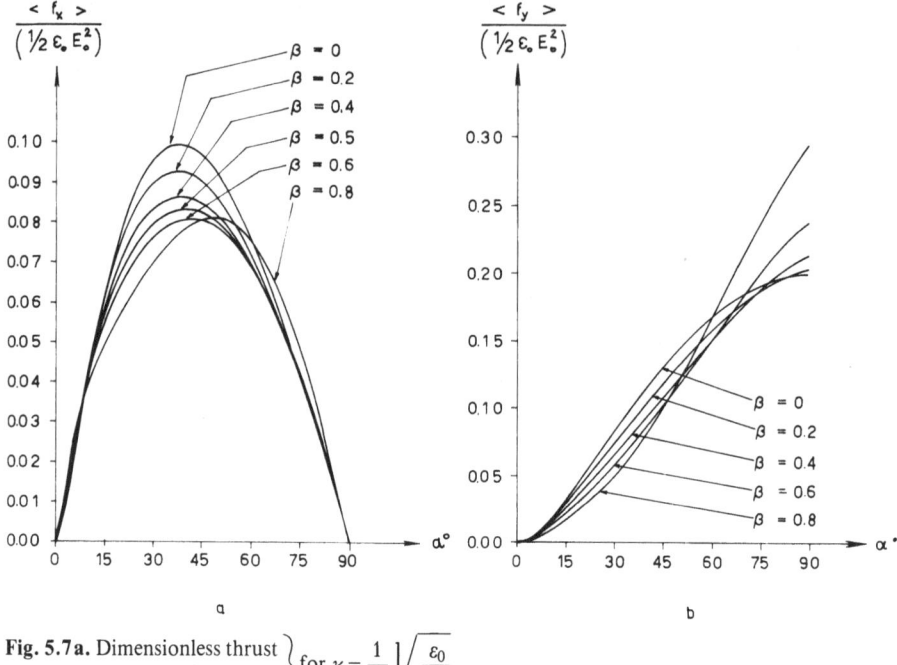

Fig. 5.7a. Dimensionless thrust
b. Expulsion forces $\left.\begin{array}{c} \\ \\ \end{array}\right\}$ for $\varkappa = \dfrac{1}{2}\sqrt{\dfrac{\varepsilon_0}{\mu_0}}$

conducting one. For greater simplicity and better clarity, the proposed analysis does not cover technological aspects, e.g., heating, evaporation [5.6], etc; it is solely intended to point out some facts pertaining to reflection, transmission and radiation pressure in the context of activated gliding matter.

5.2 Wave Propagation Inside Moving Nonmagnetic Media

The interaction mechanism of energy exchange between traveling electromagnetic waves and moving nonmagnetic media has received attention with a view to potential practical applications. To mention but some: (a) in a certain type of parametric amplifier [5.7]; (b) in modulation of electron beams, with the exchange underlying the circuit-to-beam power transfer [5.8,9]; (c) in relativistic Cerenkov generators where electron beams interact with the field of corrugated waveguides [5.10]; (d) in "electrodynamic" motors and generators [5.11], whose feasibility (as opposed to conventional magnetic devices) derives from the interaction mechanism between a dielectric rotor and a rotating field.

The present section is concerned with some of the problems involved in the last-mentioned type of wave-layer interaction, i.e. with analysis of a system which models – to a certain extent – a linear "electric" machine as opposed to its "magnetic" counterpart considered in Sect. 2.2.

5.2.1 Basic Assumptions and Statement of Problem

The configuration to be investigated is shown in Fig. 5.8.

A pair of extremely thin, highly-conducting plane sheets (which confine the electromagnetic field) are fixed at elevations $y = \pm h_1$ in a right-handed Cartesian frame of reference x, y, z, in which the running time t is continuously recorded. An uncharged, nonconducting and nonmagnetic dielectric slab of uniform thickness $2h_2$ glides within this waveconduit along the $(+x)$-axis at a constant velocity v; the region above and below the slab is filled with an idealized medium (hereafter referred to as "air") with the electrical properties of vacuum. The electrically homogeneous and isotropic slab which has a real index of refraction n in its rest frame of reference, is assumed to extend to infinity along the $\pm z$- and $\pm x$ axes, and is taken to be covered at all times by the field-confining sheets.

The excitation of this idealized waveguide is now taken to be positioned very far away along the $(-x)$-axis (at "infinity") and is further assumed to consist of an infinite array $(\pm z)$ of perfectly synchronized generators supplying electromagnetic power at the angular frequency ω. Hence, due to the geometry of the plane sheets, on the one hand, and to extension of the excitation source along the $(\pm z)$-axis to $(\pm \infty)$, on the other hand, the z-dependence may be ignored.

With these assumptions, what is the field within the wave guide?

5.2.2 Solution of Field Problem

We subdivide the interval $-h_1 < y < +h_1$ into two distinct regions of analysis: *"air"* – for $h_2 < y < +h_1$ and $-h_1 < y < -h_2$; *dielectric slab* – for $-h_2 < y < h_2$.

A) Electromagnetic Field in "Air"

Within the "air" region, the solution is obtained by means of the Fitzgerald superpotential F which has already been introduced in Sect. 2.1.2. In the

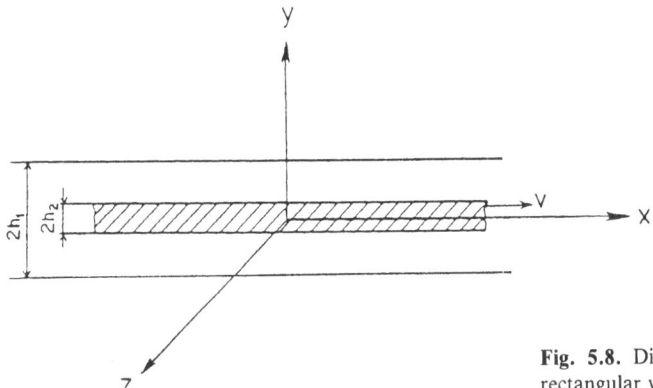

Fig. 5.8. Dielectric slab moving in rectangular waveconduit

present context, its three-dimensional vector differential equation (2.1.14) reduces to a scalar partial one, as the field is adequately described by its x-component F_x alone; hence (with the z-dependence waived as above) –

$$\frac{\partial^2 F_x}{\partial x^2} + \frac{\partial^2 F_x}{\partial y^2} = \frac{1}{c^2} \frac{\partial^2 F_x}{\partial t^2} \ . \tag{5.2.1}$$

Furthermore, using Maxwell's curl equations and (2.1.5,6), the electromagnetic field components are obtained as

$$\varepsilon_0 E_x = 0 \ , \tag{5.2.2}$$

$$\varepsilon_0 E_y = 0 \ , \tag{5.2.3}$$

$$\varepsilon_0 E_z = -\frac{1}{c^2} \frac{\partial^2 F_x}{\partial y \, \partial t} \ , \tag{5.2.4}$$

and, moreover,

$$H_x = \frac{\partial^2 F_x}{\partial y^2} \ , \tag{5.2.5}$$

$$H_y = -\frac{\partial^2 F_x}{\partial y \, \partial x} \ , \tag{5.2.6}$$

$$H_z = 0 \ . \tag{5.2.7}$$

These equations suggest the use of a potential function V, defined through

$$V \equiv \mu_0 \frac{\partial F_x}{\partial y} \ , \tag{5.2.8}$$

so that now we have

$$E_x = 0 \ , \tag{5.2.9}$$

$$E_y = 0 \ , \tag{5.2.10}$$

$$E_z = -\frac{\partial V}{\partial t} \ , \tag{5.2.11}$$

along with

$$\mu_0 H_x = +\frac{\partial V}{\partial y} \ , \tag{5.2.12}$$

$$\mu_0 H_y = -\frac{\partial V}{\partial x} \ , \tag{5.2.13}$$

$$\mu_0 H_z = 0 \ . \tag{5.2.14}$$

It is readily seen that V is nothing else but the z-component of a magnetic vector potential function V, i.e. $V \equiv V_z$ with $V = V1_z$.

The last six equations may now be condensed as

$$E = -\frac{\partial V}{\partial t} , \tag{5.2.15}$$

$$\mu_0 H = \nabla \times V . \tag{5.2.16}$$

These results could have been anticipated; as a rule, the absence of any magnetic free charges within the "air" regions is expressed by means of the well known relation

$$\nabla \cdot \mu_0 H = 0 , \tag{5.2.17}$$

so that the definition (5.2.16) of V renders (5.2.17) into an identity. With (5.2.16) taken into account, integration of the induction law yields an expression of the form

$$E = -\frac{\partial V}{\partial t} - \nabla \phi , \tag{5.2.18}$$

where ϕ is an electric scalar potential still to be determined. The advantage of the above approach becomes now clear: Eq. (5.2.15) implies from the outset that $\phi = 0$ in the present case.

We shall henceforth choose to work with the vector potential component $V \equiv V_z$, its differential equation being, in vacuum,

$$\frac{\partial^2 V_z}{\partial x^2} + \frac{\partial^2 V_z}{\partial y^2} = \frac{1}{c^2} \frac{\partial^2 V_z}{\partial t^2} . \tag{5.2.19}$$

Introducing its complex amplitude $\bar{V}_z$, which varies at the angular supply frequency ω, we have

$$V_z = \bar{V}_z e^{-i\omega t} , \tag{5.2.20}$$

i.e., within the so-called "air"

$$\frac{\partial^2 \bar{V}_z}{\partial x^2} + \frac{\partial^2 \bar{V}_z}{\partial y^2} + \frac{\omega^2}{c^2} \bar{V}_z = 0 . \tag{5.2.21}$$

In the following, the notation $\bar{K}_+$ and $\bar{K}_-$ stands for the (as yet undetermined) complex amplitudes of the two linearly independent solutions of (5.2.21); resorting to a likewise still unknown propagation constant γ, we find for the "air" interval $h_2 < y < h_1$ the solution

$$V_z = e^{-i\omega[t - \gamma(x/c)]}(\bar{K}_+ e^{i\omega\sqrt{1-\gamma^2}\,y/c} + \bar{K}_- e^{-i\omega\sqrt{1-\gamma^2}\,y/c}) , \tag{5.2.23}$$

so that the corresponding electromagnetic field components read

$$E_x = 0 , \tag{5.2.24}$$

$$E_y = 0 , \tag{5.2.25}$$

$$E_z = i\omega e^{-i\omega[t-\gamma(x/c)]}(\bar{K}_+ e^{i\omega\sqrt{1-\gamma^2}\,y/c} + \bar{K}_- e^{-i\omega\sqrt{1-\gamma^2}\,y/c}) \ , \qquad (5.2.26)$$

and

$$\mu_0 H_x = \frac{i\omega}{c}\sqrt{1-\gamma^2}\,e^{-i\omega[t-\gamma(x/c)]}$$

$$\times (\bar{K}_+ e^{i\omega\sqrt{1-\gamma^2}\,y/c} - \bar{K}_- e^{-i\omega\sqrt{1-\gamma^2}\,y/c}) \ , \qquad (5.2.27)$$

$$\mu_0 H_y = -\frac{i\omega}{c}\gamma e^{-i\omega[t-\gamma(x/c)]}$$

$$\times (\bar{K}_+ e^{i\omega\sqrt{1-\gamma^2}\,y/c} + \bar{K}_- e^{-i\omega\sqrt{1-\gamma^2}\,y/c}) \ , \qquad (5.2.28)$$

$$\mu_0 H_z = 0 \ . \qquad (5.2.29)$$

In addition to the x, y, z, t frame of reference, we now introduce a primed frame x', y', z', t' rigidly attached to the gliding dielectric slab; both frames are interconnected through the Lorentz transformation (2.1.23).

We have already mentioned phase invariance of a plane electromagnetic wave in Sect. 5.1.1 A; writing explicitly

$$\omega\frac{\gamma}{c}x + \omega\frac{\sqrt{1-\gamma^2}}{c}y + 0z - \frac{\omega}{c}ct = \text{invariant} \ , \qquad (5.2.30)$$

we may now use the well-known fact that the array

$$\left[\omega\frac{\gamma}{c}; \quad \omega\frac{\sqrt{1-\gamma^2}}{c}; \quad 0; \quad i\frac{\omega}{c} \right] \qquad (5.2.31)$$

represents a four-vector; transformation into its primed counterpart is obtained from (2.1.25), yielding

$$\omega'\gamma' = \omega\frac{\gamma-\beta}{\sqrt{1-\beta^2}}; \quad \omega'\sqrt{1-\gamma'^2} = \omega\sqrt{1-\gamma^2}; \quad \omega' = \omega\frac{1-\beta\gamma}{\sqrt{1-\beta^2}} \ .$$
$$(5.2.32)$$

These relations, see also (5.1.9,10), express here the Doppler and aberration effects, and they also directly yield the primed propagation factor γ' defined in the primed frame of reference, i.e.,

$$\gamma' = \frac{\gamma-\beta}{1-\beta\gamma} \ . \qquad (5.2.33)$$

As V_z does not change under the considered restricted Lorentz transformation [see also (5.1.11) and relevant remark], its primed counterpart V_z' reads

$$V_z' = e^{-i\omega'[t'-\gamma'(x'/c)]}(\bar{K}_+ e^{i\omega'\sqrt{1-\gamma'^2}\,y'/c} + \bar{K}_- e^{-i\omega'\sqrt{1-\gamma'^2}\,y'/c}) \ . \qquad (5.2.34)$$

Due to the relativity principle, the field equations in the primed frame of reference read

$$E' = -\frac{\partial V'}{\partial t'} - \nabla'\phi' , \tag{5.2.35}$$

$$\mu_0 H' = \nabla' \times V' . \tag{5.2.36}$$

Here the primed notation is obvious; if it could be proved that $\phi' = 0$, the primed version of (5.2.9 – 14) would then yield the field components E'_x, E'_y, E'_z, H'_x, H'_y, H'_z. Now the primed component F'_x of the Fitzgerald vector potential does not change under our restricted Lorentz transformation (Sect. 5.2.6). Resorting once more to the principle of relativity, we find again (5.2.2 – 7), but now in the primed frame of reference. Hence using a definition similar to (5.2.8), namely

$$V' \equiv \mu_0 \frac{\partial F'_x}{\partial y'} , \tag{5.2.37}$$

we find $V' \equiv V'_z$, so that recourse to the primed version of (5.2.9 – 14), which dispenses with ϕ', is justified. Hence, for the interval $h_2 < y < h_1$, we have

$$E'_x = 0 , \tag{5.2.38}$$

$$E'_y = 0 , \tag{5.2.39}$$

$$E'_z = \mathrm{i}\omega \frac{1-\beta\gamma}{\sqrt{1-\beta^2}} e^{-\mathrm{i}\omega'[t' - \gamma'(x'/c)]}$$
$$\times (\bar{K}_+ e^{\mathrm{i}\omega'\sqrt{1-\gamma^2}\, y'/c} + \bar{K}_- e^{-\mathrm{i}\omega'\sqrt{1-\gamma^2}\, y'/c}) , \tag{5.2.40}$$

along with

$$\mu_0 H'_x = \frac{\mathrm{i}\omega\sqrt{1-\gamma^2}}{c} e^{-\mathrm{i}\omega'[t' - \gamma'(x'/c)]}$$
$$\times (\bar{K}_+ e^{\mathrm{i}\omega'\sqrt{1-\gamma^2}\, y'/c} - \bar{K}_- e^{-\mathrm{i}\omega'\sqrt{1-\gamma^2}\, y'/c}) , \tag{5.2.41}$$

$$\mu_0 H'_y = -\frac{\mathrm{i}\omega}{c} \frac{\gamma-\beta}{\sqrt{1-\beta^2}} e^{-\mathrm{i}\omega'[t' - \gamma'(x'/c)]}$$
$$\times (\bar{K}_+ e^{\mathrm{i}\omega'\sqrt{1-\gamma^2}\, y'/c} + \bar{K}_- e^{-\mathrm{i}\omega'\sqrt{1-\gamma^2}\, y'/c}) , \tag{5.2.42}$$

$$\mu_0 H'_z = 0 . \tag{5.2.43}$$

Reverting to (5.2.26) and requiring $E_z = 0$ on the $y = h_1$ plane, we obtain

$$\bar{K}_+ e^{\mathrm{i}\omega\sqrt{1-\gamma^2}\, h_1/c} + \bar{K}_- e^{-\mathrm{i}\omega\sqrt{1-\gamma^2}\, h_1/c} = 0 . \tag{5.2.44}$$

This equation is identically satisfied by introducing a new complex amplitude $\bar{K}$, such that

$$\bar{K}_+ = \tfrac{1}{2}\bar{K}e^{-i\omega'\sqrt{1-\gamma^2}\,h_1/c}, \qquad \bar{K}_- = -\tfrac{1}{2}\bar{K}e^{i\omega'\sqrt{1-\gamma^2}\,h_1/c} \;. \tag{5.2.45}$$

Recalling now the invariance of V_z, on the one hand, and the equality $\omega'\sqrt{1-\gamma'^2} = \omega\sqrt{1-\gamma^2}$ on the other hand, we find at the interface $y' = h_2 + 0$ the tangential field components

$$E_z' = \omega\frac{1-\beta\gamma}{\sqrt{1-\beta^2}}\,e^{-i\omega'[t'-\gamma'(x'/c)]}\,\bar{K}\sin\left(\omega\sqrt{1-\gamma^2}\,\frac{h_1-h_2}{c}\right), \tag{5.2.46}$$

$$\mu_0 H_x' = \frac{i\omega}{c}\sqrt{1-\gamma^2}\,e^{-i\omega'[t'-\gamma'(x'/c)]}\,\bar{K}\cos\left(\omega\sqrt{1-\gamma^2}\,\frac{h_1-h_2}{c}\right). \tag{5.2.47}$$

B) Electromagnetic Field Inside the Slab

The partial differential equation for the primed component V_z' inside the slab reads

$$\frac{\partial^2 V_z'}{\partial x'^2} + \frac{\partial^2 V_z'}{\partial y'^2} = \frac{n^2}{c^2}\frac{\partial^2 V_z'}{\partial t'^2}\;. \tag{5.2.48}$$

Assuming a solution of the form

$$V_z' = e^{-i\omega'[t'-\gamma'(x'/c)]}\,f'(y')\;, \tag{5.2.49}$$

where the function $f'(y')$ is still to be determined, we obtain the ordinary differential equation

$$\frac{d^2 f'}{dy'^2} + \frac{\omega'^2}{c^2}(n^2-\gamma'^2)f' = 0\;. \tag{5.2.50}$$

The solution of (5.2.50) may exhibit a symmetric or an antisymmetric behavior; we consider the former only, i.e.,

$$V_z' = \bar{L}'e^{-i\omega'[t'-\gamma'(x'/c)]}\cos\left(\frac{\omega'}{c}y'\sqrt{n^2-\gamma'^2}\right), \tag{5.2.51}$$

where $\bar{L}'$ is again an as yet unspecified complex amplitude. The electromagnetic field components inside the slab accordingly read

$$E_x' = 0\;, \tag{5.2.52}$$

$$E_y' = 0\;, \tag{5.2.53}$$

$$E_z' = \bar{L}'i\omega\frac{1-\beta\gamma}{\sqrt{1-\beta^2}}\,e^{-i\omega'[t'-\gamma'(x'/c)]}\cos\left(\frac{\omega'}{c}y'\sqrt{n^2-\gamma'^2}\right), \tag{5.2.54}$$

$$\mu_0 H_x' = -\bar{L}'\omega\frac{1-\beta\gamma}{\sqrt{1-\beta^2}}\frac{\sqrt{n^2-\gamma'^2}}{c}e^{-i\omega'[t'-\gamma'(x'/c)]}$$

$$\times \sin\left(\frac{\omega'}{c}y'\sqrt{n^2-\gamma'^2}\right),\qquad(5.2.55)$$

$$\mu_0 H_y' = -\bar{L}'\frac{i\omega}{c}\frac{\gamma-\beta}{\sqrt{1-\beta^2}}e^{-i\omega'[t'-\gamma'(x'/c)]}$$

$$\times \cos\left(\frac{\omega'}{c}y'\sqrt{n^2-\gamma'^2}\right),\qquad(5.2.56)$$

$$\mu_0 H_z' = 0 \ .\qquad(5.2.57)$$

Field continuity at the interface $y' = h_2$ now implies, see (5.2.46, 47, 54, 55),

$$\bar{K}\sin\left(\omega\sqrt{1-\gamma^2}\frac{h_1-h_2}{c}\right) = i\bar{L}'\cos\left(\frac{\omega'}{c}h_2\sqrt{n^2-\gamma'^2}\right)\qquad(5.2.58)$$

$$\frac{i\omega}{\mu_0 c}\sqrt{1-\gamma^2}\,\bar{K}\cos\left(\omega\sqrt{1-\gamma^2}\frac{h_1-h_2}{c}\right)$$

$$= -L'\frac{\omega}{\mu_0 c}\frac{1-\beta\gamma}{\sqrt{1-\beta^2}}\sqrt{n^2-\gamma'^2}\cdot\sin\left(\frac{\omega'}{c}h_2\sqrt{n^2-\gamma'^2}\right) \ .\qquad(5.2.59)$$

Eliminating the constant complex amplitudes $\bar{K}$ and $\bar{L}'$, we obtain the equation specifying the value of the propagation factor, namely,

$$\tan\left(\omega\sqrt{1-\gamma^2}\frac{h_1-h_2}{c}\right) = \frac{\sqrt{1-\gamma^2}\sqrt{1-\beta^2}}{(1-\beta\gamma)\sqrt{n^2-\gamma'^2}}$$

$$\times\cot\left(\omega'\sqrt{n^2-\gamma'^2}\frac{h_2}{c}\right) \ .\qquad(5.2.60)$$

With (5.2.32) we have

$$\tan\left(\omega\sqrt{1-\gamma^2}\frac{h_1-h_2}{c}\right) = \sqrt{\frac{(1-\gamma^2)(1-\beta^2)}{n^2(1-\beta\gamma)^2-(\gamma-\beta)^2}}$$

$$\times\cot\left(\omega\frac{h_2}{c}\sqrt{\frac{n^2(1-\beta\gamma)^2-(\gamma-\beta)^2}{1-\beta^2}}\right),$$

$$(5.2.61)$$

and introducing the notation

$$\Omega \equiv \frac{\omega}{c} h_1, \qquad \eta \equiv \frac{h_2}{h_1}, \qquad (5.2.62)$$

we finally obtain

$$\tan\left[\Omega(1-\eta)\sqrt{1-\gamma^2}\right] = \sqrt{\frac{(1-\gamma^2)(1-\beta^2)}{n^2(1-\beta\gamma)^2-(\gamma-\beta)^2}}$$

$$\times \cot\left(\Omega\sqrt{\frac{n^2(1-\beta\gamma)^2-(\gamma-\beta)^2}{1-\beta^2}}\right). \qquad (5.2.63)$$

5.2.3 Some Kinematic Considerations

We now solve a rather specialized problem with $\eta = 1.0$, implying that the considered wave conduit is completely filled with moving dielectric material. In this case (5.2.63) yields

$$\cot\left(\Omega\sqrt{\frac{n^2(1-\beta\gamma)^2-(\gamma-\beta)^2}{1-\beta^2}}\right) = 0, \qquad (5.2.64)$$

so that

$$\Omega\sqrt{\frac{n^2(1-\beta\gamma)^2-(\gamma-\beta)^2}{1-\beta^2}} = (2k+1)\frac{\pi}{2}, \qquad k = 0,1,2,3,\ldots . \quad (5.2.65)$$

The cut-off frequency spectrum $\Omega_{c,k}$ is attained when $\gamma = 0$; hence

$$\Omega_{c,k} = (2k+1)\frac{\pi}{2}\sqrt{\frac{1-\beta^2}{n^2-\beta^2}}. \qquad (5.2.66)$$

In particular, the dimensionless cut-off circular frequency for $k = 0$, i.e., $\Omega_{c,0}$ is presented in Fig. 5.9 for the case $n = 2$; although it is symmetrical in $\pm\beta$, we have chosen to reproduce it for $\beta > 0$ only, a convention to which we adhere in the sequel.

For $k = 0$ the dependence (5.2.65) of Ω on γ may explicitly be rewritten in the form

$$\Omega = \frac{\pi}{2}\frac{\sqrt{1-\beta^2}}{1-\beta\gamma}\frac{1}{\sqrt{n^2-[(\gamma-\beta)/(1-\beta\gamma)]^2}}. \qquad (5.2.67)$$

It is reproduced in Fig. 5.10 for the region $\beta < 1/n$, and in Fig. 5.11 for $\beta \geqslant 1/n$.

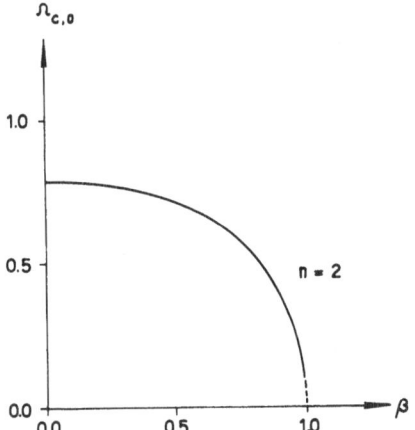

Fig. 5.9. Main cut-off frequency as function of $\beta > 0$; $n = 2$

At the (dimensionless) Cerenkov speed $\beta = 1/n$ only one value ensues for γ at $\Omega = \text{const}$; for $\beta \neq 1/n$, however, two values for γ result, yielding waves traveling along in both senses of propagation [5.12], one being "slow" and the other "fast". At very low speeds, i.e. at $\beta \simeq 0$, no such asymmetry occurs.

With the convention $\beta > 0$ in mind, we now consider the dependence of γ on γ'; to this end we rewrite (5.2.33) viewing γ' as the dependent variable, i.e. we interchange γ' and γ, while simultaneously replacing β by $(-\beta)$, i.e.,

$$\gamma = \frac{\gamma' + \beta}{1 + \beta \gamma'} \; . \tag{5.2.68}$$

For the critical value

$$\gamma'_c = -\frac{1}{\beta} \; , \tag{5.2.69}$$

the propagation factor γ increases beyond all bounds; hence, see (5.2.32), so do the primed angular frequency ω' and obviously the parameter $\Omega' \equiv \omega' h_1 / c$.

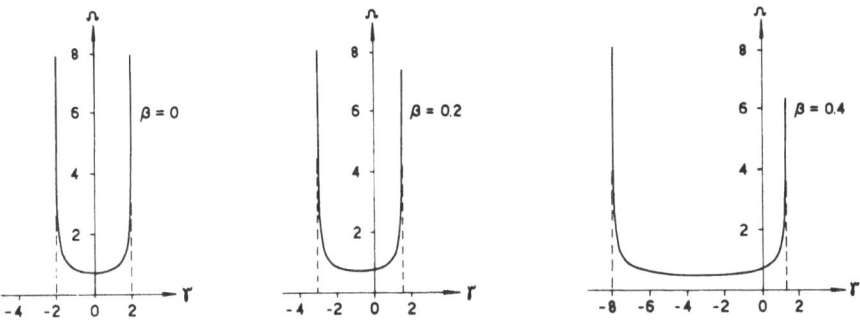

Fig. 5.10. Dimensionless angular frequency Ω as function of propagation factor γ, for $n = 2$; $\beta < 1/n$

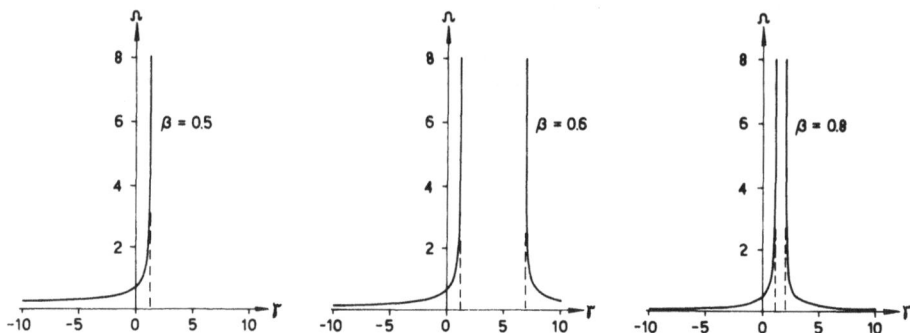

Fig. 5.11. Dimensionless angular frequency Ω as function of propagation factor γ, for $n = 2$; $\beta \geqslant 1/n$

Replacing now in (5.2.67) for $\beta = 0$ the factor γ by its primed counterpart γ', we obtain in the primed frame of reference

$$\Omega' = \frac{\pi}{2} \frac{1}{\sqrt{n^2 - \gamma'^2}} \ , \tag{5.2.70}$$

i.e., $\Omega' \to \infty$ at

$$n^2 = \gamma'^2 \equiv \gamma_c'^2 \ , \tag{5.2.71}$$

so that, in fact,

$$\beta = \frac{1}{n} \ . \tag{5.2.72}$$

The plot of γ vs. γ' is reproduced in Fig. 5.12.

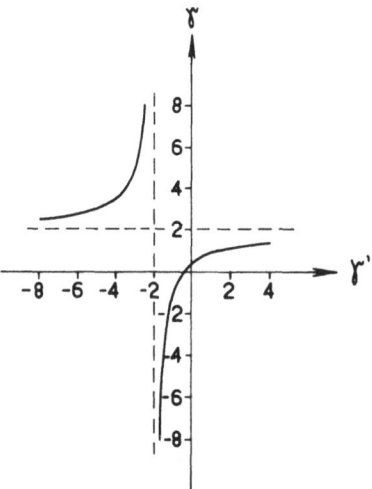

Fig. 5.12. Dependence of γ on γ'; $n = 2$

It exhibits four regions of interest:

- for $\gamma' > 0$ the sense of propagation of the waves in both frames of reference coincides with the gliding direction of the dielectric material;
- for $(-\beta) < \gamma' < 0$ the primed wave propagates along the $(-x')$-axis and the unprimed along the $(+x)$-axis;
- for $(-1/\beta) < \gamma' < (-\beta)$ the primed wave propagates along the $(-x')$-axis and the unprimed along the $(-x)$-axis;
- for $\gamma' < (-1/\beta)$ the primed wave propagates along the $(-x')$-axis and the unprimed along the $(+x)$-axis.

The above is readily extended to lossy dielectric media [5.13], including lossy waveguide walls [5.14]; we shall not, however, pursue these largely academic topics further, as our aim was only an introductory exposition.

5.2.4 Concluding Remarks

Linear as well as rotating "electric" machines have not been developed to the same extent as their "magnetic" counterparts: existence of free electric charge restricts increase of the electric field intensity E to a certain limit beyond which it breaks down, whereas in the absence of free magnetic charge there is no such restriction on H; further, stable ferromagnetic materials are quite abundant, while stable high-permittivity electric materials are scarce.

Nevertheless, such machines may come into being in the future and the present section deals with certain aspects of high frequency linear (gliding) "electric" devices.

5.2.5 Appendix

Invariance of Superpotential Component F_x

By virtue of (2.1.15) which we rewrite here, for convenience,

$$[F] = \begin{vmatrix} 0 & 0 & 0 & \dfrac{i}{c}F_x \\[2ex] 0 & 0 & 0 & \dfrac{i}{c}F_y \\[2ex] 0 & 0 & 0 & \dfrac{i}{c}F_z \\[2ex] -\dfrac{i}{c}F_x & -\dfrac{i}{c}F_y & -\dfrac{i}{c}F_z & 0 \end{vmatrix}, \qquad (5.2.73)$$

and its transform $F'_{jk} = \alpha_{jm}\alpha_{kn}F_{mn}$, see (2.1.26), we obtain

$$[F] = \begin{vmatrix} 0 & \dfrac{(v/c^2)F_y}{\sqrt{1-\beta^2}} & \dfrac{(v/c^2)F_z}{\sqrt{1-\beta^2}} & \dfrac{i}{c}F_x \\[3ex] -\dfrac{(v/c^2)F_y}{\sqrt{1-\beta^2}} & 0 & 0 & \dfrac{i}{c}\dfrac{F_y}{\sqrt{1-\beta^2}} \\[3ex] -\dfrac{(v/c^2)F_z}{\sqrt{1-\beta^2}} & 0 & 0 & \dfrac{i}{c}\dfrac{F_z}{\sqrt{1-\beta^2}} \\[3ex] -\dfrac{i}{c}F_x & -\dfrac{i}{c}\dfrac{F_y}{\sqrt{1-\beta^2}} & -\dfrac{i}{c}\dfrac{F_z}{\sqrt{1-\beta^2}} & 0 \end{vmatrix}.$$

$$(5.2.74)$$

Comparing the two arrays, we see that F_x does not change under a Lorentz transformation.

5.3 Electrodynamics of Shielding: Cylindrical Configuration and Exterior Hertz Wave

In many practical applications, electrical and electronic equipment must be protected against external interference: power lines, welding apparatus, aircraft radar, lightning, corona discharges, etc. quite often cause spurious responses in sensitive instrumentation. To avoid these undesirable occurrences, shields are employed (e.g., [5.15 – 19]); in particular, modern applications take advantage of the sealing properties of conducting materials impregnated with elastic substances. Such shields are produced in the form of very thin tubing, strips, sheeting, tapes, foils or adhesives and these permit total enclosure of the device in question so that they are increasing in popularity.

Nowadays, numerous variations of shield coatings are available, with the required electrical conduction properties being ensured by the choice of technology, namely, sintered products, interlocked metal fibers, oriented fibers conducting in a specified direction, and so on. In the present section, a mathematical model (see also [5.20 – 23]) for such an electro-magnetic shield is presented.

5.3.1 Formulation of Problem

The shield considered here is assumed to consist of a very long ("infinite") uniform metallic, nonmagnetic cylinder (Fig. 5.13) of mean radius R and thickness $\Delta \ll R$. The regions interior and exterior to the conducting tube are characterized by the electromagnetic properties of vacuum, whereas the

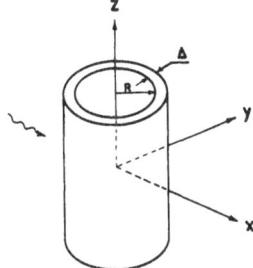

Fig. 5.13. Cylindrical shield

cylinder proper (of specific electric conductivity σ) is characterized by a conductivity $\varkappa$ given, in accordance with our regular approach, by

$$\varkappa \equiv \lim_{\substack{\sigma \to \infty \\ \varDelta \to 0}} (\sigma \varDelta) = \text{finite} \ . \tag{5.3.1}$$

A z-polarized harmonic and homogeneous plane electromagnetic wave (angular frequency ω) irradiates the cylinder. What is the overall field pattern?

5.3.2 Approach to Solution

The solution will be obtained by using the Hertz vector superpotential π (Sect. 5.3.8) formulated in a circular cylindric system of coordinates (r, α, z). Side by side with this reference system, right-handed Cartesian coordinates (x, y, z) are also introduced, the z-axes and the origins of the two frames being coincident. Furthermore, the vector π is defined so as to comprise only a z-component $\pi_z \equiv \pi$.

Resorting now to the usual notation, we obtain (Sect. 5.3.8) the following field components in Cartesian coordinates:

$$E_x = \frac{\partial^2 \pi}{\partial z \, \partial x} \ , \tag{5.3.2}$$

$$E_y = -\frac{\partial^2 \pi}{\partial z \, \partial y} \ , \tag{5.3.3}$$

$$E_z = \frac{1}{c^2} \frac{\partial^2 \pi}{\partial t^2} - \frac{\partial^2 \pi}{\partial z^2} \ , \quad \text{and} \tag{5.3.4}$$

$$H_x = -\varepsilon_0 \frac{\partial^2 \pi}{\partial t \, \partial y} \ , \tag{5.3.5}$$

$$H_y = \varepsilon_0 \frac{\partial^2 \pi}{\partial t \, \partial x} \tag{5.3.6}$$

$$H_z = 0 \ , \tag{5.3.7}$$

and similarly,

$$E_r = -\frac{\partial^2 \pi}{\partial z \, \partial r} \ , \tag{5.3.8}$$

$$E_\alpha = -\frac{1}{r}\frac{\partial^2 \pi}{\partial z \, \partial \alpha} \ , \tag{5.3.9}$$

$$E_z = \frac{1}{c^2}\frac{\partial^2 \pi}{\partial t^2} - \frac{\partial^2 \pi}{\partial z^2} \ , \tag{5.3.10}$$

along with

$$H_r = -\varepsilon_0 \frac{1}{r}\frac{\partial^2 \pi}{\partial \alpha \, \partial t} \ , \tag{5.3.11}$$

$$H_\alpha = \varepsilon_0 \frac{\partial^2 \pi}{\partial r \, \partial t} \ , \tag{5.3.12}$$

$$H_z = 0 \ , \tag{5.3.13}$$

in circular cylindrical coordinates.

In the sequel, we shall distinguish between the primary (exciting) field (superscript p) and the secondary (reaction) field (superscript s).

5.3.3 Primary Field

The steady-state primary electromagnetic wave exhibits a z-directed electric field component only, whose complex amplitude is denoted by $\bar{E}^{(p)}$; hence, see (5.3.4), the z-directed primary Hertz component $\pi^{(p)}$ reads

$$\pi^{(p)} = -\frac{c^2}{\omega^2}\bar{E}^{(p)}e^{-i\omega t}e^{i(\omega/c)x} \ , \tag{5.3.14}$$

and, accordingly, the magnetic field comprises a y-component $H_y^{(p)}$ only,

$$H_y^{(p)} = -\sqrt{\frac{\varepsilon_0}{\mu_0}}\,\bar{E}^{(p)}e^{-i\omega t}e^{i(\omega/c)x} \ . \tag{5.3.15}$$

To express $\pi^{(p)}$ in the cylindrical frame of reference, we replace at first x by $r\cos\alpha$, so that

$$\pi^{(p)} = -\frac{c^2}{\omega^2}\bar{E}^{(p)}e^{-i\omega t}e^{i(\omega/c)r\cos\alpha} \ , \tag{5.3.16}$$

and then resort to the complex Fourier series representation with coefficients c_n:

$$e^{i(\omega/c)r\cos\alpha} = \sum_{n=-\infty}^{\infty} c_n e^{in\alpha} \ . \tag{5.3.17}$$

Thus, with the dummy variable of integration α', we find

$$c_n = \frac{1}{2\pi} \int_{-\pi}^{\pi} e^{i(\omega/c)r\cos\alpha'} e^{in\alpha'} d\alpha' .$$ (5.3.18)

Now, Sommerfeld's integral representation of the Bessel function of first kind and order n, with real argument ξ, reads

$$J_n(\xi) = \frac{1}{2\pi} \int_{-\pi}^{\pi} e^{i\xi\cos\alpha'} e^{in(\alpha'-\pi/2)} d\alpha' ,$$ (5.3.19)

so that comparison of the last two expressions yields

$$c_n = e^{in(\pi/2)} J_n\left(\frac{\omega}{c} r\right) .$$ (5.3.20)

Hence, finally

$$\pi^{(p)} = -\frac{c^2}{\omega^2} \bar{E}^{(p)} \sum_{n=-\infty}^{\infty} e^{in(\pi/2)} J_n\left(\frac{\omega}{c} r\right) e^{-i(\omega t - n\alpha)}$$ (5.3.21)

5.3.4 Secondary Field

With the specified excitation, the current density in the cylindrical sheet is taken to comprise an axial component only, which by definition, is linked to the so-called "secondary" electromagnetic field. In view of the limiting process (5.3.1), the actual volume current density is replaced by the surface current density A_z; resorting to the still unknown Fourier expansion coefficients a_n, we have

$$A_z = \sum_{n=-\infty}^{\infty} a_n e^{-i(\omega t - n\alpha)} .$$ (5.3.22)

The Hertz superpotential $\pi^{(s)}$ associated with A_z must be chosen coherent in space and synchronous in time with (5.3.22); it satisfies in vacuum the homogeneous wave equation

$$\nabla^2 \pi^{(s)} - \frac{1}{c^2} \frac{\partial^2 \pi^{(s)}}{\partial t^2} = 0 , \quad r \geq R ,$$ (5.3.23)

whose solution must here, on the one hand, converge at $r = 0$, and on the other satisfy Sommerfeld's radiation condition at $r \to \infty$.

In order to formulate this solution, two − as yet unknown − series of spectral coefficients, C_n and D_n are introduced, along with the first Hankel function $H_n^{(1)}$ of order n.

Hence, by virtue of these considerations, the solution of (5.3.23) reads

$$\pi^{(s)} = \sum_{n=-\infty}^{\infty} C_n J_n \left(\frac{\omega}{c} r \right) e^{-i(\omega t - n\alpha)}, \quad r < R, \quad \text{and} \tag{5.3.24}$$

$$\pi^{(s)} = \sum_{n=-\infty}^{\infty} D_n H_n^{(1)} \left(\frac{\omega}{c} r \right) e^{-i(\omega t - n\alpha)}, \quad r > R, \tag{5.3.25}$$

so that, see (5.3.12),

$$H_\alpha^{(s)} = -\varepsilon_0 i \omega^2 \frac{1}{c} \sum_{n=-\infty}^{\infty} C_n J_n' \left(\frac{\omega}{c} r \right) e^{-i(\omega t - n\alpha)}, \quad r < R, \tag{5.3.26}$$

along with

$$H_\alpha^{(s)} = -\varepsilon_0 i \omega^2 \frac{1}{c} \sum_{n=-\infty}^{\infty} D_n H_n^{(1)'} \left(\frac{\omega}{c} r \right) e^{-i(\omega t - n\alpha)}, \quad r > R. \tag{5.3.27}$$

This field discontinuity is compensated by A_z, i.e.,

$$H_\alpha^{(s)}(R+0) - H_\alpha(R-0) = A_z, \tag{5.3.28}$$

and therefore

$$-i\omega^2 \frac{\varepsilon_0}{c} \left[D_n H_n^{(p)'} \left(\frac{\omega}{c} R \right) - C_n J_n' \left(\frac{\omega}{c} R \right) \right] = a_n. \tag{5.3.29}$$

On the other hand, continuity of the tangential electric field at $r = R$ leads to the expression

$$D_n H_n^{(1)} \left(\frac{\omega}{c} R \right) = C_n J_n \left(\frac{\omega}{c} R \right). \tag{5.3.30}$$

Combining the last two equations and resorting to the identity

$$J_n(\xi) H_n^{(1)'}(\xi) - J_n'(\xi) H_n^{(1)}(\xi) = i \frac{2}{\pi \xi}, \tag{5.3.31}$$

we obtain the spectral coefficients C_n and D_n as

$$C_n = \frac{\pi}{2} \left(\frac{\omega}{c} R \right) H_n^{(1)} \left(\frac{\omega}{c} R \right) \frac{c}{\varepsilon_0} \frac{a_n}{\omega^2}, \tag{5.3.32}$$

$$D_n = \frac{\pi}{2} \left(\frac{\omega}{c} R \right) J_n \left(\frac{\omega}{c} R \right) \frac{c}{\varepsilon_0} \frac{a_n}{\omega^2} , \tag{5.3.33}$$

whence, in turn

$$\pi^{(s)} = \frac{c}{\varepsilon_0 \omega^2} \sum_{n=-\infty}^{\infty} a_n \frac{\pi}{2} \left(\frac{\omega}{c} R \right) H_n^{(1)} \left(\frac{\omega}{c} R \right) J_n \left(\frac{\omega}{c} r \right) e^{-i(\omega t - n\alpha)} ,$$

$$r < R , \tag{5.3.34}$$

and

$$\pi^{(s)} = \frac{c}{\varepsilon_0 \omega^2} \sum_{n=-\infty}^{\infty} a_n \frac{\pi}{2} \left(\frac{\omega}{c} R \right) J_n \left(\frac{\omega}{c} R \right) H_n^{(1)} \left(\frac{\omega}{c} r \right) e^{-i(\omega t - n\alpha)} ,$$

$$r > R . \tag{5.3.35}$$

Our task is now reduced to the determination of a_n.

5.3.5 Shielding of Electric Field

The primary axial electric field component at the sheet circumference $r = R$ is given by

$$E_{zR}^{(p)} = \frac{1}{c^2} \frac{\partial^2 \pi^{(p)}}{\partial t^2} \bigg|_{r=R} = \bar{E}^{(p)} \sum_{n=-\infty}^{\infty} e^{in(\pi/2)} J_n \left(\frac{\omega}{c} R \right) e^{-i(\omega t - n\alpha)} , \tag{5.3.36}$$

while its secondary counterpart is expressed by

$$E_{zR}^{(s)} = \frac{1}{c^2} \frac{\partial^2 \pi^{(s)}}{\partial t^2} \bigg|_{r=R}$$

$$= - \sqrt{\frac{\mu_0}{\varepsilon_0}} \sum_{n=-\infty}^{\infty} a_n \frac{\pi}{2} \left(\frac{\omega}{c} R \right) H_n^{(1)} \left(\frac{\omega}{c} R \right) J_n \left(\frac{\omega}{c} R \right) e^{-i(\omega t - n\alpha)} . \tag{5.3.37}$$

Ohm's law

$$A_z = \varkappa (E_{zR}^{(p)} + E_{zR}^{(s)}) \tag{5.3.38}$$

yields the algebraic equation

$$a_n = \varkappa \left[\bar{E}^{(p)} e^{in(\pi/2)} J_n \left(\frac{\omega}{c} R \right) \right.$$

$$\left. - \sqrt{\frac{\mu_0}{\varepsilon_0}} a_n \frac{\pi}{2} \left(\frac{\omega}{c} R \right) H_n^{(1)} \left(\frac{\omega}{c} R \right) J_n \left(\frac{\omega}{c} R \right) \right] , \tag{5.3.39}$$

from which we obtain

$$a_n = \bar{E}^{(p)} \frac{\varkappa e^{in(\pi/2)} J_n\left(\frac{\omega}{c} R\right)}{1 + \varkappa \sqrt{\frac{\mu_0}{\varepsilon_0}} \frac{\pi}{2}\left(\frac{\omega}{c} R\right) H_n^{(1)}\left(\frac{\omega}{c} R\right) J_n\left(\frac{\omega}{c} R\right)} . \tag{5.3.40}$$

The primary axial electric field component along the z-axis, denoted by $E_{z0}^{(p)}$, follows directly from the definition of $\pi^{(p)}$ and reads

$$E_{z0}^{(p)} = \bar{E}^{(p)} e^{-i\omega t} . \tag{5.3.41}$$

The secondary component $E_{z0}^{(s)}$ is obtained from $\pi^{(s)}$, i.e.,

$$\begin{aligned} E_{z0}^{(s)} &= \frac{1}{c^2} \frac{\partial^2 \pi^{(s)}}{\partial t^2}\bigg|_{r=0} \\ &= -\sqrt{\frac{\mu_0}{\varepsilon_0}} \sum_{n=-\infty}^{\infty} a_n \frac{\pi}{2}\left(\frac{\omega}{c} R\right) H_n^{(1)}\left(\frac{\omega}{c} R\right) J_n(0) e^{-i(\omega t - n\alpha)} \\ &= -\sqrt{\frac{\mu_0}{\varepsilon_0}} a_0 \frac{\pi}{2}\left(\frac{\omega}{c} R\right) H_0^{(1)}\left(\frac{\omega}{c} R\right) e^{-i\omega t} . \end{aligned} \tag{5.3.42}$$

Now, from (5.3.40) we have

$$a_0 = \bar{E}^{(p)} \frac{\varkappa J_0\left(\frac{\omega}{c} R\right)}{1 + \varkappa \sqrt{\frac{\mu_0}{\varepsilon_0}} \frac{\pi}{2}\left(\frac{\omega}{c} R\right) H_0^{(1)}\left(\frac{\omega}{c} R\right) J_0\left(\frac{\omega}{c} R\right)} , \tag{5.3.43}$$

so that

$$E_{z0}^{(s)} = -\bar{E}^{(p)} \frac{\varkappa \sqrt{\frac{\mu_0}{\varepsilon_0}} \frac{\pi}{2}\left(\frac{\omega}{c} R\right) H_0^{(1)}\left(\frac{\omega}{c} R\right) J_0\left(\frac{\omega}{c} R\right)}{1 + \varkappa \sqrt{\frac{\mu_0}{\varepsilon_0}} \frac{\pi}{2}\left(\frac{\omega}{c} R\right) H_0^{(1)}\left(\frac{\omega}{c} R\right) J_0\left(\frac{\omega}{c} R\right)} e^{-i\omega t} . \tag{5.3.44}$$

The combined electric field at the z-axis is therefore obtained as

$$E_{z0} = E_{z0}^{(p)} + E_{z0}^{(s)} \equiv \bar{E}_{z0} e^{-i\omega t} \tag{5.3.45}$$

with

$$E_{z0} = \bar{E}^{(p)} \frac{1}{1 + \varkappa \sqrt{\dfrac{\mu_0}{\varepsilon_0}} \dfrac{\pi}{2} \left(\dfrac{\omega}{c} R\right) H_0^{(1)}\left(\dfrac{\omega}{c} R\right) J_0\left(\dfrac{\omega}{c} R\right)} e^{-i\omega t} . \tag{5.3.46}$$

The electric shield factor now directly follows — with the aid of the dimensionless angular frequency $\Omega \equiv (\omega/c)R$ — as the ratio

$$\left| \frac{\bar{E}^{(p)}}{\bar{E}_{z0}} \right| = \left| 1 + \varkappa \sqrt{\frac{\mu_0}{\varepsilon_0}} \frac{\pi}{2} \Omega H_0^{(1)}(\Omega) J_0(\Omega) \right| , \tag{5.3.47}$$

and the corresponding shielding effectiveness S_E is being defined as

$$S_E = 20 \log \left| \frac{\bar{E}^{(p)}}{\bar{E}_{z0}} \right| . \tag{5.3.48}$$

Let us now momentarily turn our attention to the frequency range of interest. Introducing the skin depth

$$\delta = \sqrt{\frac{2}{\omega \mu_0 \sigma}} \tag{5.3.49}$$

we find that, as opposed to (5.3.1), the fictitious limiting process

$$\lim_{\sigma \to \infty} \sigma \delta = \lim_{\sigma \to \infty} \sqrt{\frac{2\sigma}{\omega \mu_0}} \to \infty \tag{5.3.50}$$

leads to an expression increasing beyond all bounds. This implies that here

$$\Delta \ll \delta . \tag{5.3.51}$$

From a mechanical point of view this means that our considerations do not, in fact, apply to stiff shields; they are, however, meaningful for the above-mentioned elastic coatings.

The relation (5.3.51) thus yields an upper frequency limit

$$\omega \ll \frac{2}{\Delta^2 \mu_0 \sigma} . \tag{5.3.52}$$

Recently, impregnated zinc coatings $80 - 130$ μm thick have successfully been used in reducing radio-frequency interference; now from (5.3.52) we obtain, with $\Delta \simeq 100$ μm, that $f = (\omega/2\pi) \ll 1.6$ MHz; practical values pointing towards shielding effectiveness of such coatings have, however, been quoted [5.24] up to $1\,000$ MHz: it is obvious that results are highly dependent — among other factors — on geometry.

We choose here, quite arbitrarily, to focus our attention on the band up to one tenth of the limiting frequency of 1.6 MHz, i.e., up to 160 kHz; some typical applications of this band (VLF, lower LF) are radionavigation systems, low transmission rate long-range communication and standard time and frequency broadcasts. With $R \simeq 1.0$ m the dimensionless angular frequency of interest attains values up to approximately $\Omega = 3.3 \times 10^{-3}$. For such low values of the argument, limiting values of Bessel functions – in conjunction with the identity

$$\varkappa \sqrt{\frac{\mu_0}{\varepsilon_0}} \, \Omega \equiv \sigma \Delta \omega \mu_0 R \; , \qquad\qquad (5.3.53)$$

may profitably be used; the last expression in particular enhances comprehension as to how dynamics "collapses" into quasistatics when the wave interacts with a cylinder for which $\Omega \to 0$.

Some relevant results are reproduced for purposes of illustration in Fig. 5.14.

Although extension to high frequencies is here of purely academic interest, it is nevertheless tentatively being attempted on the strength of [5.24], see above, and in spite of (5.3.52); Fig. 5.15 provides some typical results indicating that at resonance frequencies, very thin shields might be completely ineffective along their axes. With $R \simeq 1.0$ m the lowest resonance frequency is in our case around 115 MHz (e.g., air-ground communications).

5.3.6 Shielding of Magnetic Field

Similar considerations apply to the magnetic field intensity; with (5.3.11, 12, 21), the primary radial and circular field components are obtained as

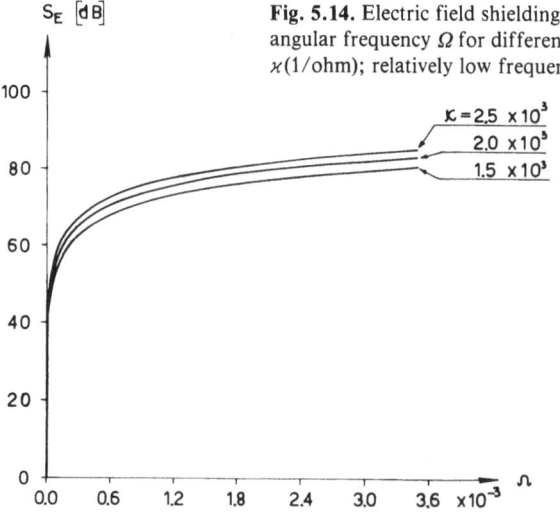

Fig. 5.14. Electric field shielding effectiveness S_E vs. dimensionless angular frequency Ω for different values of conductivity parameter $\varkappa(1/\text{ohm})$; relatively low frequency range

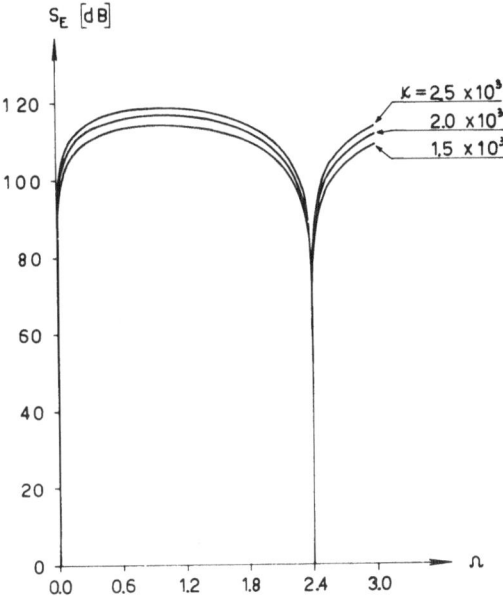

$$H_r^{(p)} = \bar{E}^{(p)} \sqrt{\frac{\varepsilon_0}{\mu_0}} \sum_{n=-\infty}^{\infty} \frac{n J_n\left(\frac{\omega}{c} r\right)}{\frac{\omega}{c} r} e^{-in(\pi/2)} e^{-i(\omega t - n\alpha)} \quad \text{and} \quad (5.3.54)$$

$$H_\alpha^{(p)} = i\bar{E}^{(p)} \sqrt{\frac{\varepsilon_0}{\mu_0}} \sum_{n=-\infty}^{\infty} J_n'\left(\frac{\omega}{c} r\right) e^{in(\pi/2)} e^{-i(\omega t - n\alpha)} . \quad (5.3.55)$$

At the axis $r = 0$, only the components pertaining to $n = \pm 1$ are non-zero, i.e.

$$H_{r0}^{(p)} = -\sqrt{\frac{\varepsilon_0}{\mu_0}} \bar{E}^{(p)} \sin \alpha\, e^{-i\omega t}, \quad r = 0 , \quad (5.3.56)$$

$$H_{\alpha 0}^{(p)} = -\sqrt{\frac{\varepsilon_0}{\mu_0}} \bar{E}^{(p)} \cos \alpha\, e^{-i\omega t}, \quad r = 0 . \quad (5.3.57)$$

These combine, along the z-axis, into the y-component $H_{y0}^{(p)}$, which reads

$$H_{y0}^{(p)} = -\sqrt{\frac{\varepsilon_0}{\mu_0}} \bar{E}^{(p)} e^{-i\omega t}$$

$$\equiv \bar{H}_{y0}^{(p)} e^{-i\omega t}, \quad r = 0 . \quad (5.3.58)$$

In the same manner, the secondary magnetic field components result from

$$H_r^{(s)} = - \sum_{n=-\infty}^{\infty} a_n \frac{\pi}{2} \left(\frac{\omega}{c} R \right) H_n^{(1)} \left(\frac{\omega}{c} R \right) \frac{n J_n \left(\frac{\omega}{c} r \right)}{\frac{\omega}{c} r} e^{-i(\omega t - n\alpha)},$$
$$(5.3.59)$$

$$H_\alpha^{(s)} = -i \sum_{n=-\infty}^{\infty} a_n \frac{\pi}{2} \left(\frac{\omega}{c} R \right) H_n^{(1)} \left(\frac{\omega}{c} R \right) J_n' \left(\frac{\omega}{c} r \right) e^{-i(\omega t - n\alpha)},$$
$$(5.3.60)$$

yielding along the z-axis the values

$$H_{r0}^{(s)} = -i a_1 \frac{\pi}{2} \left(\frac{\omega}{c} R \right) H_1^{(1)} \left(\frac{\omega}{c} R \right) \sin \alpha \, e^{-i\omega t}, \quad r = 0, \qquad (5.3.61)$$

$$H_{\alpha 0}^{(s)} = -i a_1 \frac{\pi}{2} \left(\frac{\omega}{c} R \right) H_1^{(1)} \left(\frac{\omega}{c} R \right) \cos \alpha \, e^{-i\omega t}, \quad r = 0, \qquad (5.3.62)$$

which, in turn, combine into the secondary y-component of the magnetic field, i.e.,

$$H_{y0}^{(s)} = -i a_1 \frac{\pi}{2} \left(\frac{\omega}{c} R \right) H_1^{(1)} \left(\frac{\omega}{c} R \right) e^{-i\omega t} \equiv \bar{H}_{y0}^{(s)} e^{-i\omega t}, \quad r = 0. \qquad (5.3.63)$$

Substituting a_1 from (5.3.40), we finally obtain

$$\bar{H}_{y0}^{(s)} = \sqrt{\frac{\varepsilon_0}{\mu_0}} \bar{E}^{(p)} \frac{\varkappa \sqrt{\frac{\mu_0}{\varepsilon_0}} \frac{\pi}{2} \left(\frac{\omega}{c} R \right) H_1^{(1)} \left(\frac{\omega}{c} R \right) J_1 \left(\frac{\omega}{c} R \right)}{1 + \varkappa \sqrt{\frac{\mu_0}{\varepsilon_0}} \frac{\pi}{2} \left(\frac{\omega}{c} R \right) H_1^{(1)} \left(\frac{\omega}{c} R \right) J_1 \left(\frac{\omega}{c} R \right)}.$$
$$(5.3.64)$$

Hence, the resultant magnetic field component reads

$$H_{y0} = H_{y0}^{(p)} + H_{y0}^{(s)} \equiv \bar{H}_{y0} e^{-i\omega t}, \quad r = 0, \qquad (5.3.65)$$

with

$$\bar{H}_{y0} = -\sqrt{\frac{\varepsilon_0}{\mu_0}} \bar{E}^{(p)} \frac{1}{1 + \varkappa \sqrt{\frac{\mu_0}{\varepsilon_0}} \frac{\pi}{2} \left(\frac{\omega}{c} R \right) H_1^{(1)} \left(\frac{\omega}{c} R \right) J_1 \left(\frac{\omega}{c} R \right)},$$

$$r = 0, \qquad (5.3.66)$$

so that, see (5.3.58),

$$\left| \frac{\bar{H}_{y0}^{(p)}}{\bar{H}_{y0}} \right| = \left| 1 + \varkappa \sqrt{\frac{\mu_0}{\varepsilon_0}} \frac{\pi}{2} \Omega H_1^{(1)}(\Omega) J_1(\Omega) \right| . \qquad (5.3.67)$$

For very low values of Ω we resort once more to the limiting forms of the Bessel functions, obtaining here, e.g.

$$\left| \frac{\bar{H}_{y0}^{(p)}}{\bar{H}_{y0}} \right| = \left| 1 - i \frac{1}{2} \varkappa \sqrt{\frac{\mu_0}{\varepsilon_0}} \Omega \right| . \qquad (5.3.68)$$

Shielding effectiveness defined as

$$S_H = 20 \log \left| \frac{\bar{H}_{y0}^{(p)}}{\bar{H}_{y0}} \right| , \qquad (5.3.69)$$

is reproduced in Figs. 5.16 and 17; earlier observations with regard to S_E apply here as well.

5.3.7 Concluding Remarks

Shielding effectiveness of a nonmagnetic cylindrical screen has been estimated; although a highly specialized model has been assumed, the results represent a general pattern of behavior; influence of the ratio of plane wave ($\sqrt{\mu_0/\varepsilon_0}$) to shield impedance ($1/\varkappa$) is clearly seen, as well as that of the dimensionless angular frequency Ω.

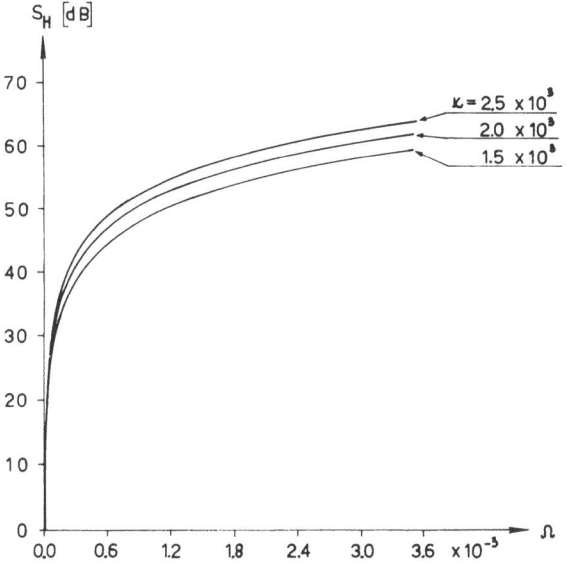

Fig. 5.16. Magnetic field shielding effectiveness S_H vs. dimensionless angular frequency Ω for different values of conductivity parameter $\varkappa$ (1/ohm); relatively low frequency range

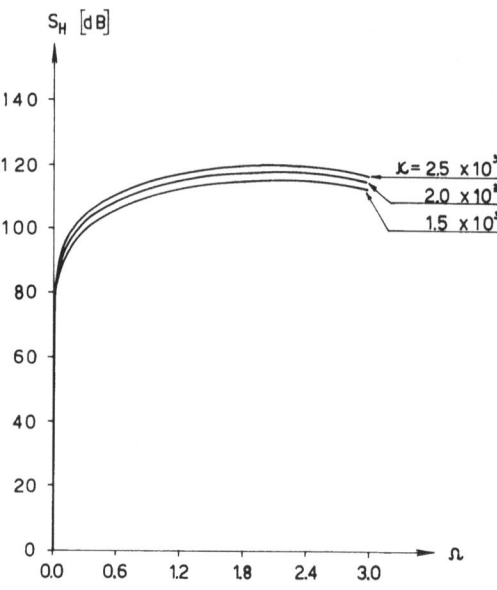

Fig. 5.17. Magnetic field shielding effectiveness S_H vs. dimensionless angular frequency Ω for different values of conductivity parameter $\varkappa$ (1/ohm); relatively high frequency range

In practice, shielding problems are often solved by assuming a vanishing tangential total electric field component across, say, the cylinder surface; in the present case this assumption was waived; circular symmetry was not implied and the fields were expanded as a function of the angle of azimuth.

We emphasize once more that the relatively high frequency region of the graphs is in the present case primarily of academic interest.

5.3.8 Appendix

The Hertz Superpotential

Starting with Maxwell's equations

$$\nabla \times H = \frac{\partial}{\partial t} \varepsilon_0 E \ , \tag{5.3.70}$$

$$\nabla \times E = -\frac{\partial}{\partial t} \mu_0 H \ , \tag{5.3.71}$$

$$\nabla \cdot \mu_0 H = 0 \ , \tag{5.3.72}$$

$$\nabla \cdot \varepsilon_0 E = 0 \ , \tag{5.3.73}$$

in vacuo, we introduce, by virtue of (5.3.72), the magnetic vector potential V such that

$$\mu_0 H = \nabla \times V \ . \tag{5.3.74}$$

Reverting to (5.3.71) we obtain

$$E = -\frac{\partial V}{\partial t} - \nabla\phi \, , \tag{5.3.75}$$

where ϕ denotes as usual the electric scalar potential. Hence (5.3.70) is now rewritten as

$$\nabla\times(\nabla\times V) = -\mu_0\varepsilon_0 \frac{\partial}{\partial t}\left(\frac{\partial V}{\partial t} + \nabla\phi\right) . \tag{5.3.76}$$

Expansion in Cartesian coordinates leads to

$$\nabla(\nabla\cdot V) - \nabla^2 V = -\frac{1}{c^2}\left(\frac{\partial^2 V}{\partial t^2} + \nabla\frac{\partial\phi}{\partial t}\right) . \tag{5.3.77}$$

With the Lorentz condition

$$\nabla\cdot V = -\frac{1}{c^2}\frac{\partial\phi}{\partial t} \tag{5.3.78}$$

we obtain on the one hand the three-dimensional homogeneous vector wave equation

$$\nabla^2 V - \frac{1}{c^2}\frac{\partial^2 V}{\partial t^2} = 0 \, , \tag{5.3.79}$$

and, on the other hand, by applying (5.3.73, 75 and 78), we obtain its scalar counterpart

$$\nabla^2\phi - \frac{1}{c^2}\frac{\partial^2\phi}{\partial t^2} = 0 \, . \tag{5.3.80}$$

The Hertz vector π is now introduced through the defining relations

$$V = -\frac{1}{c^2}\frac{\partial\pi}{\partial t} \, , \tag{5.3.81}$$

$$\phi = \nabla\cdot\pi \, , \tag{5.3.82}$$

which convert (5.3.78) into an identity. Its differential equation follows directly from (5.3.79 or 80), i.e.,

$$\nabla^2\pi - \frac{1}{c^2}\frac{\partial^2\pi}{\partial t^2} = 0 \, . \tag{5.3.83}$$

Choosing $\boldsymbol{\pi}$ so as to comprise a component $\pi_z \equiv \pi$ only, we obtain a z-component $V_z \equiv V$ for V, and thus

$$V = -\frac{1}{c^2}\frac{\partial \pi}{\partial t} \; , \tag{5.3.84}$$

$$\phi = \frac{\partial \pi}{\partial z} \; . \tag{5.3.85}$$

In view of (5.3.75)

$$E_x = -\frac{\partial^2 \pi}{\partial z \, \partial x} \; , \tag{5.3.86}$$

$$E_y = -\frac{\partial^2 \pi}{\partial z \, \partial y} \; , \tag{5.3.87}$$

$$E_z = +\frac{1}{c^2}\frac{\partial^2 \pi}{\partial t^2} - \frac{\partial^2 \pi}{\partial z^2} \; , \tag{5.3.88}$$

while (5.3.74) yields

$$\mu_0 H_x = -\frac{1}{c^2}\frac{\partial^2 \pi}{\partial t \, \partial y} \; , \tag{5.3.89}$$

$$\mu_0 H_y = \frac{1}{c^2}\frac{\partial^2 \pi}{\partial t \, \partial x} \; , \tag{5.3.90}$$

$$\mu_0 H_z = 0 \; . \tag{5.3.91}$$

In circular cylindrical coordinates

$$E_r = -\frac{\partial}{\partial r}\left(\frac{\partial \pi}{\partial z}\right) = -\frac{\partial^2 \pi}{\partial r \, \partial z} \; , \tag{5.3.92}$$

$$E_\alpha = -\frac{1}{r}\frac{\partial}{\partial \alpha}\left(\frac{\partial \pi}{\partial z}\right) = -\frac{1}{r}\frac{\partial^2 \pi}{\partial \alpha \, \partial z} \; , \tag{5.3.93}$$

$$E_z = \frac{1}{c^2}\frac{\partial^2 \pi}{\partial t^2} - \frac{\partial^2 \pi}{\partial z^2} \; , \tag{5.3.94}$$

and likewise

$$\mu_0 H_r = \frac{1}{r}\frac{\partial}{\partial \alpha}\left(-\frac{1}{c^2}\frac{\partial \pi}{\partial t}\right) = -\frac{1}{c^2}\frac{1}{r}\frac{\partial^2 \pi}{\partial \alpha \, \partial t} \; , \tag{5.3.95}$$

$$\mu_0 H_\alpha = -\frac{\partial}{\partial r}\left(-\frac{1}{c^2}\frac{\partial \pi}{\partial t}\right) = \frac{1}{c^2}\frac{\partial^2 \pi}{\partial r \, \partial t} \; , \tag{5.3.96}$$

$$\mu_0 H_z = 0 \; . \tag{5.3.97}$$

5.4 Electrodynamics of Shielding:
Spherical Configuration and Interior Fitzgerald Wave

In the preceding subsection we examined the interaction between an externally excited wave and a cylindrical configuration. However, in many electronic circuits, such as those comprising discriminator circuits, rf amplifiers, oscillators, etc., the shielding of coils and transformers is resorted to directly at the *location* of the magnetic device. Such a shield reduces the possibility of local random pickup and interference, prevents undesirable coupling and precludes mutual inductance effects; at the same time, through its own *reaction field,* the shield affects the field at the coil or transformer.

The present section is concerned with the screening effect of an electromagnetic shield and the attendant field reaction at the enclosed coil.

5.4.1 Definition of Problem

A high-frequency coil is surrounded by a non-magnetic metallic enclosure of electrical conductivity σ; the coil is excited by an imposed ac voltage of angular frequency ω. We wish to determine to what extent the magnetic field, directed along the coil axis, is reduced in these circumstances near the coil. For this purpose, we consider the following model:

- The field of the tightly wound coil is replaced by a (quasi)-homogeneous field at the origin.
- The medium inside the screen is characterized by the electromagnetic properties of vacuum.
- The enclosure is replaced (Fig. 5.18) by a metallic spherical shell of radius R and thickness Δ, for which we assume once more that $\lim_{\substack{\sigma \to \infty \\ \Delta \to 0}} (\sigma \Delta) = \varkappa$, being finite.

The proposed model yields closed-form solutions for the electromagnetic field equations and permits analytical treatment of various aspects of the problem.

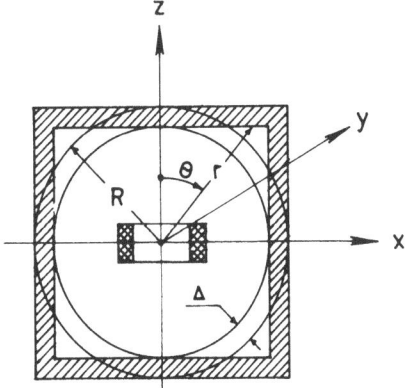

Fig. 5.18. Screened coil

5.4.2 Approach to Solution

We now introduce (Fig. 5.18), two kinds of coordinates:
- Spherical polar coordinates of radius r, polar angle θ and azimuth α;
- Right-handed Cartesian coordinates x, y, z.

The electromagnetic field is derived from a Fitzgerald vector superpotential F (Sect. 5.4.7) having a z-component only, i.e. $F = F\mathbf{1}_z$. On account of the harmonic excitation, we introduce the phasor $\bar{F}$ by means of the relation $F = \bar{F}\exp(-\mathrm{i}\,\omega t)$. The electric field components E_r, E_θ, E_α and the magnetic field components H_r, H_θ, H_α, expressed through the complex amplitudes $\bar{E}_r$, $\bar{E}_\theta$, $\bar{E}_\alpha$; $\bar{H}_r$, $\bar{H}_\theta$, $\bar{H}_\alpha$, read (Sect. 5.4.7):

$$\bar{E}_r = -\mathrm{i}\,\omega\mu_0\frac{1}{r}\frac{\partial \bar{F}}{\partial \alpha}\ , \tag{5.4.1}$$

$$\bar{E}_\theta = -\mathrm{i}\,\omega\mu_0\frac{1}{r}\frac{\partial \bar{F}}{\partial \alpha}\cot\theta\ , \tag{5.4.2}$$

$$\bar{E}_\alpha = \mathrm{i}\,\omega\mu_0\left(\sin\theta\frac{\partial \bar{F}}{\partial r}+\cos\theta\frac{\partial \bar{F}}{\partial \theta}\right)\ , \quad \text{and} \tag{5.4.3}$$

$$\bar{H}_r = \left[\frac{2}{r}\frac{\partial \bar{F}}{\partial r}+\frac{1}{r^2\sin\theta}\frac{\partial}{\partial \theta}\left(\sin\theta\frac{\partial \bar{F}}{\partial \theta}\right)+\frac{1}{r^2\sin^2\theta}\frac{\partial^2 \bar{F}}{\partial \alpha^2}\right]\cos\theta$$

$$+\sin\theta\frac{\partial}{\partial r}\left(\frac{1}{r}\frac{\partial \bar{F}}{\partial \theta}\right)\ , \tag{5.4.4}$$

$$\bar{H}_\theta = -\left(\frac{\partial^2 \bar{F}}{\partial r^2}+\frac{1}{r}\frac{\partial \bar{F}}{\partial r}+\frac{1}{r^2\sin^2\theta}\frac{\partial^2 \bar{F}}{\partial \alpha^2}\right)\sin\theta+\frac{\cos\theta}{r}\frac{\partial^2 \bar{F}}{\partial r\partial \theta}\ , \tag{5.4.5}$$

$$\bar{H}_\alpha = -\frac{1}{r}\frac{\partial^2 \bar{F}}{\partial r\partial \alpha}\cot\theta+\frac{1}{r^2}\frac{\partial^2 \bar{F}}{\partial \theta\partial \alpha}\ . \tag{5.4.6}$$

For the subsequent analysis, it is convenient to introduce once more the exciting field, due to the coil only, and the secondary field, due only to eddy-currents in the shield.

5.4.3 Exciting Field

The exciting field (superscript p) is determined by the solution of (5.4.53) in Sect. 5.4.7, applied to the primary phasor $\bar{F}^{(\mathrm{p})}$ of the Fitzgerald potential. As the medium surrounding the coil is assumed to be of vanishingly small electrical conductivity, we obtain

$$\nabla^2\bar{F}^{(\mathrm{p})}+k^2\bar{F}^{(\mathrm{p})} = 0\ , \tag{5.4.7}$$

where $k = \omega/c$.

The solution for a field decreasing with r but yielding for $r \to 0$ a *finite (quasi-)homogeneous magnetic field* directed along the z-axis (see below), is given by

$$\bar{F}^{(p)} = \bar{F}_0 \frac{\sin \dfrac{\omega}{c} r}{\dfrac{\omega}{c} r} \, , \tag{5.4.8}$$

where $\bar{F}_0$ is as yet unspecified.

We wish to emphasize that (5.4.8) is *not* the free-space solution, but rather a specialized solution, adapted to the form of the excitation located *at the origin*.

Substituting the potential $\bar{F}^{(p)}$ in (5.4.1 to 3) and (5.4.4 to 6), we obtain for the primary field phasors:

$$\bar{E}_r^{(p)} = 0 \, , \tag{5.4.9}$$

$$\bar{E}_\theta^{(p)} = 0 \, , \tag{5.4.10}$$

$$\bar{E}_\alpha^{(p)} = \bar{F}_0 i \frac{\omega^2}{c^2} \sqrt{\frac{\mu_0}{\varepsilon_0}} \frac{1}{\omega r/c} \left(\cos \omega r/c - \frac{\sin \omega r/c}{\omega r/c} \right) \sin \theta \, , \quad \text{and} \tag{5.4.11}$$

$$\bar{H}_r^{(p)} = \bar{F}_0 \left(\frac{\omega}{c} \right)^2 \frac{2}{(\omega r/c)^2} \left(\cos \omega r/c - \frac{\sin \omega r/c}{\omega r/c} \right) \cos \theta \, , \tag{5.4.12}$$

$$\bar{H}_\theta^{(p)} = \bar{F}_0 \left(\frac{\omega}{c} \right)^2 \frac{1}{\omega r/c} \left[\sin \omega r/c + \frac{\cos \omega r/c}{\omega r/c} - \frac{\sin \omega r/c}{(\omega r/c)^2} \right] \sin \theta \, , \tag{5.4.13}$$

$$\bar{H}_\alpha^{(p)} = 0 \, . \tag{5.4.14}$$

For $(\omega r/c) \to 0$, i.e., close to the origin $r = 0$, (5.4.12 and 13) yield

$$\lim_{(\omega r/c) \to 0} \bar{H}_r^{(p)} = -\bar{F}_0 \left(\frac{\omega}{c} \right)^2 \frac{2}{3} \cos \theta \, , \quad \text{and} \tag{5.4.15}$$

$$\lim_{(\omega r/c) \to 0} \bar{H}_\theta^{(p)} = -\bar{F}_0 \left(\frac{\omega}{c} \right)^2 \frac{2}{3} \sin \theta \, . \tag{5.4.16}$$

The magnetic field at the center $(r \to 0)$ is therefore (quasi-)homogeneous and directed along the negative z-axis:

$$\bar{H}_z^{(p)}(0) \mathbf{1}_z = -\bar{F}_0 \left(\frac{\omega}{c} \right)^2 \frac{2}{3} \mathbf{1}_z \, . \tag{5.4.17}$$

In line with this interpretation of the amplitude $\bar{F}_0$, (5.4.12 and 13) are recast in the form

$$\bar{H}_r^{(p)} = -\bar{H}_z^{(p)}(0)\, 3\, \frac{1}{(\omega r/c)^2}\left(\cos \omega r/c - \frac{\sin \omega r/c}{\omega r/c}\right)\cos\theta\ , \tag{5.4.18}$$

$$\bar{H}_\theta^{(p)} = -\bar{H}_z^{(p)}(0)\, \frac{3}{2}\, \frac{1}{\omega r/c}\left[\sin \omega r/c + \frac{\cos \omega r/c}{\omega r/c} - \frac{\sin \omega r/c}{(\omega r/c)^2}\right]\sin\theta\ . \tag{5.4.19}$$

Similarly, the complex electrical field component $\bar{E}_\alpha$ now reads:

$$\bar{E}_\alpha^{(p)} = -\mathrm{i}\sqrt{\frac{\mu_0}{\varepsilon_0}}\,\bar{H}_z^{(p)}(0)\, \frac{3}{2}\, \frac{1}{\omega r/c}\left(\cos \omega r/c - \frac{\sin \omega r/c}{\omega r/c}\right)\sin\theta\ . \tag{5.4.20}$$

5.4.4 Secondary Field

The secondary field (superscript s) due to an assumed circular (complex) surface current $\bar{A}_\alpha = \bar{A}_0\sin\theta$ is derived from a (complex) Fitzgerald potential $\bar{F}^{(s)}$ once more satisfying the Helmholtz equation, i.e.

$$\nabla^2 \bar{F}^{(s)} + k^2 \bar{F}^{(s)} = 0\ . \tag{5.4.21}$$

For the inner region $r < R$, a solution coherent (in r) with the primary Fitzgerald potential is chosen; introducing the as yet unspecified complex amplitude $\bar{F}_i$, we have

$$\bar{F}^{(s)} = \bar{F}_i \frac{\sin \omega r/c}{\omega r/c}\ , \qquad r < R\ . \tag{5.4.22}$$

For the outer region $r > R$, the potential is given by

$$\bar{F}^{(s)} = \bar{F}_e \frac{e^{+\mathrm{i}(\omega r/c)}}{\omega r/c}\ , \qquad r > R\ , \tag{5.4.23}$$

$\bar{F}_e$ being again an unspecified complex amplitude.

The θ-components of the complex magnetic field, as derived from the above potentials, are

$$\bar{H}_\theta = \bar{F}_i\left(\frac{\omega}{c}\right)^2 \frac{1}{\omega r/c}\left[\sin \omega r/c + \frac{\cos \omega r/c}{\omega r/c} - \frac{\sin \omega r/c}{(\omega r/c)^2}\right]\sin\theta\ , \qquad r < R\ , \tag{5.4.24}$$

$$\bar{H}_\theta = \bar{F}_e\left(\frac{\omega}{c}\right)^2 \frac{e^{+\mathrm{i}(\omega r/c)}}{\omega r/c}\left[1 + \frac{\mathrm{i}}{\omega r/c} - \frac{1}{(\omega r/c)^2}\right]\sin\theta\ , \qquad r > R\ , \tag{5.4.25}$$

whereas, for the electric field, we have

$$\bar{E}_\alpha = i\left(\frac{\omega}{c}\right)^2 \sqrt{\frac{\mu_0}{\varepsilon_0}} \, \frac{\bar{F}_i}{\omega r/c}\left(\cos\omega r/c - \frac{\sin\omega r/c}{\omega r/c}\right)\sin\theta, \quad r < R ,$$

(5.4.26)

$$\bar{E}_\alpha = i\left(\frac{\omega}{c}\right)^2 \sqrt{\frac{\mu_0}{\varepsilon_0}} \, \frac{\bar{F}_e e^{+i(\omega r/c)}}{\omega r/c}\left(+i - \frac{1}{\omega r/c}\right)\sin\theta, \quad r > R . \quad (5.4.27)$$

Continuity of the electric field at $r = R$ requires

$$\bar{F}_i\left(\cos\omega R/c - \frac{\sin\omega R/c}{\omega R/c}\right) = \bar{F}_e e^{+i(\omega R/c)}\left(+i - \frac{1}{\omega R/c}\right) . \qquad (5.4.28)$$

On the other hand, the discontinuity of the magnetic field, as obtained by applying Maxwell's first curl equation in integral form, yields

$$\bar{H}_\theta(R+0) - \bar{H}_\theta(R-0) = \bar{A}_\alpha , \quad \text{i.e.} \qquad (5.4.29)$$

$$\left(\frac{\omega}{c}\right)^2 \left\{ -F_i\frac{\sin\omega R/c + \dfrac{\cos\omega R/c}{\omega R/c} - \dfrac{\sin\omega R/c}{(\omega R/c)^2}}{\omega R/c} \right.$$

$$\left. + \bar{F}_e\frac{e^{+i(\omega R/c)}\left[1 + \dfrac{i}{\omega R/c} - \dfrac{1}{(\omega R/c)^2}\right]}{\omega R/c} \right\} = \bar{A}_0 . \qquad (5.4.30)$$

Solving (5.4.28 and 30), we obtain

$$\bar{F}_i = -\bar{A}_0\left(\frac{c}{\omega}\right)^2 e^{+i(\omega R/c)}(-i\omega R/c + 1) , \qquad (5.4.31)$$

$$\bar{F}_e = \bar{A}_0\left(\frac{c}{\omega}\right)^2\left(\frac{\omega}{c}R\cos\omega R/c - \sin\omega R/c\right) . \qquad (5.4.32)$$

Applying (5.4.17) to the amplitude $\bar{F}_i$ (instead of $\bar{F}_0$) and to the secondary magnetic field component $\bar{H}_z^{(s)}(0)$ at the center of the sphere, we have

$$\bar{H}_z^{(s)}(0) = -\bar{F}_i\left(\frac{\omega}{c}\right)^2\frac{2}{3} = +\frac{2}{3}\bar{A}_0 e^{+i(\omega R/c)}(-i\omega R/c + 1) . \qquad (5.4.33)$$

Similarly, introducing the complex amplitude $\bar{E}_\alpha^{(s)}$ of the circumferential component of the secondary field in (5.4.20) instead of $\bar{E}_\alpha^{(p)}$ and replacing $\bar{H}_z^{(p)}(0)$ by $\bar{H}_z^{(s)}(0)$ as per (5.4.33), we obtain the relation

$$\bar{E}_\alpha^{(s)}\big|_R = -i\sqrt{\frac{\mu_0}{\varepsilon_0}}\,\bar{A}_0 e^{+i(\omega R/c)}\left(-i+\frac{1}{\omega R/c}\right)$$

$$\times\left(\cos\omega R/c-\frac{\sin\omega R/c}{\omega R/c}\right)\sin\theta\;,\qquad (5.4.34)$$

holding on the spherical shell.

5.4.5 Resultant Field

The resultant complex electric field component $\bar{E}_\alpha = \bar{E}_\alpha^{(p)}+\bar{E}_\alpha^{(s)}$ is related to the surface current $\bar{A}_\alpha$ at $r=R$ by means of Ohm's law $(\sigma\Delta)\bar{E}_\alpha \equiv \varkappa\bar{E}_\alpha = \bar{A}_\alpha$, i.e.,

$$\varkappa\Bigg[-i\sqrt{\frac{\mu_0}{\varepsilon_0}}\,\bar{H}_z^{(p)}(0)\,\frac{1}{\omega R/c}\frac{3}{2}\left(\cos\omega R/c-\frac{\sin\omega R/c}{\omega R/c}\right)$$

$$-i\sqrt{\frac{\mu_0}{\varepsilon_0}}\,\bar{A}_0 e^{+i(\omega R/c)}\left(-i+\frac{1}{\omega R/c}\right)$$

$$\times\left(\cos\omega R/c-\frac{\sin\omega R/c}{\omega R/c}\right)\Bigg]\sin\theta = \bar{A}_0\sin\theta\;.\qquad (5.4.35)$$

The amplitude $\bar{A}_0$ of the surface current may therefore be related to that of the axially-directed exciting field, $\bar{H}_z^{(p)}(0)$, at the sphere center:

$$\bar{A}_0 = -\bar{H}_z^{(p)}(0)\,\dfrac{i\varkappa\sqrt{\dfrac{\mu_0}{\varepsilon_0}}\,\dfrac{3}{2}\,\dfrac{\cos\omega R/c-\dfrac{\sin\omega R/c}{\omega R/c}}{\omega R/c}}{1+i\varkappa\sqrt{\dfrac{\mu_0}{\varepsilon_0}}(1-i\omega R/c)e^{+i(\omega R/c)}\dfrac{\cos\omega R/c-\dfrac{\sin\omega R/c}{\omega R/c}}{\omega R/c}}\;.$$

$$\qquad (5.4.36)$$

Finally, the complex amplitude of the resultant z-component of the magnetic field at the center of the sphere, $\bar{H}_z(0) \equiv \bar{H}_z^{(p)}(0)+H_z^{(s)}(0)$, may now be evaluated: eliminating $\bar{H}_z^{(s)}(0)$ by means of (5.4.33, 36), and introducing the dimensionless variable $\Omega = (\omega R/c)$, we have

$$\bar{H}_z(0) = \bar{H}_z^{(p)}(0)\,\dfrac{1}{1+i\varkappa\sqrt{\dfrac{\mu_0}{\varepsilon_0}}(1-i\Omega)e^{+i\Omega}\dfrac{\cos\Omega-(\sin\Omega)/\Omega}{\Omega}}\;.\qquad (5.4.37)$$

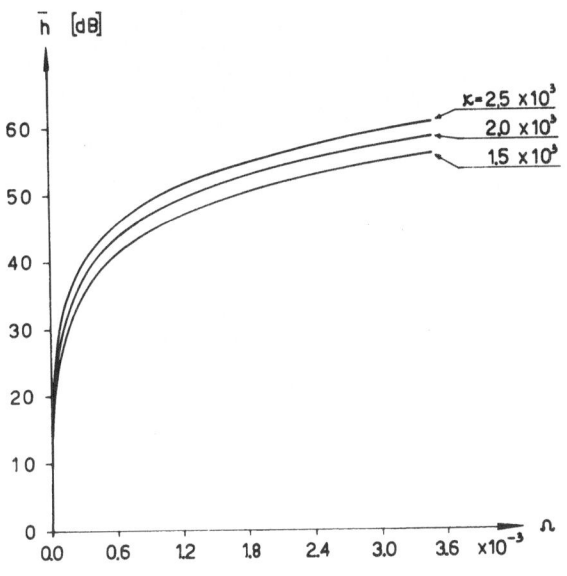

Fig. 5.19. Logarithmic field $\bar{h}$ vs. dimensionless angular frequency Ω, for different values of conductivity parameter $\varkappa\,(\text{ohm})^{-1}$; relatively low frequency range

The dimensionless logarithmic field

$$\bar{h} = 20 \log \left| \frac{\bar{H}_z^{(\mathrm{p})}(0)}{\bar{H}_z(0)} \right| , \qquad (5.4.38)$$

is plotted in Figs. 5.19 and 20. Conductivity and frequency regions have been chosen identical with those of Sect. 5.3, mainly for purposes of comparison.

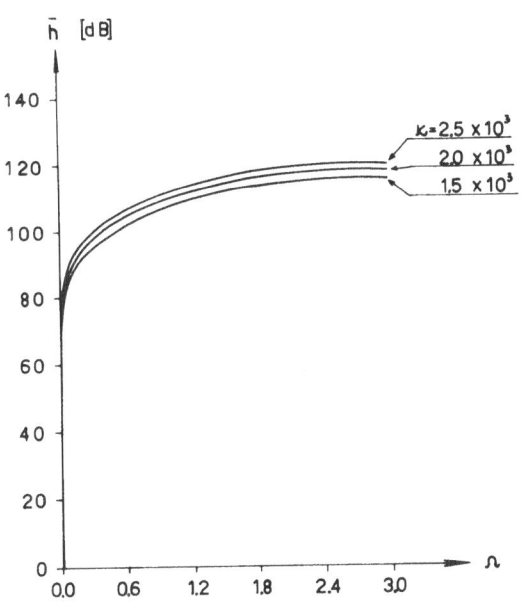

Fig. 5.20. Logarithmic field $\bar{h}$ vs. dimensionless angular frequency Ω, for different values of conductivity parameter $\varkappa\,(\text{ohm})^{-1}$; relatively high frequency range

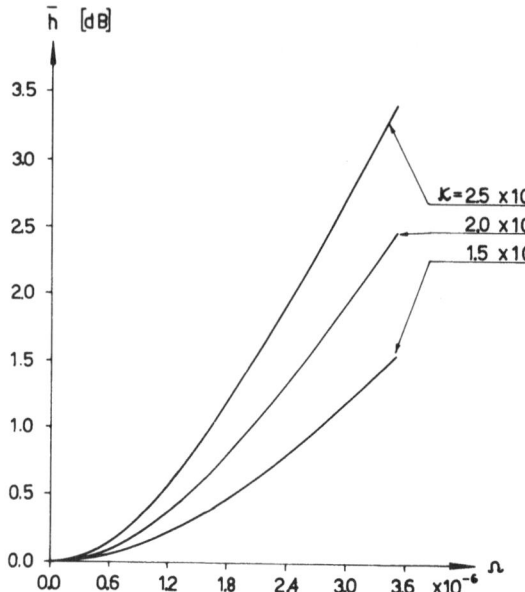

Fig. 5.21. Logarithmic field $\bar{h}$ vs. dimensionless angular frequency Ω, for different values of conductivity parameter $\varkappa\,(\text{ohm})^{-1}$; very low frequency range

Once more the high frequency range is of academic interest; in the low frequency range (5.4.37) is replaced by

$$\bar{H}_z(0) = \bar{H}_z^{(p)}(0)\,\frac{1}{1 - i\,3^{-1}\varkappa\sqrt{\mu_0\Omega/\varepsilon_0}}\ . \tag{5.4.39}$$

Very low frequency behavior is reproduced in Fig. 5.21.

5.4.6 Concluding Remarks

A quantitative approach in determining the effects of a non-magnetic shield on an enclosed coil was outlined. It appears that resonance tends to set in for certain values of the dimensionless angular frequency; one should, however, be aware that this result is here again mainly quantitative, as the actual screen has been replaced by a fictitious sphere with extremely high ("infinite") conductivity and thickness Δ much smaller than the skin depth $\delta = \sqrt{2/\omega\mu_0\sigma}$; the relevant remarks at end of Sect. 5.3.5 apply therefore here as well.

5.4.7 Appendix

The Fitzgerald Superpotential

Hitherto we have considered the Fitzgerald superpotential either from the view point of dynamics and in the absence of conductivity (Sect. 2.1.2), or alternatively, in the presence of conductivity but under conditions of quasi-statics (Sect. 2.2.8). We now consider this vector function in the presence of *both* dynamics and conductivity.

Thus, in a stationary, homogeneous and isotropic medium of electrical conductivity σ, permeability μ_0 and dielectric constant ε_0, Maxwell's equations read:

$$\nabla \times H = \sigma E + \varepsilon_0 \frac{\partial E}{\partial t} \; , \tag{5.4.40}$$

$$\nabla \times E = -\mu_0 \frac{\partial H}{\partial t} \; . \tag{5.4.41}$$

The continuity equation

$$\nabla \cdot \left(\sigma E + \varepsilon_0 \frac{\partial E}{\partial t} \right) = 0 \; , \tag{5.4.42}$$

is satisfied identically if an electric vector potential C is introduced through the relation

$$\nabla \times C = \sigma E + \varepsilon_0 \frac{\partial E}{\partial t} \; . \tag{5.4.43}$$

Resorting also to a magnetic scalar potential χ, we obtain from (5.4.40 and 43)

$$H = C - \nabla \chi \; . \tag{5.4.44}$$

Applying the curl operator to the expression $[\sigma E + \varepsilon_0 (\partial E / \partial t)]$, and recalling (5.4.41), we have

$$\nabla \times \left(\sigma E + \varepsilon_0 \frac{\partial E}{\partial t} \right) = -\mu_0 \frac{\partial}{\partial t} \left(\sigma H + \varepsilon_0 \frac{\partial H}{\partial t} \right) \; , \tag{5.4.45}$$

or, by (5.4.43, 44),

$$\nabla \times (\nabla \times C) = -\mu_0 \frac{\partial}{\partial t} \left[\sigma (C - \nabla \chi) + \varepsilon_0 \frac{\partial}{\partial t} (C - \nabla \chi) \right] \; . \tag{5.4.46}$$

Expressed in Cartesian coordinates, this last expression reads

$$\nabla (\nabla \cdot C) - \nabla^2 C = -\mu_0 \left[\sigma \frac{\partial C}{\partial t} - \sigma \nabla \frac{\partial \chi}{\partial t} + \varepsilon_0 \frac{\partial^2 C}{\partial t^2} - \varepsilon_0 \nabla \frac{\partial^2 \chi}{\partial t^2} \right] \; . \tag{5.4.47}$$

Imposing now the condition

$$\nabla \cdot C = \mu_0 \left(\sigma \frac{\partial \chi}{\partial t} + \varepsilon_0 \frac{\partial^2 \chi}{\partial t^2} \right) \; , \tag{5.4.48}$$

we obtain the vector differential equation

$$\nabla^2 C = \mu_0 \sigma \frac{\partial C}{\partial t} + \frac{1}{c^2} \frac{\partial^2 C}{\partial t^2} \ . \tag{5.4.49}$$

The condition (5.4.48) is identically satisfied by the vector function F, introduced through the relations

$$C = \mu_0 \sigma \frac{\partial F}{\partial t} + \frac{1}{c^2} \frac{\partial^2 F}{\partial t^2} \ , \tag{5.4.50}$$

$$\chi = \nabla \cdot F \ . \tag{5.4.51}$$

The vector differential equation for F is obtained from the last two expressions along with (5.4.44) the latter being subject to the requirement

$$\nabla \cdot (\mu_0 H) = 0 \ , \tag{5.4.52}$$

and it reads

$$\nabla^2 F = \mu_0 \sigma \frac{\partial F}{\partial t} + \frac{1}{c^2} \frac{\partial^2 F}{\partial t^2} \ . \tag{5.4.53}$$

The preceding sections showed that consistent results are obtained if the vector F is directed along the z-axis; representing now $F = F 1_z$ through the complex amplitude $\bar{F}$, i.e.,

$$F = \bar{F} e^{-i\omega t} \ , \tag{5.4.54}$$

(5.4.53) yields

$$\nabla^2 \bar{F} + \left(i \omega \mu_0 \sigma + \frac{\omega^2}{c^2} \right) \bar{F} = 0 \ . \tag{5.4.55}$$

Resorting once more to (5.4.50) the phasor $\bar{C}$, representing the z-component of the electric vector potential, reads

$$\bar{C} = - \left(i \omega \mu_0 \sigma + \frac{\omega^2}{c^2} \right) \bar{F} \ . \tag{5.4.56}$$

We next express the various field components in spherical coordinates (r, θ, α). The superpotential being directed by choice along the z-axis, we obtain from its "physical components" F_r, F_θ, F_α the phasors,

$$\bar{F}_r = \bar{F} \cos \theta, \quad \bar{F}_\theta = - \bar{F} \sin \theta, \quad \bar{F}_\alpha = 0 \ , \tag{5.4.57}$$

and therefore, on account of (5.4.56) the "physical components" of the phasors representing the electric vector potential are given by

$$\bar{C}_r = -\left(i\omega\mu_0\sigma + \frac{\omega^2}{c^2}\right)\bar{F}\cos\theta \ ,$$

$$\bar{C}_\theta = \left(i\omega\mu_0\sigma + \frac{\omega^2}{c^2}\right)\bar{F}\sin\theta, \qquad \bar{C}_\alpha = 0 \ .$$

(5.4.58)

Taking advantage of (5.4.44) in phasor representation and resorting to (5.4.51, 58) we obtain the complex amplitudes $\bar{H}_r$, $\bar{H}_\theta$ and $\bar{H}_\alpha$ of the magnetic field intensity

$$\bar{H}_r = \left[\frac{2}{r}\frac{\partial\bar{F}}{\partial r} + \frac{1}{r^2\sin\theta}\frac{\partial}{\partial\theta}\left(\sin\theta\frac{\partial\bar{F}}{\partial\theta}\right) + \frac{1}{r^2\sin^2\theta}\frac{\partial^2\bar{F}}{\partial\alpha^2}\right]\cos\theta$$

$$+ \sin\theta\frac{\partial}{\partial r}\left(\frac{1}{r}\frac{\partial\bar{F}}{\partial\theta}\right) \ ,$$

(5.4.59)

$$\bar{H}_\theta = -\left(\frac{\partial^2\bar{F}}{\partial r^2} + \frac{1}{r}\frac{\partial\bar{F}}{\partial r} + \frac{1}{r^2\sin^2\theta}\frac{\partial^2\bar{F}}{\partial\alpha^2}\right)\sin\theta$$

$$+ \frac{\cos\theta}{r}\frac{\partial^2\bar{F}}{\partial r\partial\theta} \ ,$$

(5.4.60)

$$\bar{H}_\alpha = -\frac{1}{r}\frac{\partial^2\bar{F}}{\partial r\partial\alpha}\cot\theta + \frac{\partial^2\bar{F}}{\partial\theta\partial\alpha}\frac{1}{r^2} \ .$$

(5.4.61)

As regards the electrical field, (5.4.43 and 56) similarly yield its complex representation

$$\bar{E}_r = -i\omega\mu_0\frac{1}{r}\frac{\partial\bar{F}}{\partial\alpha} \ ,$$

(5.4.62)

$$\bar{E}_\theta = -i\omega\mu_0\frac{1}{r}\frac{\partial\bar{F}}{\partial\alpha}\cot\theta \ ,$$

(5.4.63)

$$\bar{E}_\alpha = i\omega\mu_0\left(\sin\theta\frac{\partial\bar{F}}{\partial r} + \frac{1}{r}\cos\theta\frac{\partial\bar{F}}{\partial\theta}\right) \ .$$

(5.4.64)

References

Chapter 1

1.1 A. Sommerfeld: *Electrodynamics* (Academic, New York 1952) p. 281
1.2 L. D. Landau, E. M. Lifshitz: *The Classical Theory of Fields* (Pergamon, Oxford 1971) pp. 256 – 257
1.3 E. Schmutzer: *Relativistische Physik* (Akademische Verlagsgesellschaft, Leipzig 1968) pp. 386 – 404
1.4 J. Van Bladel: *Relativity and Engineering* (Springer, Berlin 1984) pp. 264 – 266
1.5 N. Rosen, D. Schieber: Am. J. Phys. **50**, 974 – 975 (1982)
1.6 R. M. Fano, L. J. Chu, R. B. Adler: *Electromagnetic Fields, Energy and Forces* (Wiley, New York 1960) pp. 376 – 418
1.7 B. D. H. Tellegen: Am. J. Phys. **30**, 650 – 652 (1962)
1.8 P. Penfield Jr., H. A. Haus: *Electrodynamics of Moving Media* (MIT Press, Cambridge, MA 1967) p. 244
1.9 H. A. Haus, P. Penfield Jr.: Phys. Lett. **26 A**, 412 – 413 (1968)
1.10 H. A. Haus, P. Penfield Jr.: Physica **42**, 447 – 454 (1969)
1.11 H. A. Haus, D. Schieber: Electrodynamics Memo No. 34, MIT (1973)
1.12 C. Møller: *The Theory of Relativity* (Clarendon, Oxford 1972) pp. 201 – 224
1.13 J. R. Melcher: *Continuum Electromechanics* (MIT Press, Cambridge, MA 1981) p. 2.5
1.14 R. B. Adler, L. J. Chu, R. M. Fano: *Electromagnetic Energy Transmission and Radiation* (Wiley, New York 1960) p. 29
1.15 F. Ollendorff: *Elektronik des Einzelelektrons* (Springer, Wien 1955) p. 43
1.16 A. A. Andronov, A. A. Vitt, S. E. Chaikin: *Theory of Oscillators* (Pergamon, Oxford 1966) pp. 36 – 39
1.17 F. Ollendorff: Arch. Elektrotechnik **56**, 1 – 11 (1974); and **59**, 133 – 140 (1977)
1.18 F. Ollendorff: *Erdströme* (Birkhäuser, Basel 1969) p. 327
1.19 F. Ollendorff: *Grundlagen der Kristallelektronik* (Springer, Wien 1966) p. 163
1.20 C. W. Michels, L. A. Patterson: Phys. Rev. **60**, 589 – 592 (1941)
1.21 N. Rosen: Am. J. Phys. **20**, 161 – 164 (1952)
1.22 J. Fischer: *Elektrodynamik*, (Springer, Berlin, Heidelberg 1976) p. 412

Chapter 2

2.1 A. Sommerfeld: *Electrodynamics* (Academic, New York 1952) p. 286
2.2 A. Sommerfeld: *Mechanics of Deformable Bodies* (Academic, New York 1950) p. 132
2.3 L. D. Landau, E. M. Lifshitz: *Electrodynamics of Continuous Media* (Pergamon, Oxford 1960) p. 208
2.4 R. P. Feynman, R. B. Leighton, M. Sands: *The Feynman Lectures on Physics*, Vol. 2 (Addison-Wesley, Reading, MA 1975) p. 17 – 2
2.5 C. Møller: *The Theory of Relativity* (Clarendon, Oxford 1972) p. 214
2.6 L. D. Landau, E. M. Lifshitz: *The Classical Theory of Fields* (Pergamon, Oxford 1971) pp. 256 – 257
2.7 W. G. V. Rosser: *Classical Electromagnetism via Relativity* (Butterworths, London 1968) p. 157
2.8 A. Einstein: *Albert Einstein, Philosopher Scientist*, Vol. I, ed. by P. A. Schilpp (Open Court Publishing Co., La Salle, IL 1969) p. 63

2.9   T. Schlomka, G. Schenkel: Ann. Phys. **6**, 51 – 62 (1949)
2.10   F. Ollendorff: Arch. Elektr. **44**, 80 – 94 (1959)
2.11   R. Becker, F. Sauter: *Theorie der Elektrizität*, 19th ed. (Teubner, Stuttgart 1969) Vol. I, p. 264
2.12   E. Jahnke, F. Emde: *Tables of Functions* (Dover, New York 1945)
2.13   E. R. Laithwaite: Nature **4831**, 811 – 813 (1962)
2.14   E. R. Laithwaite: *Induction Machines for Special Purposes* (Newnes, London 1966)
2.15   E. R. Laithwaite: *Linear Electric Motors* (Mills & Boon, London 1971)
2.16   F. Tölke: *Praktische Funktionenlehre* (Springer, Berlin, Heidelberg 1967)
2.17   M. Abramovitz, I. S. Stegun: *Handbook of Mathematical Functions* (Dover, New York 1965) p. 608
2.18   F. Ollendorff: *Berechnung Magnetischer Felder* (Springer, Wien 1952) p. 228

Chapter 3
3.1   I. S. Gradshteyn, I. M. Ryzhik: *Tables of Integrals, Series and Products* (Academic, New York, London 1965)
3.2   L. V. Bewley: *Traveling Waves on Transmission Systems* (Dover, New York 1951) p. 210
3.3   G. A. Campbell: Bell Syst. Tech. J. **2**, 1 – 30 (1923)
3.4   O. Mayr: ETZ **36**, 1352 – 1355; **38**, 1436 – 1440 (1925)
3.5   J. R. Carson: Bell Syst. Tech. J. **5**, 539 – 554 (1926)
3.6   G. Haberland: Z. angew. Math. Mech. **5**, 366 – 379 (1926)
3.7   A. Sommerfeld: Ann. Phys. **28**, 665 – 736 (1909)
3.8   H. Weyl: Ann. Phys. **60**, 481 – 500 (1919)
3.9   M. J. O. Strutt: Ann. Phys. **1**, 721 – 771 (1929)
3.10   G. H. Brown, R. F. Lewis, J. Epstein: Proc. IRE **25**, 753 – 787 (1937)
3.11   J. R. Wait: Proc. IRE **46**, 1539 – 1541 (1958)
3.12   M. Krakowski: Arch. Elektrotech. **20**, 107 – 123 (1971)
3.13   M. Krakowski, G. S. Szymanski: Proc. IEE **126**, 139 – 140 (1979)
3.14   A. Timotin, A. Tugulea, C. Fluerasu: Revue Roum. Sci. Techn.-Electrotechn. Energ. **11**, 307 – 327 (1966)
3.15   M. Stafl: Acta Tech. **5**, 124 – 142 (1960)
3.16   F. Ollendorff: *Erdströme* (Birkhäuser, Basel 1969) pp. 211 – 228
3.17   D. Müller-Hillebrand: ETZ-A **82**, 232 – 249 (1961)
3.18   F. Ollendorff: Arch. Elektrotech. **52**, 137 – 147 (1968)
3.19   M. Abramovitz, I. A. Stegun: *Handbook of Mathematical Functions* (Dover, New York 1965) No. 5.2.12; 5.2.13, p. 232
3.20   A. D. Daly, S. P. Knight, M. Caulton, R. Ekholdt: IEEE Trans. MTT-**15**, 713 – 721 (1967)
3.21   M. Caulton, S. P. Knight, A. D. Daly: IEEE Trans. ED-**15**, 459 – 466 (1968)
3.22   M. Caulton: Microwave J. **13**, 51 – 58 (1970)
3.23   S. C. Aitchison, R. Davies, D. I. Higgins, R. S. Longley, H. B. Newton, F. J. Wells, C. J. Williams: IEEE Trans. MTT-**12**, 928 – 937 (1971)
3.24   E. R. De Brecht: IEEE Trans. MTT-**20**, 41 – 48 (1972)
3.25   M. Caulton: Lumped Elements in Microwave Integrated Circuits, in *Advances in Microwaves*, ed. by L. Young and H. Sobol (Academic, New York 1974) pp. 143 – 202
3.26   S. J. Joshi, R. J. Cockril, A. J. Turner: IEEE Trans. ED-**28**, 158 – 162 (1981)
3.27   H. Q. Tserng, H. M. Macksey: IEEE Trans. ED-**28**, 163 – 165 (1981)
3.28   L. R. Van Tuyl: IEEE Trans. ED-**28**, 166 – 170 (1981)
3.29   H. Q. Tserng, M. H. Macksey, R. S. Nelson: IEEE Trans. ED-**28**, 183 – 190 (1981)
3.30   F. E. Terman: *Radio Engineer Handbook* (McGraw-Hill, New York 1943) p. 58
3.31   W. F. Grover: *Inductance Calculations* (Dover, New York 1946) p. 167
3.32   J. M. C. Dukes: *Printed Circuits, Their Design and Application* (MacDonald, London 1961) pp. 129 – 130
3.33   H. G. Dill: *Electronic Design* (Hayden, New York 1964) No. 4, pp. 52 – 60

Chapter 4
4.1   F. Ollendorff: *Erdströme* (Birkhäuser, Basel 1969) p. 319 and 331
4.2   F. Ollendorff: ETZ, **83**, 573 – 580 (1962)
4.3   I. S. Gradshteyn, I. M. Ryzhik: *Tables of Integrals, Series and Products* (Academic, New York 1965)
4.4   F. Nechleba: Arch. Elektrotech. **53**, 309 – 315 (1970)
4.5   D. Schieber: EE Publication No. 256, Technion (1975)
4.6   H. Haas: Arch. Elektrotech. **61**, 89 – 96 (1979)
4.7   L. D. Landau, E. M. Lifshitz: *Electrodynamics of Continuous Media* (Pergamon, Oxford 1960) pp. 248 – 250
4.8   P. Frank, R. v. Mises: *Die Differential und Integralgleichungen der Mechanik und Physik* (Dover, New York 1961) p. 566
4.9   E. Jahnke, F. Emde: *Tables of Functions with Formulae and Curves* (Dover, New York 1945) p. 6
4.10 L. Brillouin: *Wave Propagation and Group Velocity* (Academic, New York 1960) p. 11
4.11 A. B. Pippard: Proc. Roy. Soc. Lond. **A216**, 547 – 568 (1953)
4.12 F. London: *Superfluids* (Wiley, New York 1950) Vol. I, p. 29
4.13 R. E. Matick: *Transmission Lines for Digital and Communication Networks* (McGraw-Hill, New York 1969) p. 254
4.14 M. Abramovitz, I. Stegun: *Handbook of Mathematical Functions* (Dover, New York 1968) No. 9.1.21, p. 360
4.15 A. Sommerfeld: *Mechanics of Deformable Bodies* (Academic, New York 1950) p. 183
4.16 H. Lamb: *Hydrodynamics* (Dover, New York 1945) pp. 459 – 460

Chapter 5
5.1   A. Ashkin, J. M. Dziezic: Appl. Phys. Lett. **19**, 283 – 285 (1971)
5.2   A. Ashkin: Sci. Am. **226**, 63 – 71 (1972)
5.3   A. Ashkin, J. M. Dziedzic: Appl. Phys. Lett. **24**, 586 – 588 (1974)
5.4   A. Ashkin, J. M. Dziedzic: Science **187**, 1073 – 1075 (1975)
5.5   A. Ashkin, J. M. Dziedzic: Phys. Rev. Lett. **38**, 1351 – 1354 (1977)
5.6   F. B. Bunkin, A. M. Prokhorov: Sov. Phys. Usp. **19**, 561 – 572 (1977)
5.7   J. R. Pierce: J. Appl. Phys. **30**, 1341 – 1346 (1959)
5.8   L. R. Walker: J. Appl. Phys. **26**, 1031 – 1033 (1955)
5.9   R. J. Briggs: J. Appl. Phys. **35**, 3268 – 3272 (1964)
5.10 S. Y. Galuzo: Radio Eng. and Electron Phys. **27**, 113 – 117 (1982)
5.11 S. D. Choi, D. A. Dunn: Proc. IEEE **59**, 737 – 748 (1971)
5.12 J. A. Kong: *Theory of Electromagnetic Waves* (Wiley, New York 1975) p. 147, p. 296
5.13 F. Ollendorff: ZAMP **26**, 611 – 618 (1975)
5.14 F. Ollendorff, D. Schieber: Arch. Elektr. **61**, 209 – 213 (1979)
5.15 C. Nuccio, D. Planthaber: IBM Tech. Disclosure Bull. **20**, 281 (1977)
5.16 E. M. Honig Jr.: IEEE Trans. EC-**19**, 377 – 382 (1977)
5.17 G. Franceschetti: IEEE Trans. EC-**21**, 335 – 348 (1979)
5.18 B. Rashkow: Electron. Design **9**, 88 – 94 (1979)
5.19 A. E. Booth, G. T. Hodgson: "Electromagnetic Pulse (EMP)", IEEE Int. Symp. on Electromagnetic Compatibility, Boulder, Colorado (1981)
5.20 R. W. P. King, C. W. Harrison Jr.: IRE Trans. AP-9, 166 – 170 (1961)
5.21 L. Hannakam: ETZ-A **87**, 227 – 232 (1966)
5.22 H. Kaden: Frequenz **23**, 159 – 164 (1969)
5.23 J. R. Wait: Canadian J. Phys. **51**, 209 – 218 (1973)
5.24 P. F. Gerbosi: Indus. Res./Develop. **21**, 119 – 121 (1979)

Acknowledgement

The kind permission of the Journals listed below to dráw upon my papers published in them is gratefully acknowledged.

Proc. IEE (London)
Vol. 119, 1499 – 1503 (1972)
Vol. 120, 1519 – 1520 (1973)
Vol. 121, 117 – 122 (1974)

Proc. I.E.E.E.
Vol. 61, 647 – 656 (1973)

Trans. I.E.E.E. on Electromagnetic Compatibility
Vol. EMC-15, 12 – 16 (1973)

Trans. I.E.E.E. on Magnetics
Vol. MAG-11, 948 – 953 (1975)

Journal of the Franklin Institute
Vol. 155, 249 – 261 (1973)
Vol. 296, 15 – 31 (1973)
Vol. 302, 293 – 312 (1976)
Vol. 308, 163 – 169 (1979)
Vol. 310, 119 – 129 (1980)
Vol. 310, 271 – 280 (1980)
Vol. 314, 231 – 242 (1982)
Vol. 314, 243 – 262 (1982)
Vol. 315, 149 – 164 (1983)
Vol. 316, 241 – 259 (1983)

Archiv für Elektrotechnik
Vol. 63, 111 – 115 (1981)
Vol. 68, 155 – 159 (1985)
Vol. 68, 175 – 181 (1985)
Vol. 68, 305 – 312 (1985)

Subject Index